AF388903

The IX Iberian Congress of
Ichthyology

The IX Iberian Congress of Ichthyology

Editor

Alberto Teodorico Correia

MDPI • Basel • Beijing • Wuhan • Barcelona • Belgrade • Manchester • Tokyo • Cluj • Tianjin

Editor
Alberto Teodorico Correia
Centro Interdisciplinar de Investigação Marinha e Ambiental
Portugal

Editorial Office
MDPI
St. Alban-Anlage 66
4052 Basel, Switzerland

This is a reprint of Proceedings published online in the open access journal *Biology and Life Sciences Forum* (ISSN 2673-9976) (available at: https://www.mdpi.com/2673-9976/13/1).

For citation purposes, cite each article independently as indicated on the article page online and as indicated below:

LastName, A.A.; LastName, B.B.; LastName, C.C. Article Title. *Journal Name* **Year**, *Volume Number*, Page Range.

ISBN 978-3-0365-6479-1 (Hbk)
ISBN 978-3-0365-6480-7 (PDF)

Cover image courtesy of Alberto Teodorico Correia

Contents

About the Editor

Alberto Teodorico Correia

Alberto Teodorico Correia has a doctorate and a Habilitation degree in Aquatic Sciences. He is an Associate Professor at the Fernando Pessoa University and is the Principal Investigator at the Interdisciplinary Centre for Marine and Environmental Research. He is a recognized specialist in the field of the ecology and ecotoxicology of fishes. He has about 100 international peer-review publications, has supervised several MSc and PhD students, and regularly participates in scientific projects.

Preface to "The IX Iberian Congress of Ichthyology"

The present volume contains selected manuscripts presented at the IX Iberian Congress of Ichthyology, which was organized by the Interdisciplinary Centre of Marine and Environmental Research (CIIMAR) and held in Porto, Portugal from 20th to 23rd June 2022.

The conference was attended by 123 participants, mainly from Portugal and Spain, and presented 77 regular oral communications and 55 posters. The congress was organized in parallel with oral presentations and poster exhibitions and included six plenary keynote lectures delivered by world renowned researchers.

A selection of 92 manuscripts subjected to blind peer review was considered for publication as the Special Issue presented here. Not all the subject areas of the congress are equally covered by this special volume, but the selected papers cover a wide range of topics such as i) Phylogeny, Systematics, and Genetics, ii) Morphology, Ontogeny, and Palaeontology, iii) Migration, Habitat Use, and Connectivity, iv) Physiology, Endocrinology, and Toxicology, v) Non-native and Invasive Species, vi) Fisheries and Aquaculture, vii) Fish Behaviour and Welfare, and viii) Climate Changes and Extreme Environments. Additionally, three special symposia have been organized: Invasaqua, Fish otoliths: current applications, methodological prerequisites and future challenges (Second Edition) and Fish Genetics: From Sanger to Next (First Edition).

We wish to acknowledge all the members of the congress' scientific committee that have assisted us in assuring the scientific quality of these selected papers. We also acknowledge the support of the various institutions and sponsors who enabled the organization of this event, namely the Fundação Eng. António de Almeida (FEEA), the Iberian Ichthyological Society (SIBIC), and Innovasea-Nautilus-Oceanica.

Alberto Teodorico Correia
Editor

Abstract

Contribution to the Conservation Genetics of an Endangered Fish, Endemic to the Spanish Mediterranean Coast: The Spanish Toothcarp, *Apricaphanius iberus* (Valenciennes, 1846) [†]

Tessa Lynn Nester [1,*,‡], Alfonso López-Solano [1], Silvia Perea [2] and Ignacio Doadrio [1]

[1] Department of Biodiversity and Evolutionary Biology, National Museum of Natural Sciences, Spanish National Research Council (CSIC), c/José Gutiérez Abascal 2, 28006 Madrid, Spain; alfonso.lopez@mncn.csic.es (A.L.-S.); doadrio@mncn.csic.es (I.D.)

[2] Department of Zoology, Instituto de Biología, Universidad Nacional Autónoma de México, Tercer Circuito Exterior S/N. C.P., Mexico City 04510, Mexico; silvia.perea@st.ib.umna.mx

*   Correspondence: tessa@mncn.csic.es

†   Presented at the IX Iberian Congress of Ichthyology, Porto, Portugal, 20–23 June 2022.

‡   Presenting author (Oral presentation).

Abstract: The Spanish toothcarp, *Apricaphanius iberus* (Valenciennes, 1846), is a small, endangered fish species, endemic to the coastal waters, interior lakes and salt marshes of the Spanish Mediterranean coast. Anthropogenic-driven factors such as agricultural exploitation and urbanization have progressively degraded its natural habitats, fragmentating its populations and driving the Spanish toothcarp to the brink of extinction. In this study, we aimed to improve our understanding of the Spanish toothcarp's genetic diversity and population structure using single-nucleotide polymorphisms (SNPs) to conduct a genetic analysis of 101 individuals from 12 of its populations along the Mediterranean coast of the Iberian Peninsula. We analyzed the phylogenetic relationships and genetic structure of its populations, as well as their migration rates. Our results showed that the genetic diversity values were similar and relatively moderate across populations, except the northernmost population, Aigüamolls (Girona), which was the most genetically differentiated, although the individuals belonging to this population presented the lowest amount of genetic differentiation among each other. Significant migration was not detected and F_{ST} values were fairly high between populations, indicating levels of differentiation and genetic isolation that were attributable to fragmentation. One of the northernmost populations, Albuixech (Valencia), and the southernmost population, Adra (Almería), comprised a sister group, possibly indicating Adra's factitious origin. The results from our study enabled us to define eight Operational Conservation Units (OCUs) that should be implemented into current and future conservation programs aimed at keeping the Spanish toothcarp from disappearing completely from the wild.

Keywords: *Apricaphanius iberus*; endangered species; conservation genetics; fragmented populations; single nucleotide polymorphisms (SNPs); Operational Conservation Units (OCUs)

Citation: Nester, T.L.; López-Solano, A.; Perea, S.; Doadrio, I. Contribution to the Conservation Genetics of an Endangered Fish, Endemic to the Spanish Mediterranean Coast: The Spanish Toothcarp, *Apricaphanius iberus* (Valenciennes, 1846). *Biol. Life Sci. Forum* **2022**, *13*, 1. https://doi.org/10.3390/blsf2022013001

Academic Editor: Alberto Teodorico Correia

Published: 1 June 2022

Publisher's Note: MDPI stays neutral with regard to jurisdictional claims in published maps and institutional affiliations.

Author Contributions: Conceptualization, T.L.N., S.P. and I.D.; methodology, T.L.N., A.L.-S., S.P. and I.D.; software, T.L.N. and S.P.; validation, I.D.; formal analysis, T.L.N., A.L-S., S.P. and I.D.; investigation, T.L.N. and A.L.-S.; resources, T.L.N. and A.L.-S.; data curation, T.L.N., A.L.-S., S.P. and I.D.; writing—original draft preparation, T.L.N.; writing—review and editing, T.L.N., A.L.-S., S.P. and I.D.; visualization, T.L.N.; supervision, T.L.N. and I.D.; project administration, I.D.; funding acquisition, I.D. All authors have read and agreed to the published version of the manuscript.

Funding: This research was funded by Ministerio de Ciencia e Innovación (PID2019-103936GB-C22).

Institutional Review Board Statement: Not applicable.

Informed Consent Statement: Not applicable.

Data Availability Statement: The data from this study will be stored in the Genbank database.

Conflicts of Interest: The authors declare no conflict of interest.

Abstract

Molecular Taxonomy of Deep-Sea Swallowers of the Family Chiasmodontidae [†]

Jesús Portas [1], Rafael Bañón [2,‡], José Luis del Río [3], David Barros-García [4] and Alejandro de Carlos [1,*]

[1] Facultade de Bioloxía, Universidade de Vigo, 36310 Vigo, Spain; jesusportasfndz@gmail.com
[2] Grupo de Estudio do Medio Mariño (GEMM), 15960 Ribeira, Spain; anoplogaster@yahoo.es
[3] Instituto Español de Oceanografía, Centro Oceanográfico de Vigo, 36390 Vigo, Spain; joseluis.delrio@ieo.es
[4] CIIMAR—Interdisciplinary Centre of Marine and Environmental Research, 4450-159 Matosinhos, Portugal; davbarros1985@gmail.com
*   Correspondence: adcarlos@uvigo.es; Tel.: +34-986-812576
†   Presented at the IX Iberian Congress of Ichthyology, Porto, Portugal, 20–23 June 2022.
‡   Presenting author (Poster presentation).

Abstract: The taxonomy of deep-sea fishes remains a little explored and sometimes unclear subject, as is the case of the family Chiasmodontidae. In multiple fishery assessment surveys carried out between 2017 and 2021 in the Western North Atlantic, around 50 specimens of this family were caught and tentatively named onboard as *Chiasmodon niger*. To test this preliminary identification, based on taxonomic and geographical criteria, DNA barcodes of the 5′ end of the mitochondrial *COI* gene were developed for 37 specimens. In this way it was possible to observe the polymorphisms between sequences and to calculate the gene distance between each of them. GenBank and BOLD repositories were also used for presumptive identification based on the percentage of similarity between sequences. In addition, the sequences of the 37 specimens were compared with all the barcode records of the COI-5P region of chiasmodontids found in the BOLD database. From the comparison between all the sequences and the data analyses performed on them, two main conclusions were reached. The first is that all specimens captured in the Western North Atlantic area were confirmed to belong to the species *Chiasmodon niger*. The other conclusion that can be drawn is that the databases present major problems with regard to this family. The identification of the 37 sample specimens allowed the detection of misidentifications of some sequences extracted from the repositories, as well as possible synonymies, as is the case with *Chiasmodon niger* and *Chiasmodon harteli*. This highlights taxonomic conflicts that should be addressed as a subject for future studies.

Keywords: barcoding; *Chiasmodon*; *COI*; identification; mtDNA

Citation: Portas, J.; Bañón, R.; del Río, J.L.; Barros-García, D.; de Carlos, A. Molecular Taxonomy of Deep-Sea Swallowers of the Family Chiasmodontidae. *Biol. Life Sci. Forum* **2022**, *13*, 13. https://doi.org/10.3390/blsf2022013013

Academic Editor: Alberto Teodorico Correia

Published: 2 June 2022

Publisher's Note: MDPI stays neutral with regard to jurisdictional claims in published maps and institutional affiliations.

Author Contributions: Conceptualization, R.B. and A.d.C.; methodology, R.B., J.P. and A.d.C.; formal analysis, R.B., D.B.-G. and A.d.C.; resources, R.B., J.L.d.R. and A.d.C.; data curation, J.P. and A.d.C.; writing—original draft preparation, J.P, R.B. and A.d.C.; funding acquisition, J.L.d.R. and A.d.C. All authors have read and agreed to the published version of the manuscript.

Funding: A.d.C. was funded by the GRC Program from the Xunta de Galicia, Spain, grant code ED431C 2019/28.

Institutional Review Board Statement: Not applicable.

Informed Consent Statement: Not applicable.

Data Availability Statement: DNA sequences and their corresponding metadata, such as specimen photographs, capture coordinates and dates, are part of the project "Fishes of the Family Chiasmodontidae" (code CHIAS), deposited in the Barcode of Life database (www.boldsystems.org).

Conflicts of Interest: The authors declare no conflict of interest.

biology and life sciences forum

Abstract

Genetic Structure of Meagre (*Argyrosomus regius*) in Portugal: Implications for Fisheries Management [†]

Rita Almeida [1,*,‡], **Catarina S. Mateus** [1], **Maria Judite Alves** [2,3], **João P. Marques** [4], **Joana Pereira** [1], **Nuno Prista** [5], **Henrique Cabral** [6], **Pedro R. Almeida** [1,7] and **Bernardo R. Quintella** [4,8,*]

[1] MARE—Marine and Environmental Sciences Centre, Universidade de Évora, 7002-554 Évora, Portugal; cspm@uevora.pt (C.S.M.); joana_gomespereira@hotmail.com (J.P.); pmra@uevora.pt (P.R.A.)
[2] MUHNAC—Museu Nacional de História Natural e da Ciência, 1250-102 Lisboa, Portugal; mjalves@museus.ulisboa.pt
[3] Natural History and Systematics (NHS) Research Group of cE3c—Centre for Ecology, Evolution and Environmental Changes, Universidade de Lisboa, 1749-016 Lisboa, Portugal
[4] MARE—Marine and Environmental Sciences Centre, Faculdade de Ciências, Universidade de Lisboa, 1749-016 Lisboa, Portugal; jpmarques@fc.ul.pt
[5] Department of Aquatic Resources, Institute of Marine Research, Swedish University of Agricultural Sciences, 453 30 Lysekil, Sweden; nuno.prista@slu.se
[6] INRAE—Institut National de Recherche Pour L'agriculture, L'alimentation et L'environnement, UR EABX, 33612 Cestas, France; henrique.cabral@inrae.fr
[7] Departamento de Biologia, Escola de Ciência e Tecnologia, Universidade de Évora, 7002-554 Évora, Portugal
[8] Departamento de Biologia Animal, Faculdade de Ciências, Universidade de Lisboa, 1749-016 Lisboa, Portugal
[*] Correspondence: ritapalmeida93@gmail.com (R.A.); bsquintella@fc.ul.pt (B.R.Q.)
[†] Presented at the IX Iberian Congress of Ichthyology, Porto, Portugal, 20–23 June 2022.
[‡] Presenting author (Oral communication).

Citation: Almeida, R.; Mateus, C.S.; Alves, M.J.; Marques, J.P.; Pereira, J.; Prista, N.; Cabral, H.; Almeida, P.R.; Quintella, B.R. Genetic Structure of Meagre (*Argyrosomus regius*) in Portugal: Implications for Fisheries Management. *Biol. Life Sci. Forum* **2022**, *13*, 16. https://doi.org/10.3390/blsf2022013016

Academic Editor: Alberto Teodorico Correia

Published: 6 June 2022

Publisher's Note: MDPI stays neutral with regard to jurisdictional claims in published maps and institutional affiliations.

Abstract: The meagre *Argyrosomus regius* (Asso, 1801) is a marine migratory species with a wide distribution range encompassing the north-eastern and central-eastern Atlantic Ocean, the Mediterranean Sea, and the western Black Sea. *A. regius* is one of the largest overexploited sciaenids, being a valuable resource for aquaculture and fisheries along its distribution range. The Iberian Peninsula is considered an intermediate area between two genetically distinct groups of *A. regius* populations, one in the north-eastern Atlantic Ocean and one in the eastern Mediterranean Sea. The current knowledge on the population dynamics and distribution of this species has been derived from commercial and recreational fishery catches; therefore, little is known about the importance of the Iberian Peninsula for the species' management and conservation. The aim of this study is to evaluate the *A. regius* population genetic structure along the Portuguese coast taking into consideration the north-eastern Atlantic region. To achieve this goal, the genetic diversity, differentiation, populational structure and demographic history of *A. regius* populations along the Atlantic coast were analyzed using 15 microsatellite loci. The detected populational structure indicates that *A. regius* species in Portugal are divided into two distinct stocks, one across the Portuguese western coast, possibly related to the Tagus spawning and nursery area, and another one on the southern coast. This study reveals the need for *A. regius*-specific fishery management plans in Portugal and underlines the importance of considering the genetic structure of *A. regius* populations when delineating such management plans.

Keywords: *Argyrosomus regius*; demographic history; gene flow; microsatellites; population structure; stock identification

Author Contributions: Conceptualization, B.R.Q., C.S.M. and M.J.A.; methodology, C.S.M. and M.J.A.; Sampling: J.P.M., R.A. and H.C.; Laboratorial analysis: R.A. and J.P.; Data analysis: R.A. and C.S.M.; Resources: B.R.Q. and M.J.A.; writing—original draft preparation, R.A., C.S.M., B.R.Q. and M.J.A.; writing—review and editing, R.A., C.S.M., B.R.Q., M.J.A., N.P., J.P.M., H.C., P.R.A., J.P.;

supervision, C.S.M. and B.R.Q.; project administration, B.R.Q.; funding acquisition, B.R.Q., C.S.M., M.J.A., P.R.A. and N.P. All authors have read and agreed to the published version of the manuscript.

Funding: This research was funded by National Funds through FCT (Foundation for Science and Technology) via the project "MIGRACORV—Integrated approach to study the movement dynamics of the meagre *Argyrosomus regius*" (PTDC/BIA-BMA/30517/2017) and MARE's strategic programme (UID/MAR/04292/2020).

Institutional Review Board Statement: Not applicable.

Informed Consent Statement: Not applicable.

Conflicts of Interest: The authors declare no conflict of interest.

biology and life sciences forum

Abstract

Atlantic Bonito (*Sarda sarda*) Genomic ddRadSeq Analysis Confirms Population Differentiation across Northeast Atlantic and Mediterranean Locations—Implications for Fishery Management [†]

Judith Ollé and Jordi Viñas *,[‡]

Laboratori d'Ictiologia Genètica, Departament de Biologia, Universitat de Girona, 17003 Girona, Catalonia, Spain; judith.olle@udg.edu
* Correspondence: jordi.vinas@udg.edu
† Presented at the IX Iberian Congress of Ichthyology, Porto, Portugal, 20–23 June 2022.
‡ Presenting author (Poster presentation).

Citation: Ollé, J.; Viñas, J. Atlantic Bonito (*Sarda sarda*) Genomic ddRadSeq Analysis Confirms Population Differentiation across Northeast Atlantic and Mediterranean Locations—Implications for Fishery Management. *Biol. Life Sci. Forum* **2022**, *13*, 23. https://doi.org/10.3390/blsf2022013023

Academic Editor: Alberto Teodorico Correia

Published: 6 June 2022

Publisher's Note: MDPI stays neutral with regard to jurisdictional claims in published maps and institutional affiliations.

Abstract: Atlantic bonito (*Sarda sarda*) is an epipelagic migratory species of the family Scombridae and it is usually classified under the common name of small tuna. This species is heavily targeted by artisanal and commercial fisheries in the northeast Atlantic and Mediterranean. However, no management directive has been proposed for this species. In 2019, ICCAT (International Commission for the Conservation of Atlantic tuna) identified the Atlantic bonito as a critical species to be studied, due to the lack of knowledge in several key biological aspects such as growth, reproduction, and stock structure. In a preliminary genetic analysis of more than 600 individuals from the Northeast Atlantic and Mediterranean, using a single molecular marker (mitochondrial DNA control region), we detected a significant genetic differentiation between two groups of locations. One group with two locations within the Mediterranean (Spain and Tunisia) and one location in the Atlantic near to the strait of Gibraltar (Portugal); and another group comprising locations along the African coast (Morocco, Mauritania, Senegal, and Côte d'Ivoire). A new analysis, using a genomic approach (ddRadseq analysis) confirm the same populations structure. Population genetic analysis of 95 individuals from seven locations recovered more than 8500 SNPs with a coverage greater than 20X (mean = 27.9; SD = 16.8) and confirmed that the genetic discontinuity is not placed in the Strait of Gibraltar. The locations from north and south Portugal were grouped together with the two locations within the Mediterranean Sea (Spain and Tunisia) and clearly differentiated from the rest of locations in the African coast (Morocco, Senegal, and Côte d'Ivoire). These results updated the genetic population structure of Atlantic bonito, and they can be used as starting point to infer a proper management regulation for this relevant commercial species.

Keywords: Atlantic bonito; *Sarda sarda*; population genomic structure; ddRadseq; fishery management; genomic analysis; northeast Atlantic; Mediterranean

Author Contributions: J.O. and J.V. conceived the study, analyzed the data and write the manuscript. All authors have read and agreed to the published version of the manuscript.

Funding: This research was funded by ICCAT.

Institutional Review Board Statement: Not applicable.

Informed Consent Statement: Not applicable.

Data Availability Statement: Not applicable.

Conflicts of Interest: The authors declare no conflict of interest.

biology and life sciences forum

Abstract

Complete Mitochondrial Genome of the Spanish Toothcarp, *Apricaphanius iberus* (Valenciennes, 1846) (Actinopterygii, Aphaniidae) and Its Phylogenetic Position within the Cyprinodontiformes Order [†]

Alfonso López-Solano [1,*,‡], Silvia Perea [2], Ignacio Doadrio [1] and Tessa Lynn Nester [1]

[1] Departamento de Biodiversidad y Biología Evolutiva, MNCN/CSIC C/José Gutiérez Abascal 2, 28006 Madrid, Spain; doadrio@mncn.csic.es (I.D.); tessa@mncn.csic.es (T.L.N.)

[2] Departamento de Zoología, Instituto de Biología, Universidad Nacional Autónoma de México, Tercer Circuito Exterior S/N. C.P., Ciudad de México 04510, Mexico; sperea2@gmail.com

[*] Correspondence: alfonso.lopez@mncn.csic.es

[†] Presented at the IX Iberian Congress of Ichthyology, Porto, Portugal, 20–23 June 2022.

[‡] Presenting author (Poster presentation).

Citation: López-Solano, A.; Perea, S.; Doadrio, I.; Nester, T.L. Complete Mitochondrial Genome of the Spanish Toothcarp, *Apricaphanius iberus* (Valenciennes, 1846) (Actinopterygii, Aphaniidae) and Its Phylogenetic Position within the Cyprinodontiformes Order. *Biol. Life Sci. Forum* **2022**, *13*, 29. https://doi.org/10.3390/blsf2022013029

Academic Editor: Alberto Teodorico Correia

Published: 6 June 2022

Publisher's Note: MDPI stays neutral with regard to jurisdictional claims in published maps and institutional affiliations.

Abstract: The Spanish Toothcarp (*Apricaphanius iberus*) is a small fish species endemic to the eastern coast of the Iberian Peninsula. Its area of distribution includes a variety of salt and freshwater habitats that experience large fluctuations in temperature and salinity throughout the year. The Spanish Toothcarp belongs to the Cyprinodontiformes order and to the family Aphaniidae. It is currently considered "Endangered" (category IUCN: EN), i.e., facing a very high risk of extinction in the wild. The genetics of *A. iberus* are not well known since most studies have only evaluated the genetic structure of the species under a conservation framework in order to identify its potential conservation units. In this study, the whole mitochondrial genome of *A. iberus* was obtained for the first time in the context of an *A. iberus* genome reference sequencing. The mitogenome was reconstructed and aligned against 83 other cyprinodontiformes and two outgroup taxa. From this, a phylogenetic reconstruction was created using PartitionFinder to test the best evolutionary model for both the 13 protein coding genes and two non-coding ribosomal genes. Following this, a phylogenetics analysis was conducted using two methods: Maximum-Likelihood approximation (IQTree) including bootstrap branching support and Bayesian inference (MrBayes) applying the partitions models obtained. By doing so, it was possible to find the molecular position of *A. iberus* within the Cyprinodontiformes order. The results showed *A. iberus* grouped together with *Orestias ascotanensis* and both species formed a sister group with North American and Caribbean cyprinodontiformes. These new data will be valuable for a better understanding of the evolution of *A. iberus* and will be highly useful for future genetic studies.

Keywords: mitogenome; *Apricaphanius iberus*; cyprinodontiformes; phylogeny

Author Contributions: Conceptualization, A.L.-S., S.P. and I.D.; methodology, A.L.-S., T.L.N., S.P. and I.D.; software, A.L.-S., T.L.N. and S.P.; validation, S.P. and I.D.; formal analysis, A.L.-S., T.L.N., S.P. and I.D.; investigation, A.L.-S., T.L.N. and S.P.; resources, A.L.-S.; data curation, S.P. and I.D.; writing—original draft preparation, A.L.-S.; writing—review and editing, S.P., T.L.N. and I.D.; visualization, A.L.-S. and T.L.N.; supervision, S.P. and I.D.; project administration, I.D.; funding acquisition, I.D. All authors have read and agreed to the published version of the manuscript.

Funding: This research was funded by Ministerio de Ciencia e Innovación (PID2019-103936GB-C22).

Institutional Review Board Statement: Not applicable.

Informed Consent Statement: Not applicable.

Data Availability Statement: The data from this study will be stored in the Genbank database.

Conflicts of Interest: The authors declare no conflict of interest.

 *biology and life sciences forum*

Abstract

Distribution Update of the Endemic Vettonian Spined Loach *Cobitis vettonica* Doadrio & Perdices, 1997, in Portugal [†]

Anabel Perdices [1,*,‡] and Manuela Coelho [2]

1 Department of Biodiversity and Evolutionary Biology, Museo Nacional de Ciencias Naturales Madrid (MNCN, CSIC), 28006 Madrid, Spain
2 Centre for Ecology, Evolution and Environmental Changes (c3Ec), Universidade de Ciências de Lisboa, 1749-016 Lisbon, Portugal; mmcoelho@ciencias.ulisboa.pt
* Correspondence: aperdices@mncn.csic.es
† Presented at the IX Iberian Congress of Ichthyology, Porto, Portugal, 20–23 June 2022.
‡ Presenting author (Poster presentation).

Abstract: Current knowledge of the Iberian freshwater fish fauna distributions need to be updated. Species with localized distributions in places with difficult access are more susceptible to being overlooked. This situation is critically important for endemic species with restricted distribution and potentially threatened status, which should be included in the national Red List. Since the beginning of the XXI century, more than 15 new Iberian endemic freshwater fish species have been described or reassigned, with only half of them exhibiting some threatened criteria. Despite the updated distribution of the widespread species *C. paludica* and *C. vettonica* in Spain, the distribution of *C. vettonica* remains unknown in Portugal in spite of some indications, suggesting the presence of this species in Portugal. *Cobitis vettonica* Doadrio & Perdices, 1997, found in the Alagón river, a tributary of the Tagus in Spain, and was also found in the Águeda river, a tributary of the Duero drainage (Spain). In this study, we reported the Vettonian spined loach also known as the Alagón spined loach (*Cobitis vettonica*), for the first time in Portugal. We found *C. vettonica* specimens in three tributaries of the Tagus in Portugal: Erges, Aravil and Ponsul, in ten localities. These three Portuguese rivers are adjacent to the Alagón river in Spain (the type locality of *C. vettonica*). We found *C. vettonica* in sympatry with *C. paludica* in six localities of these three tributaries. *Cobitis vettonica* is endangered in the Spanish Red List, and its threatened status must be defined for Portuguese waters.

Keywords: Alagón spined loach; endemic species; endangered species; Iberian Peninsula; distribution

Citation: Perdices, A.; Coelho, M. Distribution Update of the Endemic Vettonian Spined Loach *Cobitis vettonica* Doadrio & Perdices, 1997, in Portugal. *Biol. Life Sci. Forum* **2022**, 13, 31. https://doi.org/10.3390/blsf2022013031

Academic Editor: Alberto Teodorico Correia

Published: 6 June 2022

Publisher's Note: MDPI stays neutral with regard to jurisdictional claims in published maps and institutional affiliations.

Author Contributions: Conceptualization, A.P. and M.C.; formal analysis, A.P.; writing—review and editing—A.P. All authors have read and agreed to the published version of the manuscript.

Funding: The study was funded by FCT project FRH/BPD/14637/2003.

Institutional Review Board Statement: Not applicable.

Informed Consent Statement: Not applicable.

Data Availability Statement: Not applicable.

Conflicts of Interest: The authors declare no conflict of interest.

Abstract

The Family Ariidae (Siluriformes) in the New World: Composition and Species Concentration Areas [†]

Arturo Acero Pizarro [‡]

Instituto para el Estudio de las Ciencias del Mar (Cecimar), Universidad Nacional de Colombia sede Caribe, El Rodadero, Santa Marta 111321, Colombia; aacerop@unal.edu.co

† Presented at the IX Iberian Congress of Ichthyology, Porto, Portugal, 20–23 June 2022.

‡ Presenting author (Oral communication).

Abstract: Ariidae (sea catfishes) is one of the two catfish (Siluriformes) families that occur frequently in tropical marine and estuarine waters. There are at least 130 ariid species worldwide, but when samples are carefully studied, they tend to yield cryptic new species, given the group´s extremely conservative morphology. Mouth brooding is one of the synapomorphies of the family and is, consequently, responsible for the strong speciation within the group. Sea catfishes comprise two subfamilies: Galeichthyinae, including *Galeichthys peruvianus*, which live mainly in South Africa; *G. peruvianus* is endemic to Peru; and is, therefore, an interesting biogeographic enigma. The subfamily Ariinae, on the other hand, occurs along the family distribution, inhabiting continental tropical shores influenced by freshwater and with abundant mangrove development. The closing of the Tethys Sea sharply separated the Ariinae into two geographic groups that do not share any genus. The Old World, including African, Asian, and Australian–New Guinean waters, comprises almost 30 genera and 60 species. Eleven genera and at least 70 species are known from the New World. Seven genera and forty-one species occur in the Tropical Eastern Pacific, while from the Caribbean, eight genera and thirty species are known. The main New World regions where sea catfish species are concentrated are the Panama Bight, including Southern Caribbean, and the area between the mouths of the Orinoco and Amazon rivers.

Keywords: sea catfishes; biogeography; Tropical Eastern Pacific

Citation: Acero Pizarro, A. The Family Ariidae (Siluriformes) in the New World: Composition and Species Concentration Areas. *Biol. Life Sci. Forum* **2022**, *13*, 33. https://doi.org/10.3390/blsf2022013033

Academic Editor: Alberto Teodorico Correia

Published: 6 June 2022

Funding: This research received no external funding.

Institutional Review Board Statement: Not applicable.

Informed Consent Statement: Not applicable.

Data Availability Statement: Not applicable.

Conflicts of Interest: The author declares no conflict of interest.

 *biology and life sciences forum*

Abstract

Genetic and Morphological Identification of Formalin Fixed Larval Fishes: How Long Is Too Long? [†]

Anthony G. Miskiewicz [1,2,*,‡], Sharon A. Appleyard [3], Safia Maher [3], Ana Lara-Lopez [4], Paloma Matis [1,5], Stewart Fielder [6] and Iain Suthers [1,5]

[1] School of Biological, Earth & Environmental Sciences, University of New South Wales, Sydney, NSW 2052, Australia; paloma.matis@sims.org.au (P.M.); i.suthers@unsw.edu.au (I.S.)

[2] Ichthyology Department, Australian Museum Research Institute, Australian Museum, Sydney, NSW 2010, Australia

[3] CSIRO Australian National Fish Collection, National Research Collections Australia, Hobart, TAS 7000, Australia; sharon.appleyard@csiro.au (S.A.A.); safia.maher@csiro.au (S.M.)

[4] Institute for Marine and Antarctic Studies, University of Tasmania, Hobart, TAS 7000, Australia; ana.lara@utas.edu.au

[5] Sydney Institute of Marine Science, Chowder Bay Road, Mosman, NSW 2088, Australia

[6] NSW Department of Primary Industries, Port Stephens Fisheries Institute, Taylors Beach, NSW 2316, Australia; stewart.fielder@dpi.nsw.gov.au

[*] Correspondence: tonymisk@yahoo.com

[†] Presented at the IX Iberian Congress of Ichthyology, Porto, Portugal, 20–23 June 2022.

[‡] Presenting author (Poster presentation).

Citation: Miskiewicz, A.G.; Appleyard, S.A.; Maher, S.; Lara-Lopez, A.; Matis, P.; Fielder, S.; Suthers, I. Genetic and Morphological Identification of Formalin Fixed Larval Fishes: How Long Is Too Long? *Biol. Life Sci. Forum* **2022**, *13*, 69. https://doi.org/10.3390/blsf2022013069

Academic Editor: Alberto Teodorico Correia

Published: 8 June 2022

Publisher's Note: MDPI stays neutral with regard to jurisdictional claims in published maps and institutional affiliations.

Abstract: Identification of larvae of fish is usually based on assembling a developmental series of wild-collected larvae, usually formalin preserved, using pigmentation patterns, morphology and fin meristics. However, for many species, larvae are still undescribed, or there are only limited descriptions of larvae development. Formalin fixation of larval fish was previously thought to prevent genetic sequencing compared to ethanol-preserved larvae. In this poster, we detail the results of an integrative taxonomic approach based on morphology, imaging and DNA barcoding of the mitochondrial (mtDNA) cytochrome c oxidase subunit (COI) gene. We used this approach in both cultured yellow tail kingfish, *Seriola lalandi* and wild-sourced fish larvae fixed in 5% formalin. DNA barcoding and genetic species identification were 100% successful in *S. lalandi* fixed in formalin for up to 6 months, while barcoding of wild-caught fish larvae enabled species identification of 93% for up to 8-week formalin-fixed specimens. While COI genetic identifications from the in-field experiments were patchier than the controlled experiments, our study highlights the possibility of recovering suitable DNA from formalin-fixed larvae for up to six months. This was achieved by applying DNA extraction methods that use de-cross-linking steps and species identification based on both full-length reference and mini-barcodes. Our study provides a practical framework for undertaking both morphological and genetic identification to document the larval development of previously undescribed species from historic and current formalin-fixed samples collected around southern Australia.

Keywords: fish larvae; southern Australia; cytochrome c oxidase subunit I (COI) barcoding; formalin fixation

Author Contributions: Conceptualization, A.G.M., S.A.A., A.L.-L., P.M., S.F. and I.S.; methodology, A.G.M., S.A.A., A.L.-L., P.M. and I.S.; validation, A.G.M., S.A.A. and S.M.; formal analysis, S.A.A.; investigation, A.G.M., S.A.A., A.L.-L. and S.F.; resources, S.A.A. and S.F.; data curation, A.G.M., S.A.A. and S.M.; writing—original, draft preparation, S.A.A.; review and editing, A.G.M., S.A.A., S.M., A.L.-L., P.M., S.F. and I.S. All authors have read and agreed to the published version of the manuscript.

Funding: This research was funded by the Commonwealth Scientific Industrial Research Organisation, Sydney Institute of Marine Science and co-funding from the University of New South Wales, NSW Recreational Fishing Trrust and Australia's Integrated Marine Observing System (IMOS). Data was sourced from IMOS–IMOS is enabled by the National Collaborative Research Infrastructure Strategy (NCRIS).

Institutional Review Board Statement: Larval spawning and handling were undertaken at the Port Stephens research institute under the NSW Department of Primary Industries (Fisheries), Animal Care and Ethics Permits 93-1 and 93-3 for Larval Fish Rearing and Marine Fish Breeding respectively. Larval fish collections at the IMOS national reference stations were collected under ethics permit UNSW ACEC 19/96B.

Informed Consent Statement: Not applicable.

Data Availability Statement: The data presented in this study are available in https://doi.org/10.1016/j.jembe.2022.151763 (accessed on 27 May 2022).

Conflicts of Interest: The authors declare no conflict of interest.

biology and life sciences forum

Abstract

Evidence of High Levels of Gene Flow in a Widely Distributed Catadromous Species: The Thin-Lippedgrey Mullet [†]

Esmeralda Pereira [1,*,‡], Catarina Sofia Mateus [1], Maria Judite Alves [2], Rita Almeida [1], Joana Pereira [1], Bernardo Ruivo Quintella [3,4] and Pedro Raposo de Almeida [1,5]

[1] MARE—Centro de Ciências do Mar e do Ambiente, Universidade de Évora, 7002-554 Évora, Portugal; cspm@uevora.pt (C.S.M.); ritapalmeida93@gmail.com (R.A.); joana_gomespereira@hotmail.com (J.P.); pmra@uevora.pt (P.R.d.A.)
[2] Museu Nacional de História Natural e da Ciência & Centre for Ecology, Evolution and Environmental Changes (cE3c), University of Lisbon, Rua da Escola Politécnica 58, 1250-102 Lisbon, Portugal; mjalves@museus.ulisboa.pt
[3] MARE—Centro de Ciências do Mar e do Ambiente, Faculdade de Ciências, Universidade de Lisboa, 1749-016 Lisboa, Portugal; bsquintella@fc.ul.pt
[4] Departamento de Biologia Animal, Faculdade de Ciências, Universidade de Lisboa, 1749-016 Lisboa, Portugal
[5] Departamento de Biologia, Escola de Ciências e Tecnologia, Universidade de Évora, 7002-554 Évora, Portugal
* Correspondence: ecdp@uevora.pt
† Presented at the IX Iberian Congress of Ichthyology, Porto, Portugal, 20–23 June 2022.
‡ Presenting author (oral communication).

Citation: Pereira, E.; Mateus, C.S.; Alves, M.J.; Almeida, R.; Pereira, J.; Quintella, B.R.; Almeida, P.R.d. Evidence of High Levels of Gene Flow in a Widely Distributed Catadromous Species: The Thin-Lippedgrey Mullet. *Biol. Life Sci. Forum* **2022**, *13*, 91. https://doi.org/10.3390/blsf2022013091

Academic Editor: Alberto Teodorico Correia

Published: 15 June 2022

Publisher's Note: MDPI stays neutral with regard to jurisdictional claims in published maps and institutional affiliations.

Abstract: The thin-lipped grey mullet *Chelon ramada* (Risso, 1827) is a catadromous species that is distributed along the Northeast Atlantic, from the Norwegian coastline down to Mauritania, on the African coast (20–60° N, 18° E–42° W), and displays diverse patterns of habitat use and migratory behaviors. This widely distributed species is observed in large shoals throughout coastal areas and in brackish and freshwater environments, yet no previous studies have addressed the population's genetic structure. To study the patterns of genetic variation, gene flow and connectivity in the *C. ramada* distribution range), 457 fin clips sampled from 14 locations (Portuguese coast, Bay of Biscay, North seas, Celtic sea, Western Mediterranean and Eastern Mediterranean) were genotyped using 11 microsatellite DNA markers. No significant genetic differentiation among locations or geographic clustering of samples was observed, which points towards the existence of a unique genetic group. The results suggest strong gene flow from the Western Mediterranean to the Portuguese coast (Nm = 1) and vice versa (Nm = 0.87). The Portuguese coast has displayed the highest values of gene flow with all the sampling sites ([0.4–0.6]) whereas Northeast Atlantic coast and Eastern Mediterranean maintained symmetrical lower values of gene flow that ranged between [0.20–0.30]. The present study provides evidence that high levels of gene flow are maintained within the distribution range, contributing to the existence of a panmictic population.

Keywords: catadromy, panmixia; genetic structure; connectivity; dispersion; microsatellites; *Chelon ramada*

Funding: This work was also supported by National Funds through FCT—Foundation for Science and Technology via the project UIDB/04292/2020, by the FCT doctoral grants cofunded by FSE though the "Programa Operacional regional Alentejo" attributed to E. Pereira (SFRH/BD/121042/2016) as well as through the individual contract attributed to Catarina S. Mateus within the project "EVOLAMP—Genomic footprints of the evolution of alternative life histories in lampreys" (PTDC/BIA-EVL/30695/2017), and Bernardo R.Quintella (2020.02413.CEECIND).

Institutional Review Board Statement: Not applicable.

Informed Consent Statement: Not applicable.

Data Availability Statement: Not applicable.

Conflicts of Interest: The authors declare no conflict of interest.

biology and life sciences forum

Abstract

Genetic Structure and Diversity of Brown Trout (*Salmo trutta* L.) in Portugal [†]

Joana Pereira [1,*,‡], Sara Silva [1], Pedro R. Almeida [1,2], Carlos M. Alexandre [1], Rita Almeida [1], Andreia Domingues [1], Maria Judite Alves [3] and Catarina S. Mateus [1,*]

[1] MARE—Marine and Environmental Sciences Centre/ARNET—Aquatic Research Network Universidade de Évora, 7002-554 Évora, Portugal; ssrs@uevora.pt (S.S.); pmra@uevora.pt (P.R.A.); cmea@uevora.pt (C.M.A.); ritapalmeida93@gmail.com (R.A.); afdomingues@uevora.pt (A.D.)

[2] Departamento de Biologia, Escola de Ciência e Tecnologia, Universidade de Évora, 7002-554 Évora, Portugal

[3] MUHNAC—Museu Nacional de História Natural e da Ciência & Natural History and Systematics (NHS) Research Group, CE3C—Centre for Ecology, Evolution and Environmental Changes, Universidade de Lisboa, 1749-016 Lisboa, Portugal; mjalves@museus.ulisboa.pt

* Correspondence: joana_gomespereira@hotmail.com (J.P.); cspm@uevora.pt (C.S.M.)

† Presented at the IX Iberian Congress of Ichthyology, Porto, Portugal, 20–23 June 2022.

‡ Presenting author (oral communication).

Abstract: Population genetic studies have been extensively used as tools for the management and conservation of salmonid species and related habitats. The brown trout, *Salmo trutta* (Linnaeus 1758), is one of the most studied species within its family, and is frequently used as a population model. It can have a highly complex and variable life history, often presenting a migratory ecotype (i.e., sea trout), and it is considered an indicator of the quality of aquatic ecosystems. Moreover, it has a high socioeconomic value for commercial and recreational fishing. The destruction or alteration of aquatic habitats, over-exploitation, exotic species and climate change are some of the factors that threaten the sustainability of the species in Portugal. To analyse the genetic structure, gene flow and connectivity patterns among Portuguese brown trout populations, we sampled approximately 392 individuals from 15 sites across the distribution of the species in its national territory. DNA fingerprinting methodologies were carried out to determine the structural differences between populations, using a set of microsatellite loci developed for salmonids. The overall results suggest significant genetic differences between the populations sampled. This study has enabled a breakthrough in understanding the genetic structure of *Salmo trutta* populations in the southern limit of the species' global distribution, assessing the impact of natural and human factors on the genetic structure of its populations, and consequently developing mitigation measures for the effective management and conservation of the species.

Keywords: brown trout; microsatellites; genetic structure; management and conservation

Citation: Pereira, J.; Silva, S.; Almeida, P.R.; Alexandre, C.M.; Almeida, R.; Domingues, A.; Alves, M.J.; Mateus, C.S. Genetic Structure and Diversity of Brown Trout (*Salmo trutta* L.) in Portugal. *Biol. Life Sci. Forum* **2022**, *13*, 104. https://doi.org/10.3390/blsf2022013104

Academic Editor: Alberto Teodorico Correia

Published: 16 June 2022

Publisher's Note: MDPI stays neutral with regard to jurisdictional claims in published maps and institutional affiliations.

Author Contributions: Conceptualization, C.M.A., C.S.M., P.R.A. and M.J.A.; methodology, C.S.M. and M.J.A.; Sampling: S.S., R.A., A.D. and J.P.; Laboratorial analysis: J.P. and R.A.; Data analysis: J.P., R.A. and C.S.M.; Resources: C.S.M., P.R.A., C.M.A. and M.J.A.; writing—original draft preparation, J.P., C.S.M. and C.M.A.; writing—review and editing, J.P., C.S.M., C.M.A., M.J.A., R.A., S.S., A.D. and P.R.A.; supervision, C.S.M. and C.M.A.; project administration, C.S.M.; funding acquisition, C.M.A., C.S.M. and P.R.A. All authors have read and agreed to the published version of the manuscript.

Funding: This research was funded by INTERREG ATLANTIC AREA via the project DiadES—Assessing and enhancing ecosystem services provided by diadromous fish in a climate change context (EAPA_18/2018)). Support was also provided by the Portuguese Foundation for Science and Technology (FCT) through the strategy plan for MARE (Marine and Environmental Sciences Centre), via project UIDB/04292/2020, and under the project LA/P/0069/2020 granted to the Associate Laboratory ARNET. FCT also supported this study through the individual contract attributed to

Carlos M. Alexandre within the project CEECIND/02265/2018, and the PhD scholarships attributed to Sara Silva (2021.05558.BD) and Andreia Domingues (2021.05644.BD).

Institutional Review Board Statement: Not applicable.

Informed Consent Statement: Not applicable.

Conflicts of Interest: The authors declare no conflict of interest.

Abstract

The Present Genetic Structure of Vendace (*Coregonus albula* (L.)): Populations in Latvian Lakes as a Result of Its Management and Conservation [†]

Jelena Oreha and Natalja Škute *,[‡]

Institute of Life Sciences and Technology, Daugavpils University, LV-5401 Daugavpils, Latvia; jelena.oreha@du.lv
* Correspondence: natalja.skute@du.lv; Tel.: +371-29185513
† Presented at the IX Iberian Congress of Ichthyology, Porto, Portugal, 20–23 June 2022.
‡ Presenting author (Poster presentation).

Abstract: The European vendace (*Coregonus albula* (L.)) is often treated as one of the glacial relicts of the animal world, and it is a widespread species in the waters of the Holarctic. Together with whitefish and salmonid, vendace belongs to a list of economically valuable fish species. In Latvian lakes, its share in the fishery is not big, and it is insignificant and unstable. The objectives of this study were to reveal the genetic structure and variability of the vendace gene pool in Latvian Lakes and to evaluate the influence of vendace translocations on the flourishment of a population and following a decline. The present research used nine microsatellite markers to study vendace populations from nine Latvian Lakes. The indices of genetic variation include the following: number and frequency of alleles at a locus, occurrence of private alleles, observed and expected heterozygosity levels, and the richness of alleles and private alleles were determined. Additionally, the genetic structure and differentiation of the populations were accessed. Investigated vendace populations have a high level of expected heterozygosity, with a high mean allelic richness and private allelic richness in Lake Riču suggesting that this vendace population is indigenous. Three clustering methods reveal similar groupings in three genetic groups. At the present time, the vendace populations of the investigated Latvian lakes seem to be a "mixture" of several populations and, therefore, may not be fully indigenous. The level of genetic variability differs among the studied populations. Such differences may be caused by the consequences of the translocations and genetic drift, which influence allele frequencies in different ways and could be driven by some environmental factors. The results of the study allow the acceptance of each of the studied populations as a different management unit and prompts the development of an optimal strategy for their effective conservation and management.

Keywords: population genetics; fish transfer; indigenous population; divergence; translocation

Citation: Oreha, J.; Škute, N. The Present Genetic Structure of Vendace (*Coregonus albula* (L.)): Populations in Latvian Lakes as a Result of Its Management and Conservation. *Biol. Life Sci. Forum* **2022**, *13*, 110. https://doi.org/10.3390/blsf2022013110

Academic Editor: Alberto Teodorico Correia

Published: 16 June 2022

Publisher's Note: MDPI stays neutral with regard to jurisdictional claims in published maps and institutional affiliations.

Author Contributions: Methodology, Validation, Formal analysis, Investigation, Data Curation, Writing—Original Draft, Visualization, J.O.; Conceptualization, Resources, Review & Editing, Supervision, Project administration, funding acquisition, N.Š. All authors have read and agreed to the published version of the manuscript.

Funding: This research received no external funding.

Institutional Review Board Statement: Not applicable.

Informed Consent Statement: Not applicable.

Data Availability Statement: The datasets generated during and/or analyzed during the current study are available from the corresponding author on a reasonable request.

Conflicts of Interest: The authors declare no conflict of interest.

biology and life sciences forum

Abstract

Endemic Sicilian Brown Trout Endangered by Hatchery Introgression and Low Gene Diversity [†]

Nuria Sanz [1,*,‡], **Antonino Duchi** [2] **and Monica Giampiccolo** [3]

[1] Department of Biology, Universitat de Girona, 17003 Girona, Spain
[2] Giordano Bruno 8, 97100 Ragusa, Italy; carrua@tin.it
[3] Via Saragat, 97100 Ragusa, Italy; dott.mgiampiccolo@gmail.com
[*] Correspondence: nuria.sanz@udg.edu
[†] Presented at the IX Iberian Congress of Ichthyology, Porto, Portugal, 20–23 June 2022.
[‡] The presenting author (Poster Presentation).

Abstract: Brown trout (*Salmo trutta*) populations living in Mediterranean peninsulas and the surrounding islands belong to old evolutionary lineages that persisted during Quaternary glaciations. Many of these populations inhabit marginal areas along the south limit of the distribution of the species, where they face hard environmental conditions (drought, high temperatures and sudden shifts in water flow) that can get worse by anthropogenic activities and climate change. In islands, this vulnerable situation is exacerbated by geographical isolation. Sicilian trout remain only in the south-eastern part of the island and, based on their morphological characteristics, they have been recognized as a distinct species named *Salmo cetti*. Despite their genetic singularity among other Italian trout (a unique Italian native trout of Atlantic origin), the mitochondrial DNA haplotypes that were found in Sicily are clearly related with the brown trout Atlantic lineages from North Africa and the Iberian Peninsula. In the present study, brown trout in four rivers from north-eastern Sicily were genetically analysed. Based on the genotyping of mitochondrial (control region) and nuclear (LDHC, GP1, GP14, GP31, GP34, GP37, GP38, GP42, SS2 and SL) genes, this study aimed to: evaluate the impact of past stocking practices in natural populations; estimate mitochondrial and nuclear gene diversity; and reconstruct the phylogenetic relationships of Sicilian trout. The initial results showed that only trout from the Tellessimo River remain free from hatchery introgression. Gene diversity was low in most populations, and both mitochondrial and nuclear phylogenetic reconstruction related Sicilian trout with old Atlantic lineages.

Keywords: introgression; mitochondrial DNA; phylogenetics; *Salmo trutta*; *Salmo cetti*

Citation: Sanz, N.; Duchi, A.; Giampiccolo, M. Endemic Sicilian Brown Trout Endangered by Hatchery Introgression and Low Gene Diversity. *Biol. Life Sci. Forum* **2022**, *13*, 133. https://doi.org/10.3390/blsf2022013133

Academic Editor: Alberto Teodorico Correia

Published: 20 June 2022

Publisher's Note: MDPI stays neutral with regard to jurisdictional claims in published maps and institutional affiliations.

Author Contributions: Conceptualization: A.D. and N.S.; sampling: A.D. and M.G.; laboratory and software; N.S.; validation: A.D., N.S. and M.G.; formal analysis: N.S.; writing—original draft preparation: N.S.; review and editing: A.D. and N.S.; funding acquisition, A.D. and N.S. All authors have read and agreed to the published version of the manuscript.

Funding: This study was partially funded by PO FEAMP 2014/2020 Misura a titolarità 2.51.

Institutional Review Board Statement: Ethical review and approval were waived for this study and not applicable.

Informed Consent Statement: Not applicable.

Data Availability Statement: All data is available under request.

Conflicts of Interest: The authors declare no conflict of interest.

Abstract

Analysis of the Efficiency of the Taxonomic Identification of Small Fishes Using Artificial Neural Networks: A CASE Study of the Ichthyofauna of the Carajás Mountain (Pará—Northern Brazil) [†]

Lays C. L. Nogueira [1], Rafael Schroeder [1,2,‡], Rodrigo Sant'Ana [1,*] and Antônio C. Beaumord [3]

[1] Laboratório de Estudos Marinhos Aplicados, Escola do Mar, Ciência e Tecnologia, Universidade do Vale do Itajaí (UNIVALI), Rua Uruguai, 458-Centro, Itajaí 88302-901, Brazil; layslemes.nogueira@gmail.com (L.C.L.N.); schroeder@edu.univali.br (R.S.)
[2] Centro Interdisciplinar de Investigação Marinha e Ambiental (CIIMAR), Terminal de Cruzeiros do Porto de Leixões, Avenida General Norton de Matos S/N, 4550-208 Matosinhos, Portugal
[3] Laboratório de Estudos de Impactos Ambientais, Escola do Mar, Ciência e Tecnologia, Universidade do Vale do Itajaí (UNIVALI), Rua Uruguai, 458-Centro, Itajaí 88302-901, Brazil; beaumord@univali.br
* Correspondence: rsantana@univali.br
† Presented at the IX Iberian Congress of Ichthyology, Porto, Portugal, 20–23 June 2022.
‡ Presenting author (Poster presentation).

Citation: Nogueira, L.C.L.; Schroeder, R.; Sant'Ana, R.; Beaumord, A.C. Analysis of the Efficiency of the Taxonomic Identification of Small Fishes Using Artificial Neural Networks: A CASE Study of the Ichthyofauna of the Carajás Mountain (Pará—Northern Brazil). *Biol. Life Sci. Forum* **2022**, *13*, 102. https://doi.org/10.3390/blsf2022013102

Academic Editor: Alberto Teodorico Correia

Published: 16 June 2022

Publisher's Note: MDPI stays neutral with regard to jurisdictional claims in published maps and institutional affiliations.

Abstract: The development of techniques that assist in the processes of taxonomic identification is of utmost importance, considering the scarcity of specialists and literature available in remote and diverse areas. Environmental studies such as the Biodiversity Monitoring Program of the Carajás National Forest in northern Brazil (FLONA de Carajás—PA, 6°6′29″ S, 50°18′16″ W) face challenges in this regard. These challenges include the particularity of the morphological and evolutionary characteristics of the fauna, present in a very diverse area of intense anthropogenic intervention by the use of resources of economic interest. Thus, this work sought to analyze the efficiency of using Artificial Neural Networks (ANN), more specifically the "XCeption" algorithm, configured for the taxonomic identification of samples captured during this monitoring program. These samples were previously identified using traditional taxonomic identification keys. The taxa *Aequidens tetramerus*, *Astyanax abramis*, *Bryconops* spp., *Knodus* spp., and *Moenkhausia* spp. were used. After capturing the images, the content was assigned to different folders, named "Training" and "Test". This procedure seeks to quantify the model's ability to classify data characteristically different from that presented in the training base. The accuracy results obtained during the training phase of the algorithm used, executed in about 16 hours, were 98% for the Training phase and 92% for the Validation phase, with some categories presenting better prediction results, such as classes 4 (100%) and 2 (85%). The testing phase, executed in about 1 hour, obtained an accuracy value of 78.26%, with a 95% confidence interval (63.64–89.05%) and a *kappa* of 70%. The applied methodology presented high accuracy, configuring itself as an important tool for identifying fish species in extremely diverse and remote environments.

Keywords: deep learning; morphology; cichlidae; characidae; iguanodectidae; FLONA de Carajás (PA)

Author Contributions: Conceptualization, A.C.B. and R.S. (Rodrigo Sant'Ana); methodology, R.S. (Rodrigo Sant'Ana) and L.C.L.N.; formal analysis, R.S. (Rodrigo Sant'Ana) and L.C.L.N.; investigation, A.C.B., L.C.L.N., R.S. (Rodrigo Sant'Ana) and R.S. (Rafael Schroeder); resources, A.C.B.; data curation, A.C.B.; writing—original draft preparation, L.C.L.N.; writing—review and editing, A.C.B., L.C.L.N., R.S. (Rodrigo Sant'Ana) and R.S. (Rafael Schroeder); supervision, A.C.B. and R.S. (Rodrigo Sant'Ana); project administration, A.C.B. All authors have read and agreed to the published version of the manuscript.

Funding: This research received no external funding.

Institutional Review Board Statement: Not applicable.

Informed Consent Statement: Not applicable.

Data Availability Statement: The data that support the findings of this study are available from the corresponding author upon reasonable request.

Conflicts of Interest: The authors declare no conflict of interest.

Abstract

Fish Assemblage Response to Eutrophic-Mediated Environmental Stress Events in the Mar Menor Coastal Lagoon (SE of Spain) [†]

Antonio Zamora-López [*,‡], **Adrián Guerrero-Gómez, Mar Torralva, José Manuel Zamora-Marín, Antonio Guillén-Beltrán, Patricio López-Martínez de la Plaza and Francisco José Oliva-Paterna**

Department of Zoology and Physical Anthropology, Faculty of Biology, University of Murcia, 30100 Murcia, Spain; adrian.guerrero@um.es (A.G.-G.); torralva@um.es (M.T.); josemanuel.zamora@um.es (J.M.Z.-M.); antonio.guillenb@um.es (A.G.-B.); patricio.lopez.martinez.plaza@gmail.com (P.L.-M.d.l.P.); fjoliva@um.es (F.J.O.-P.)
* Correspondence: antonio.zamora2@um.es; Tel.: +34-677-809-208
† Presented at the IX Iberian Congress of Ichthyology, Porto, Portugal, 20–23 June 2022.
‡ Presenting author: (Poster).

Abstract: In the last two decades, the extensive development of intensive irrigated agriculture around the Mar Menor coastal lagoon (SE Iberian Peninsula) has disturbed the characteristics of this initially oligotrophic ecosystem. The organic pollution that flows into the lagoon from agriculture and, to a lesser extent, from urban wastewater, has triggered the eutrophication of the system, resulting in dystrophic crises and mass mortality events of aquatic fauna. In order to explore temporal changes in the ecological quality of the shallow areas of the Mar Menor, fish assemblage parameters have been obtained and integrated into a multimetric fish index (modified-EMFI). A total of 18 sampling sites around the lagoon perimeter were seasonally assessed. Seasonal sampling was carried out in three two-years periods: (a) the reference period (2002–2004); (b) the critical eutrophic first period (2018–2019); and (c) the critical eutrophic second period with multispecies mass mortality events (2020–2021). The effect of sampling-site confinement (three levels of water-renewal time) and the degree of anthropic pressure on shallow areas were evaluated. Despite the magnitude of the eutrophication impact on the lagoon, the shallow areas' ecological quality (according to the modified-EMFI) did not show a drastic drop. Nevertheless, significantly lower values were detected in the critical eutrophic second period. The level of confinement showed effects on the ecological quality of the shallow areas, although the effects associated with the degree of anthropic pressure at site-approach were not significant. These results suggest that the shallow areas of the lagoon, beyond their usual function as recruitment areas for juvenile fish, could be playing a complementary role as fish refuge habitats and buffering against the long-term eutrophication process that is affecting the lagoon. However, fish assemblages in shallow areas could be undergoing significant changes in their structure and composition. We understand that the loss of the taxonomic and functional integrity of the shallow-area fish assemblage has repercussions on the dynamics of the fish populations present throughout the lagoon, and can compromise the recovery of this ecosystem.

Keywords: fish assemblage; shallow areas; ecological quality; multimetric fish index; eutrophication

Citation: Zamora-López, A.; Guerrero-Gómez, A.; Torralva, M.; Zamora-Marín, J.M.; Guillén-Beltrán, A.; López-Martínez de la Plaza, P.; Oliva-Paterna, F.J. Fish Assemblage Response to Eutrophic-Mediated Environmental Stress Events in the Mar Menor Coastal Lagoon (SE of Spain). *Biol. Life Sci. Forum* **2022**, *13*, 5. https://doi.org/10.3390/blsf2022013005

Academic Editor: Alberto Teodorico Correia

Published: 2 June 2022

Publisher's Note: MDPI stays neutral with regard to jurisdictional claims in published maps and institutional affiliations.

Author Contributions: Conceptualization, A.Z.-L., M.T. and F.J.O.-P.; methodology, A.Z.-L., M.T. and F.J.O.-P.; software, A.G.-G.; validation, A.Z.-L., A.G.-G and P.L.-M.d.l.P.; formal analysis, A.Z.-L. and A.G.-G.; investigation, A.Z.-L., J.M.Z.-M. and A.G.-G.; resources, A.Z.-L. and A.G.-G.; data curation, A.Z.-L. and A.G.-B.; writing—original draft preparation, A.Z.-L.; writing—review and editing, J.M.Z.-M., M.T., A.G.-G. and F.J.O.-P.; visualization, J.M.Z.-M.; supervision, M.T. and F.J.O.-P.; project administration, M.T.; funding acquisition, M.T. and F.J.O.-P. All authors have read and agreed to the published version of the manuscript.

Funding: Part of this research was supported by the Environmental Service and Mar Menor and Service of the Autonomous Government of Murcia (Spain). Proyect (04812210001) "Study and spatio-temporal monitoring of priority faunal populations (ichthyofauna and the invasive species Callinectes sapidus): current situation in shallow and intertidal areas of the Mar Menor."

Institutional Review Board Statement: Not applicable.

Informed Consent Statement: Not applicable.

Data Availability Statement: The data presented in this study are available within the downloadable reports on this website https://canalmarmenor.carm.es/ciencia/estudios-de-investigacion/.

Conflicts of Interest: The authors declare no conflict of interest.

Abstract

Marine Recreational Fisheries in the Canary Islands: State of Knowledge, Preliminary Results [†]

Alberto Bilbao-Sieyro [1,*,‡], Yeray Pérez-González [1], Pablo Martín-Sosa [2], José Juan Castro-Hernández [3], David Jiménez-Alvarado [3] and José J. Pascual-Fernández [4]

1   GMR Canarias, S.A.U., 35259 Ingenio, Spain; yeraypg@gmrcanarias.com
2   Instituto Español de Oceanografía (CNIEO-CSIC), Centro Oceanográfico de Canarias, 38180 Santa Cruz de Tenerife, Spain; pablo.martin-sosa@ieo.csic.es
3   Instituto Universitario Ecoaqua, ULPGC, 35017 Las Palmas de Gran Canaria, Spain; jose.castro@ulpgc.es (J.J.C.-H.); david.jimenez@ulpgc.es (D.J.-A.)
4   Instituto Universitario de Investigación Social y Turismo (ISTUR), Universidad de La Laguna, Apartado 456, 38205 La Laguna, Spain; jpascual@ull.edu.es
*   Correspondence: albertobs@gmrcanarias.com; Tel.: +34-928-385-078
†   Presented at the IX Iberian Congress of Ichthyology, Porto, Portugal, 20–23 June 2022.
‡   Presenting author (Oral communication).

Citation: Bilbao-Sieyro, A.; Pérez-González, Y.; Martín-Sosa, P.; Castro-Hernández, J.J.; Jiménez-Alvarado, D.; Pascual-Fernández, J.J. Marine Recreational Fisheries in the Canary Islands: State of Knowledge, Preliminary Results. *Biol. Life Sci. Forum* **2022**, *13*, 7. https://doi.org/10.3390/blsf2022013007

Academic Editor: Alberto Teodorico Correia

Published: 2 June 2022

Publisher's Note: MDPI stays neutral with regard to jurisdictional claims in published maps and institutional affiliations.

Abstract: According to the data collection framework (DFC, Regulation (EU) 2017/1004) for marine recreational fisheries (MRF), the compilation of biological, environmental, technical and socioeconomic data is essential. However, the MRF in the Canary Islands does not have official catch statistics even though there are around 90,000 recreational fishers unequally distributed among the eight islands along 1500 km of coastline. Nevertheless, this has not been an obstacle for the research groups that have made notable efforts to infer the role of MRF in Canary Island fishery. We found 54 research references (2000–2021) in this regard. The studies were classified according to Pita et al. (2020). Most of the references (22) corresponded to peer-reviewed articles. The key results were mainly related to catch and effort estimates (26) and there were few publications referring to dissemination (3) and legislation (1). The data collection techniques included: face to face interviews, on-site data gathering and telephonic and online surveys. Is noteworthy that the catch/effort data is directly taken from fishers in only three publications (peer-reviewed). On a regional basis, only one work considers all modalities (coast, boat and spearfishing) and the other two are about spearfishing with significant differences regarding total catch estimates. As many authors have pointed out, most or all these methodologies are based on non-probabilistic samples or are specific to one island. Consequently, inferences must be taken with caution. The research carried out so far has been opportunistic (research groups) and with a short-term vision (administrations) to respond to specific needs. However, as in any fishery assessment, a long-term vision is necessary. Since January 2021 an MRF working group was created to contribute to the development of DFC at the national level. It would be a big step if the research groups could detail a roadmap with unified methodologies. In this way, the management of public resources would be more efficient. This effort should be led by the two fishing administrations (national and regional) of the Canary Islands. This work was developed within the framework of the PLASMAR+ Project (MAC2/1.1a/347), co-financed by the European Regional Development Fund (EDRF) and INTERREG V-A Spain-Portugal MAC 2014-2020.

Keywords: marine recreational fisheries; state of knowledge; Canary Islands

---

Author Contributions: Conceptualization, A.B.-S., Y.P.-G., P.M.-S., J.J.C.-H., D.J.-A. and J.J.P.-F.; methodology, A.B.-S.; validation, A.B.-S., Y.P.-G., P.M.-S., J.J.C.-H., D.J.-A. and J.J.P.-F.; formal analysis A.B.-S., Y.P.-G., P.M.-S., J.J.C.-H., D.J.-A. and J.J.P.-F.; investigation, A.B.-S., Y.P.-G., P.M.-S., J.J.C.-H., D.J.-A. and J.J.P.-F.; writing—original draft preparation, A.B.-S.; writing—review and editing, A.B.-S., Y.P.-G., P.M.-S., J.J.C.-H., D.J.-A. and J.J.P.-F. All authors have read and agreed to the published version of the manuscript.

Funding: This research was co-funded by the European Regional Development Fund (EDRF) and INTERREG V-A Spain-Portugal MAC 2014-2020.

Institutional Review Board Statement: Not applicable.

Informed Consent Statement: Not applicable.

Data Availability Statement: Not applicable.

Conflicts of Interest: The authors declare no conflict of interest.

biology and life sciences forum

Abstract

Another Tool for Chondrichthyan Ex Situ Conservation: First-Time *Chimaera monstrosa* Sperm Cryopreservation [†]

Pablo García-Salinas, Victor Gallego and Juan F. Asturiano *[,‡]

Grupo de Acuicultura y Biodiversidad, Instituto de Ciencia y Tecnología Animal, Universitat Politècnica de València, 46022 Valencia, Spain; pablo.g.salinas@outlook.com (P.G.-S.); vgalbiach@ualg.pt (V.G.)
* Correspondence: jfastu@dca.upv.es; Tel.: +34-96-387-93-85
† Presented at the IX Iberian Congress of Ichthyology, Porto, Portugal, 20–23 June 2022.
‡ Presenting author (Poster presentation).

Abstract: Chondrichthyans, which comprises elasmobranchs (sharks and rays) and holocephalans (chimaeras), are one of the most endangered group of vertebrates on the planet. Ex situ conservation programs, such as captive breeding, are tools that can be used to improve the status of some of the most sensitive species belonging to these groups. However, the use of reproductive techniques is necessary when planning sustainable breeding programs in controlled environments. In recent years, our group has described the protocols necessary to achieve viable sperm extraction and its cryopreservation in 13 species. However, the use of these techniques in the holocephalan group has not yet been explored. Here, the process of obtaining viable sperm in a holocephalan species, *Chimaera monstrosa*, is presented for the first time. The sperm was obtained from animals recovered from bottom trawling bycatch. It was possible to recover sperm from both males (n = 3), using cannulation and abdominal massage, and females (n = 2), by directly accessing their oviductal glands. Sufficient sperm was obtained from the males to apply cryopreservation protocols developed for elasmobranchs. For this purpose, the sperm was diluted in an extender for elasmobranchs (1 sperm:9 extender) previously developed by our group. The cryopreservation of sperm was achieved through the addition of different cryoprotectants to the extender: methanol, dimethyl sulfoxide (DMSO) and fresh egg yolk. Samples were frozen inside a Styrofoam box using vapor of liquid nitrogen and preserved in liquid nitrogen. Sperm quality was assessed by studying motility and membrane integrity post-thawing. The initial motility and membrane integrity values were close to 54%. The best post-thawing motility values were obtained with a combination of 5% DMSO, 5% methanol and 10% egg yolk, which induced motility values close to 25% and membrane integrity values close to 24%. This is the first time that sperm from this group of animals has been cryopreserved, expanding our knowledge on their reproductive biology and the tools available for their conservation.

Keywords: holocephalan; sperm extraction; cryopreservation; sharks; rays; assisted reproduction techniques

Citation: García-Salinas, P.; Gallego, V.; Asturiano, J.F. Another Tool for Chondrichthyan Ex Situ Conservation: First-Time *Chimaera monstrosa* Sperm Cryopreservation. *Biol. Life Sci. Forum* **2022**, *13*, 9. https://doi.org/10.3390/blsf2022013009

Academic Editor: Alberto Teodorico Correia

Published: 2 June 2022

Publisher's Note: MDPI stays neutral with regard to jurisdictional claims in published maps and institutional affiliations.

Author Contributions: Conceptualization, P.G.-S. and J.F.A.; methodology, P.G.-S., V.G.; writing—review and editing, P.G.-S. and J.F.A. All authors have read and agreed to the published version of the manuscript.

Funding: This research received no external funding.

Institutional Review Board Statement: All the samples were obtained from bycatch dead animals.

Informed Consent Statement: Not applicable.

Data Availability Statement: Experimental data can be demanded to the corresponding author.

Conflicts of Interest: The authors declare no conflict of interest.

biology and life sciences forum

Abstract

Microplastics Contamination of Large Pelagic Fish in the Open Atlantic Ocean [†]

Rúben Pereira [1,2,*,‡], Sabrina M. Rodrigues [1,2], Diogo Silva [1,2], Vânia Freitas [1], C. Marisa R. Almeida [1], António Camilo [3], Susana Barbosa [4], Eduardo Silva [4] and Sandra Ramos [1]

[1] Interdisciplinary Centre of Marine and Environmental Research (CIIMAR), University of Porto, 4450-208 Porto, Portugal; sabrinardmagalhaes@gmail.com (S.M.R.); silva.diogom@gmail.com (D.S.); vpfreitas@ciimar.up.pt (V.F.); calmeida@ciimar.up.pt (C.M.R.A.); ssramos@ciimar.up.pt (S.R.)

[2] Abel Salazar Institute of Biomedical Sciences (ICBAS), University of Porto, 4050-313 Porto, Portugal

[3] Portuguese Navy's Research Centre (CINAV), 2810-145 Almada, Portugal; antoniocamilo@gmail.com

[4] Institute for Systems and Computer Engineering, Technology and Science (INESCTEC), 4200-465 Porto, Portugal; susana.a.barbosa@inesctec.pt (S.B.); eduardo.silva@inesctec.pt (E.S.)

* Correspondence: ruben.pereira12@gmail.com

† Presented at the IX Iberian Congress of Ichthyology, Porto, Portugal, 20–23 June 2022.

‡ Presenting author (Poster presentation).

Citation: Pereira, R.; Rodrigues, S.M.; Silva, D.; Freitas, V.; Almeida, C.M.R.; Camilo, A.; Barbosa, S.; Silva, E.; Ramos, S. Microplastics Contamination of Large Pelagic Fish in the Open Atlantic Ocean. *Biol. Life Sci. Forum* **2022**, *13*, 11. https://doi.org/10.3390/blsf2022013011

Academic Editor: Alberto Teodorico Correia

Published: 2 June 2022

Publisher's Note: MDPI stays neutral with regard to jurisdictional claims in published maps and institutional affiliations.

Abstract: Fish are one of the most important components of the ocean, playing relevant ecological roles and providing several ecosystem services. Large migratory fish, such as tunas, mackerels and amberjacks, can function as valuable indicators of ocean health, since they are top predators and are exposed to several anthropogenic pressures, including pollution from different sources. Microplastics (MPs), small plastic particles (<5 mm), are ubiquitous throughout the world, occurring even in habitats with little anthropogenic pressure such as open sea waters. Taking advantage of the unique fish collection gathered by the NRP Sagres Crew during the 2020 Circumnavigation Expedition, biological samples of fish gastrointestinal tract and gills were collected and properly stored on board for further laboratorial analyses to assess MP contamination, using optimized protocols. MPs were characterized in terms of shape and color, and polymers were identified using FTIR. Seven fish were collected across the middle of the Atlantic Ocean, between the continents of Africa and South America, and along the South American coast. Three *Acanthocybium solandri*, two *Seriola lalandi*, one *Thunnus abacares* and one *Coryphaena* spp. were collected, with the total length ranging from 54 cm to 127 cm, and total weight from 1 kg to 11 kg, including adults and one juvenile (*S. lalandi*). A total of 124 MPs were observed in the gastrointestinal tract and gills, including 72% of fibers and 28% particles. Rayon was the most abundant polymer (25%), and a large majority MPs were blue (85%). Fibers were mainly Rayon (34%) and blue, while particles were mostly composed of polypropylene (71%). MPs were found in all fish, with an average of 18 ± 11 MPs per fish. In all sampled fish, both organs presented MPs with a mean number of 9 ± 5 MPs in the gills and 9 ± 6 MPs in the gut. These results demonstrate the ubiquitous occurrence of MPs throughout the world, even in remote areas such as the open Atlantic Ocean, and in top predators which are more prone to bioaccumulate pollutants. This study reinforces the need for further research regarding plastic pollution and MP contamination in species from higher trophic levels.

Keywords: microplastics; fish; pollution; bioaccumulation; open waters; Atlantic Ocean

Funding: This research was partially funded by Ocean3R (NORTE-01-0145-FEDER-000064) and ATLANTIDA (ref. NORTE-01-0145-FEDER-000040) projects, supported by the Norte Portugal Regional Operational Programme (NORTE 2020) under the PORTUGAL 2020 Partnership Agreement and through the European Regional Development Fund (ERDF). Additionally, FCT is awarded by the Strategic Funding UIDB/04423/2020 and UIDP/04423/2020 through national funds provided by FCT and ERDF, Ph.D. fellowship to R.P. (2021.04850.BD) S.M.R. (SFRH/BD/145736/2019), D.S. (2020.06088.BD) and a research contract to S.R. (DL57/2016/CP1344/CT0020).

Institutional Review Board Statement: Not applicable.

Informed Consent Statement: Not applicable.

Data Availability Statement: Not applicable.

Conflicts of Interest: The authors declare no conflict of interest.

Abstract

Fishers' Experiential Knowledge about Biological Traits of Commercial Marine Species in the Azores [†]

Régis Santos [1,*,‡], Morgan Casal-Ribeiro [1,2], Ualerson Iran Peixoto [1,2], Wendell Medeiros-Leal [1], Ana Nova-Pabon [1] and Mário Pinho [1,2]

[1] Okeanos-UAc Instituto de Investigação em Ciências do Mar, Universidade dos Açores, Rua Prof. Dr. Frederico Machado, 4, 9900-138 Horta, Portugal; morgan.rd.ribeiro@uac.pt (M.C.-R.); ualerson.ip.silva@uac.pt (U.I.P.); wendell.mm.silva@uac.pt (W.M.-L.); ana.mn.pabon@uac.pt (A.N.-P.); mario.rr.pinho@uac.pt (M.P.)

[2] IMAR Instituto do Mar, Universidade dos Açores, Rua Prof. Dr. Frederico Machado, 4, 9900-138 Horta, Portugal

* Correspondence: regisvinicius@gmail.com; Tel.: +351-292-200-400

† Presented at the IX Iberian Congress of Ichthyology, Porto, Portugal, 20–23 June 2022.

‡ Presenting author (Poster presentation).

Citation: Santos, R.; Casal-Ribeiro, M.; Peixoto, U.I.; Medeiros-Leal, W.; Nova-Pabon, A.; Pinho, M. Fishers' Experiential Knowledge about Biological Traits of Commercial Marine Species in the Azores. *Biol. Life Sci. Forum* **2022**, *13*, 15. https://doi.org/10.3390/blsf2022013015

Academic Editor: Alberto Teodorico Correia

Published: 6 June 2022

Publisher's Note: MDPI stays neutral with regard to jurisdictional claims in published maps and institutional affiliations.

Abstract: Fishers' experiential knowledge (FEXK) is increasingly recognized as a potential source of information on fishery resources, particularly in data-deficient scenarios. In the Azores, fisheries are classified as small-scale because around 60% of the vessels are smaller than nine meters in length and target several species. Local communities benefit from this activity in various ways, including cultural aspects, employment, and food security. However, for some commercially exploited resources, information on biological and ecological aspects remains poorly known. This study analysed the potential of small-scale FEXK to supply information on the biology and ecology of commercially important marine species, including veined squid *Loligo forbesii*, blue jack mackerel *Trachurus picturatus*, and blackspot seabream *Pagellus bogaraveo*. We conducted interviews with fishers on the bio-ecological characteristics of these species and reviewed the scholarly literature on the same subject. The findings indicated that fishers thoroughly understand species movement, growth, diet, and reproductive aspects. Our results confirm that FEXK might be a valuable source of ecological characteristics for these species, hence assisting in managing fisheries in the Azores.

Keywords: local ecological knowledge; small-scale fisheries; demersal; population aspects; Azores

Author Contributions: R.S. conceived the original idea, performed the analysis and interpretation of data, and wrote the manuscript. M.C.-R. and U.I.P. helped to perform the interviews. W.M.-L., A.N.-P. and M.P. provided critical feedback. R.S. supervised the project. All authors have read and agreed to the published version of the manuscript.

Funding: This work is part of the PESCAz project (ref. MAR–01.03.02–FEAMP–0039) financed by the European Maritime and Fisheries Fund (EMFF) through the Regional Government of the Azores under the MAR2020 operational program.

Institutional Review Board Statement: No ethical approval was required as this study is non-experimental.

Informed Consent Statement: Informed consent was obtained from all subjects involved in the study.

Data Availability Statement: The data underlying this article will be shared upon reasonable request to the corresponding author.

Conflicts of Interest: The authors declare no conflict of interest.

Abstract

The How, the Why, and the What of a New Freshwater Fish Dataset of Individual Body Size in the Iberian Peninsula [†]

Ignasi Arranz [1,*,‡], Rafael Miranda [2] and Lluís Benejam [3]

[1] Laboratoire Evolution et Diversité Biologique, Université Toulouse III-Paul Sabatier, 31062 Toulouse, France
[2] Biodiversity Data Analytics and Environmental Quality (BEQ) Group, Institute of Biodiversity and Environment (BIOMA), Universidad de Navarra, 31009 Pamplona, Spain; rmiranda@unav.es
[3] Aquatic Ecology Group, University of Vic—Central University of Catalonia, 08500 Vic, Spain; lluis.benejam@uvic.cat
* Correspondence: ignasiarranz@gmail.com
† Presented at the IX Iberian Congress of Ichthyology, Porto, Portugal, 20–23 June 2022.
‡ Presenting author (Oral communication).

Abstract: Never before have there been so many ecological datasets available. The organization, accessibility, and collection of ecological datasets create more opportunities among researchers to find new and original questions, perform strong analytical analyses, and gain reliability among the results obtained through alternative methodologies. In addition, increasing the number of ecological datasets in ecoregions and organisms more susceptible to global change may be particularly useful, in order to improve the understanding of ecological processes and mechanisms. Freshwater fish communities in the Iberian Peninsula show unique biological characteristics and high levels of endemicity, but they are sadly impaired by human impacts, including nutrient pollution, landscape changes, and climate warming. Despite previous efforts in creating publicly archived datasets of Iberian fish, most of them are based on specific geographical areas or species-centric approaches, including species occurrences and relative abundances. The individual body size of fish is the simplest metric to collect from the field but it has strong implications on the most physiological, ecological, and evolutionary processes of fish. Thus, a complete raw dataset covering broad spatial and temporal gradients of freshwater fish in the Iberian Peninsula, with information on the fish length and weight, would be a key step if we want to synthesize our current knowledge of Iberian fish trends in space and time. In this talk, we reflect on our own opinion on the how, the why, and the what of having such a dataset. We point out that a new open fish dataset of individual body size would fulfill its potential to advance the knowledge of freshwater fish in the Iberian Peninsula, and develop efficient management plans using size-based approaches, such as the community size spectrum or functional size diversity. We also emphasize that environmental factors (e.g., altitude, ecosystem size) that need to be integrated with the information of the biotic communities. Finally, we provide several suggestions to build this dataset by searching for synergies among researchers and practitioners to create a single unifying standard dataset of fish individual body size across freshwater habitats in the Iberian Peninsula.

Keywords: community size structure; data acquisition; fish assemblage; freshwater habitat; spatiotemporal gradient

Citation: Arranz, I.; Miranda, R.; Benejam, L. The How, the Why, and the What of a New Freshwater Fish Dataset of Individual Body Size in the Iberian Peninsula. *Biol. Life Sci. Forum* **2022**, *13*, 18. https://doi.org/10.3390/blsf2022013018

Academic Editor: Alberto Teodorico Correia

Published: 6 June 2022

Publisher's Note: MDPI stays neutral with regard to jurisdictional claims in published maps and institutional affiliations.

Author Contributions: Conceptualization, I.A.; writing—original draft preparation, I.A.; writing—review and editing, I.A., R.M. and L.B.; funding acquisition, L.B. All authors have read and agreed to the published version of the manuscript.

Funding: This research was funded by the Spanish Ministry of Science, Innovation and Universities, grant number RTI2018-095363-B-I00.

Institutional Review Board Statement: Not applicable.

Informed Consent Statement: Not applicable.

Data Availability Statement: Not applicable.

Conflicts of Interest: The authors declare no conflict of interest.

biology and life sciences forum 

Abstract

Reproduction Techniques Applied to Chondrichthyans Conservation [†]

Pablo García-Salinas, Victor Gallego and Juan F. Asturiano *,[‡]

Grupo de Acuicultura y Biodiversidad, Instituto de Ciencia y Tecnología Animal, Universitat Politècnica de València, 46022 Valencia, Spain; pablo.g.salinas@outlook.com (P.G.-S.); vgalbiach@ualg.pt (V.G.)

* Correspondence: jfastu@dca.upv.es; Tel.: +34-96-387-9385
† Presented at the IX Iberian Congress of Ichthyology, Porto, Portugal, 20–23 June 2022.
‡ Presenting author (Poster presentation).

Citation: García-Salinas, P.; Gallego, V.; Asturiano, J.F. Reproduction Techniques Applied to Chondrichthyans Conservation. *Biol. Life Sci. Forum* **2022**, *13*, 19. https://doi.org/10.3390/blsf2022013019

Academic Editor: Alberto Teodorico Correia

Published: 6 June 2022

Publisher's Note: MDPI stays neutral with regard to jurisdictional claims in published maps and institutional affiliations.

Abstract: Chondrichthyan fishes, which comprise sharks, rays, and chimaeras, are one of the most threatened groups of vertebrates. Given this situation, one possible strategy for the protection of these species could be the use of ex situ conservation projects. However, to develop sustainable ex situ conservation programs, captive breeding techniques, such as sperm extraction and its preservation, should be used. Two main obstacles must be overcome to develop these techniques: first, the lack of knowledge and the scarce previous work focused on the conservation of gametes from these animals; secondly, the peculiarities of the reproductive anatomy of each particular species. Through a detailed description of their reproductive anatomy, we have been able to develope the best techniques to obtain viable sperm from 17 species. Extraction has been performed in both live and dead animals, using cannulation, abdominal massage, and dissection. Exceptionally, we have even been able to recover viable sperm from the reproductive tract of females. Moreover, we have formulated artificial seminal plasma that can be used as an extender to maintain sperm motility for 36 days at 4 °C. By supplementing this extender with different combinations of cryoprotectants, i.e., methanol, dimethyl sulfoxide (DMSO), and fresh egg yolk, we were able to successfully cryopreserve (for the first time in most of these species) the sperm of 14 chondrichthyan species. Sperm samples were frozen inside a styrofoam box using the vapour of liquid nitrogen and were preserved in liquid nitrogen. The sperm quality was assessed by studying the motility and membrane integrity post thawing, demonstrating its effectiveness in the 14 species tested. In rays, the use of 10% DMSO or 10% methanol rendered post-thawing motility values higher than 40%. In sharks and the chimaera species, the best post-thawing motility values were obtained with a combination of 5% DMSO, 5% methanol and 10% egg yolk, which induced mean values close to 35%. All this information broadens our knowledge on the reproductive techniques that can be applied to chondrichthyans, laying the foundations for the first cryobanks for their sperm.

Keywords: anatomy; sperm extraction; cryopreservation; sharks; rays; chimaeras; assisted reproduction techniques

Author Contributions: Conceptualization, P.G.-S. and J.F.A.; sampling, P.G.-S. and V.G.; methodology, P.G.-S. and V.G.; writing—review and editing, P.G.-S. and J.F.A. All authors have read and agreed to the published version of the manuscript.

Funding: Study partially funded by the Fundación Biodiversidad (PRCV00683). P.G.-S. had a contract from the European Union through the Operational Program of the European Social Fund (ESF) of the Comunitat Valenciana 2014–2020 ACIF 2018 (ACIF/2018/147). V.G. had a postdoc contract from the MICIU, Programa Juan de la Cierva-Incorporación (IJCI-2017-34200).

Institutional Review Board Statement: The study was conducted according to the guidelines of the Declaration of Helsinki, and approved by the Animal Care and Welfare Committee of Fundación Oceanogràfic (Project reference: OCE-16-19 on 1 August 2020).

Informed Consent Statement: Not applicable.

Data Availability Statement: Experimental data can be demanded to the corresponding author.

Conflicts of Interest: The authors declare no conflict of interest.

biology and life sciences forum

Abstract

LIFE RESQUE ALPYR: Restoration of Aquatic Ecosystems in Protected Areas of the Alps and Pyrenees [†]

Marc Ventura [1,*], Rocco Tiberti [2], Teresa Buchaca [1], Josep M. Ninot [3], Aaron Pérez-Haase [3] and Quim Pou-Rovira [4,5,‡]

[1] Integrative Freshwater Ecology Group, Centre for Advanced Studies of Blanes (CEAB-CSIC), 17300 Blanes, Spain; buch@ceab.csic.es
[2] Department of Earth and Environmental Sciences, University of Pavia, 27100 Pavia, Italy; rocco.tiberti@gmail.com
[3] Department of Evolutionary Biology, Ecology and Environmental Sciences, University of Barcelona, 08028 Barcelona, Spain; jninot@ub.edu (J.M.N.); aaronperez@ub.edu (A.P.-H.)
[4] Sorelló, Estudis al Medi Aquatic SL, 17003 Girona, Spain; quim.pou@sorello.net
[5] Associació la Sorellona, 17003 Girona, Spain
[*] Correspondence: ventura@ceab.csic.es
[†] Presented at the IX Iberian Congress of Ichthyology, Porto, Portugal, 20–23 June 2022.
[‡] Presenting author (Poster presentation).

Citation: Ventura, M.; Tiberti, R.; Buchaca, T.; Ninot, J.M.; Pérez-Haase, A.; Pou-Rovira, Q. LIFE RESQUE ALPYR: Restoration of Aquatic Ecosystems in Protected Areas of the Alps and Pyrenees. *Biol. Life Sci. Forum* **2022**, *13*, 22. https://doi.org/10.3390/blsf2022013022

Academic Editor: Alberto Teodorico Correia

Published: 2 June 2022

Publisher's Note: MDPI stays neutral with regard to jurisdictional claims in published maps and institutional affiliations.

Abstract: In alpine biogeographic regions, aquatic and semi-aquatic habitats are important biodiversity reservoirs and habitats for species of community interest, but they are often threatened by multiple factors. The conservation state of protected habitats and species in the EU is expected to worsen as long as no actions or conservation strategies are implemented. LIFE RESQUE ALPYR aims to restore mountain aquatic habitats by improving the conservation of several target habitats/species in four Nature 2000 sites from the alpine biogeographical regions of the Pyrenees (in northeastern Iberian Peninsula) and the Alps (in northwestern Italy). The target habitats include eleven aquatic and semi-aquatic habitats, of which five are prioritary: high mountain lakes (HCIs 3110 and 3130), alpine and subalpine grasslands, heaths and meadows (HCIs 4020*, 6230*, 6410, and 6520), mires (HCIs 7110*, 7140, 7230, and 91D0*) and petrifying springs (HCI 7220*). The target species include native amphibians found either in both areas (*Rana temporaria*) or solely in the Pyrenees (*Euproctus asper* and *Alytes obstetricans*); the semi-aquatic mammal *Galemys pyrenaicus* living in Pyrenean streams and lakes; and seven insectivorous bats, including *Barbastella barbastellus*, *Myotis myotis*, and *Plecotus macrobullaris*, which are present in the Pyrenees and the Alps, and *Rhinolophus hipposideros*, *Myotis blythii*, *Myotis bachsteinii*, and *Nyctalus lasiopterus* from the Pyrenees. The target habitats and most of the target species have naturally fragmented distributions, occurring in small areas of the European alpine biogeographic zone, and are affected by anthropogenic pressures. The introduction of trout or minnows in most alpine lakes caused the disappearance of native amphibians and invertebrates at local and landscape scales, indirectly affecting aquatic mammals and terrestrial species, such as bats, that rely on aquatic insects for feeding. Minnows can also cause the strong eutrophication of lakes, leading to drastic habitat degradation. The affected habitats and species are HCIs 3110 and 3130, *R. temporaria*, *E. asper*, *A. obstetricans*, *G. pyrenaicus*, *R. hipposideros*, *P. macrobullaris*, *B. barbastellus*, *M. myotis*, *M. blythii*, *M. bachsteinii*, and *N. lasiopterus*. The proposed actions and methods with regard to fish species involve the experimental eradication of non-native fish in high mountain lakes by means of both chemical (rotenone) and mechanical methods (traps, nets, and electrofishing). The project will provide data regarding replicable and exportable conservation actions and will increase awareness of pertinent conservation issues among stakeholders and the public. In addition, the project will promote the transfer of its background data and results to conservation authorities concerned with other European high mountain areas.

Keywords: ecological restoration; non-native fish eradication; Nature 2000; Pyrenees; Alps; LIFE RESQUE ALPYR

Author Contributions: All authors have contributed equally to the present work and have read and agreed to the published version of the manuscript.

Funding: This research was funded by the European Comission grant number LIFE20.

Institutional Review Board Statement: Not applicable.

Informed Consent Statement: Informed consent was obtained from all subjects involved in the study.

Data Availability Statement: Not applicable.

Conflicts of Interest: The authors declare no conflict of interest.

Abstract

Modelling the Distribution of Freshwater Fish Species to Update Natura 2000 Standard Data Forms in Spain [†]

Mónica Lanzas [1,2,*,‡], Jorge R. Sánchez-González [3,4], Frederic Casals [1,3,4], Felipe Morcillo [3,5], Francisco Guil [6] and Virgilio Hermoso [1,7]

1. Centre de Ciència i Tecnologia Forestal de Catalunya (CTFC), Crta Sant Llorenç de Morunys km 2, 25280 Solsona, Spain; frederic.casals@udl.cat (F.C.); virgilio.hermoso@gmail.com (V.H.)
2. TRAGSATEC, Tecnologías Y Servicios Agrarios, S.A., 28006 Madrid, Spain
3. Sociedad Ibérica de Ictiología (SIBIC), 31008 Navarra, Spain; jorge.sanchez@udl.cat (J.R.S.-G.); fmorcill@ucm.es (F.M.)
4. Department de Ciència Animal, Universitat de Lleida (UdL), 25003 Lleida, Spain
5. Grupo Investigación Ecología, Cambio Global y Restauración de Ecosistemas (GI-UCM-910314), Universidad Complutense de Madrid (UCM), 28040 Madrid, Spain
6. S.G. Biodiversidad Terrestre y Marina, Ministerio para la Transición Ecológica y el Reto Demográfico, 28003 Madrid, Spain; fguil@miteco.es
7. Dpto. Biología Vegetal y Ecología, Universidad de Sevilla (US), 41004 Sevilla, Spain
* Correspondence: lanzas.monica@gmail.com
† Presented at the IX Iberian Congress of Ichthyology, Porto, Portugal, 20–23 June 2022.
‡ Presenting author (Oral communication).

Citation: Lanzas, M.; Sánchez-González, J.R.; Casals, F.; Morcillo, F.; Guil, F.; Hermoso, V. Modelling the Distribution of Freshwater Fish Species to Update Natura 2000 Standard Data Forms in Spain. *Biol. Life Sci. Forum* **2022**, *13*, 24. https://doi.org/10.3390/blsf 2022013024

Academic Editor: Alberto Teodorico Correia

Published: 6 June 2022

Publisher's Note: MDPI stays neutral with regard to jurisdictional claims in published maps and institutional affiliations.

Abstract: Freshwater systems are among the most threatened ecosystems worldwide, and fish species inhabiting them are increasingly endangered by different pressures. One of the most important tools in the European Union (EU) to halt this decline is the Natura 2000 network (N2000). The Habitat Directive (HD) includes freshwater habitats and 39 native fish species from Spain considered of Community interest. Here, we evaluate the degree of spatial coverage of freshwater fish in the N2000 network in Spain, in accordance with reporting needs for the HD. Each N2000 site needs to provide estimates of occupancy as part of the Standard Data Forms, that could be outdated or incomplete. Updated information should help enhance conservation of freshwater fish species in Spain. We compiled a dataset with 10,000 field observations for 60 species and a dataset of environmental predictors including climate, topography, and land cover variables. We then used BIOMOD2 for modelling the spatial distribution of 40 freshwater fish species, 28 of them included in the HD. We then translated these distributions into two presence-absence maps: a maximum potential distribution, and a minimum potential distribution, and used them to measure the degree of coverage of species in the N2000. We found that, on average, up to 30% and 35% of the maximum and minimum potential distribution, respectively, of freshwater fish species were covered under N2000. However, there were differences between species, only a quarter of the species had at least 40% of its minimum potential distribution under N2000. For instance, *Cobitis calderoni* had a coverage of its minimum potential distribution inside N2000 under the 10% compared with *Parachondrostoma turiense* or *Pseudochondrostoma polylepis* that reached the 70% of coverage. The spatial coverage of species under N2000 and its consideration in the policy framework could help to ensure its conservation and to better monitor its conservation status over time. The information presented here could help prioritizing conservation measures inside N2000 for freshwater ecosystems, and to designate new Special Areas of Conservations to fill some of the gaps identified in this study, as part of the objectives of the new EU Biodiversity Strategy.

Keywords: species distribution models; conservation; Natura 2000 network

Author Contributions: All authors have participated in the conceptualization, validation, investigation an writing—review and editing of the manuscript. Methodology was developed by V.H., F.C., F.M., J.R.S.-G. and M.L. The formal analysis: M.L. and V.H. Writing—original draft preparation: V.H., J.R.S.-G. and M.L. All authors have read and agreed to the published version of the manuscript.

Funding: This research was funded by the Spanish Ministry for Ecological Transition and the Demographic Challenge (MITERD) through the project title "Propuesta metodológica para la estimación de la distribución de las especies de peces continentales de España y su incorporación en los Furmularios Normalizados de Datos de la Red Natura2000". VH was funded by the Junta de Andalucía through an Emergia contract (EMERGIA20_00135). M.L. was funded by the Secretaria d'Universitats i Recerca of Generalitat de Catalunya and European Social Found (ESF) through a FI predoctoral contract (2020 FI_B1 00128) and by TRAGSATEC S.A.

Institutional Review Board Statement: Not applicable.

Informed Consent Statement: Not applicable.

Data Availability Statement: Information is available upon request and will be soon published in the Spanish Ministry for the Ecological Transition webpage.

Conflicts of Interest: The authors declare no conflict of interest.

biology and life sciences forum

Abstract

Size-at-Age, Growth and Condition of a Trout (*Salmo trutta* L.) Population from Central Portugal [†]

Ana R. Ribeiro [1,*,‡], Sara Silva [1], Catarina S. Mateus [1], Bernardo R. Quintella [2], Pedro R. Almeida [1,3] and Carlos M. Alexandre [1]

[1] MARE—Centro de Ciências do Mar e do Ambiente, Universidade de Évora, 7004-516 Évora, Portugal; ssrs@uevora.pt (S.S.); cspm@uevora.pt (C.S.M.); pmra@uevora.pt (P.R.A.); cmea@uevora.pt (C.M.A.)

[2] MARE—Centro de Ciências do Mar e do Ambiente, Departamento de Biologia Animal, Faculdade de Ciências da Universidade de Lisboa, 1749-016 Lisboa, Portugal; bsquintella@fc.ul.pt

[3] Departamento de Biologia, Escola de Ciências e Tecnologia, Universidade de Évora, 7004-516 Évora, Portugal

[*] Correspondence: anacarita_ribeiro@hotmail.com; Tel.: +351-934372088

[†] Presented at the IX Iberian Congress of Ichthyology, Porto, Portugal, 20–23 June 2022.

[‡] Presenting author (Poster presentation).

Abstract: The trout (*Salmo trutta* L.) is a salmonid species that is endemic to European rivers, which has a significant interest from ecological, environmental quality, and socioeconomic (i.e., recreational fisheries) perspectives. Much has been written about its bioecology around the world but there is a lack of information about the populations inhabiting the southern limit of its natural distribution. This study aims to learn more about trout biology in the Mondego River basin, Central Portugal, an area where the species is highly searched for in angling activities, and in need of effective measures for its sustainable management and conservation that consider the specificities of this highly plastic species. From electric fishing campaigns, complemented with the help of local fly-fishing anglers, we captured 104 trout in the Mondego River basin. From these, we collected length and weight data, scale samples for age, and life-history determination, and all the trout were tagged with PIT tags to be identified if recaptured. Observed length varied between 95 mm and 560 mm, and five age groups (1+ to 5+) were identified, showing that the population is well structured. The results also show that growth varies considerably throughout the species ontogeny with a larger annual increment between the 3+ and 4+ classes. Age-at-length relationships that were found for the population from Mondego River are considerably different from others that were obtained for the Iberian Peninsula. The weight–length relationship shows a negative allometric growth. The mean value of Fulton's condition factor that was calculated for the population was 1.04, indicating the poor condition of the fish, and there was no significant variability between different age classes when considering this parameter. The present study provided basic information about trout population biology at the southern limit of its distribution which is useful to allow comparisons with other populations from other regions, being essential to help fishery biologists and managers developing appropriate and adapted management and conservation strategies for this species.

Keywords: bioecology; condition factor; weight–length relationship; Mondego River

Citation: Ribeiro, A.R.; Silva, S.; Mateus, C.S.; Quintella, B.R.; Almeida, P.R.; Alexandre, C.M. Size-at-Age, Growth and Condition of a Trout (*Salmo trutta* L.) Population from Central Portugal. *Biol. Life Sci. Forum* **2022**, *13*, 25. https://doi.org/10.3390/blsf2022013025

Academic Editor: Alberto Teodorico Correia

Published: 6 June 2022

Publisher's Note: MDPI stays neutral with regard to jurisdictional claims in published maps and institutional affiliations.

Author Contributions: Conceptualization, C.M.A., P.R.A. and B.R.Q.; methodology, C.M.A., B.R.Q., S.S. and A.R.R.; formal analysis, C.M.A., S.S. and A.R.R.; investigation, C.M.A., S.S. and A.R.R.; resources, C.M.A. and P.R.A.; data curation, A.R.R. and S.S.; writing—original draft preparation, A.R.R. and S.S.; writing—review and editing, C.M.A., P.R.A., B.R.Q. and C.S.M.; supervision, C.M.A. and B.R.Q.; project administration, C.S.M., C.M.A. and P.R.A.; funding acquisition, P.R.A., C.M.A. and C.S.M. All authors have read and agreed to the published version of the manuscript.

Funding: This study was conducted under the scope of the project "DiadES—Assessing and enhancing ecosystem services provided by diadromous fish in a climate change context" (EAPA_18/2018), funded by the INTERREG Atlantic Area program. Funding was also provided by the Portuguese Science Foundation through the strategy plan for MARE (Marine and Environmental Sciences Centre), via project UIDB/04292/2020, through the individual contracts attributed to Carlos M. Alexandre (CEECIND/02265/2018) and Bernardo R. Quintella (2020.02413.CEECIND) and throught a PhD scholarship to Sara Silva (2021.05558.BD).

Institutional Review Board Statement: Not applicable.

Informed Consent Statement: Not applicable.

Data Availability Statement: Not applicable.

Conflicts of Interest: The authors declare no conflict of interest.

Abstract

Application of eDNA Metagenomics to Describe Freshwater Fish Communities [†]

Manuel Curto [1,*,‡], Ana Veríssimo [2], Filipe Ribeiro [1], Carlos D. Santos [3], Sissel Jentoft [4], Judite Alves [5,6] and Hugo F. Gante [5,6,7,8]

[1] Marine and Environmental Sciences Center, Faculty of Sciences, University of Lisbon, 1749-016 Lisbon, Portugal; fmvribeiro@gmail.com
[2] Research Center in Biodiversity and Genetic Resources, 4485-661 Vairao, Portugal; verissimoac@gmail.com
[3] Centro de Estudos do Ambiente e do Mar, Faculdade de Ciências, Universidade de Lisboa, 1749-016 Lisboa, Portugal; cdsantos@fc.ul.pt
[4] Centre for Ecological and Evolutionary Synthesis (CEES), Department of Biosciences, University of Oslo, NO-0371 Oslo, Norway; sissel.jentoft@ibv.uio.no
[5] Center for Ecology, Evolution and Environmental Changes, Universidade de Lisboa, 1250-102 Lisboa, Portugal; mjalves@museus.ulisboa.pt (J.A.); hugo.gante@kuleuven.be (H.F.G.)
[6] Museu Nacional de História Natural e da Ciência, Universidade de Lisboa, 1250-102 Lisboa, Portugal
[7] Department of Biology, KU Leuven, 3000 Leuven, Belgium
[8] Royal Museum for Central Africa, Section Vertebrates, 3080 Tervuren, Belgium
[*] Correspondence: macurto@fc.ul.pt; Tel.: +351-933-971-577
[†] Presented at the IX Iberian Congress of Ichthyology, Porto, Portugal, 20–23 June 2022.
[‡] Presenting author (Oral communication).

Citation: Curto, M.; Veríssimo, A.; Ribeiro, F.; Santos, C.D.; Jentoft, S.; Alves, J.; Gante, H.F. Application of eDNA Metagenomics to Describe Freshwater Fish Communities. *Biol. Life Sci. Forum* **2022**, *13*, 27. https://doi.org/10.3390/blsf2022013027

Academic Editor: Alberto Teodorico Correia

Published: 6 June 2022

Publisher's Note: MDPI stays neutral with regard to jurisdictional claims in published maps and institutional affiliations.

Abstract: Freshwater fish communities are highly threatened by human activities, making it necessary to establish methodologies able to efficiently monitor possible change. Recently, environmental DNA (eDNA) has shown to be one of the most promising sources of biodiversity information, especially when combined with high throughput sequencing approaches, such as DNA metabarcoding, allowing a simplified and efficient description of fish communities. However, metabarcoding relies on PCR amplification of eDNA, which has some limitations, such as priming biases, exclusion of shorter target eDNA fragments, limited taxonomic range, and limited taxonomic information in a single barcode. Alternatively, metagenomics sequences all eDNA fragments present in a sample without enriching for a specific taxonomic group, locus, or fragment length, overcoming these limitations. So far, eDNA metagenomics has mostly been implemented to describe microorganism communities and its applicability to large metazoans, such as fishes, is still understudied. In the current study, we test the power of this method to describe freshwater fish communities using water samples from the Ave and the Tagus rivers. Metagenomics is able to describe the whole freshwater biome by detecting taxa across the tree of life. When compared with eDNA metabarcoding, eDNA metagenomics provides more information in two aspects: First, a higher detection ability for fish species represented by low abundant and highly degraded DNA. Second, it differentiates between local and transported eDNA, which is not possible with eDNA metabarcoding. Our results also show a higher detectability for taxa represented by a whole genome sequence in the reference database. Therefore, some of the diversity is still being missed, since not all organisms have a genome available. Efforts are being made to produce genomic resources for all eukaryotes, meaning that in the future metagenomic data produced now will still be useful for the description of new diversity and characterization of community change.

Keywords: eDNA shotgun sequencing; eDNA metabarcoding; sequence length; multi-loci

---

Author Contributions: All authors participated in the conceptualization of the project. M.C., F.R., and H.F.G. conducted filed work; M.C., A.V. conducted the laboratory work under the supervision of H.F.G. and J.A.; and data analysis was done by M.C., C.D.S., and H.F.G. with extensive support of S.J., M.C. wrote the first draft of the manuscript and all authors contributed to improve it. All authors have read and agreed to the published version of the manuscript.

Funding: This work was funded by the Portuguese Foundation for Science and Technology (FCT) though the project PTDC/BIA-CBI/31644/2017.

Institutional Review Board Statement: Not relevant for this study.

Informed Consent Statement: Not applicable.

Data Availability Statement: All sequence data were deposit in public databases.

Conflicts of Interest: The authors present no conflict of interests.

biology and life sciences forum

Abstract

Population Dynamics of Two Resident Fish Species with Contrasting Habitat Preferences: Temporal Changes over Eutrophic Crises in the Mar Menor Coastal Lagoon (SE Iberian Peninsula) [†]

Adrián Guerrero-Gómez [*,‡], Antonio Zamora-López, Francisco José Oliva-Paterna, José Manuel Zamora-Marín, Patricio López-Martínez de la Plaza and Mar Torralva

Department of Zoology and Physical Anthropology, Faculty of Biology, University of Murcia, 30100 Murcia, Spain; antonio.zamora2@um.es (A.Z.-L.); fjoliva@um.es (F.J.O.-P.); josemanuel.zamora@um.es (J.M.Z.-M.); patricio.lopez.martinez.plaza@gmail.com (P.L.-M.d.l.P.); torralva@um.es (M.T.)
* Correspondence: adrian.guerrero@um.es
† Presented at the IX Iberian Congress of Ichthyology, Porto, Portugal, 20–23 June 2022.
‡ Presenting author (Poster presentation).

Citation: Guerrero-Gómez, A.; Zamora-López, A.; Oliva-Paterna, F.J.; Zamora-Marín, J.M.; López-Martínez de la Plaza, P.; Torralva, M. Population Dynamics of Two Resident Fish Species with Contrasting Habitat Preferences: Temporal Changes over Eutrophic Crises in the Mar Menor Coastal Lagoon (SE Iberian Peninsula). *Biol. Life Sci. Forum* **2022**, *13*, 32. https://doi.org/10.3390/blsf2022013032

Academic Editor: Alberto Teodorico Correia

Published: 6 June 2022

Publisher's Note: MDPI stays neutral with regard to jurisdictional claims in published maps and institutional affiliations.

Abstract: The Mar Menor coastal lagoon is a paradigmatic example of a transitional system under high human disturbance, which in the last years has resulted in occasional fish kill due to extreme anoxic conditions. Here, we assess the abundance temporal trend of two conservation-concern resident fish species with contrasting habitat preferences: *Pomatoschistus marmoratus*, associated with sandy open bottoms, and *Syngnathus abaster*, associated with vegetated bottoms. Both species have been historically abundant in the shallow areas of the Mar Menor. Temporal change was related to three dystrophic crisis events, leading to a lagoon-scale mortality of deep seagrass meadow before spring 2016 and two fish kills occurring in autumn 2019 and summer 2021. After deep seagrass meadow mortality (2017/19), *S. abaster* showed an increased abundance (annual mean: 5 ind/100 m^2 to 23 ind/100 m^2), whereas abundance of *P. marmoratus* decreased (annual mean: 37 ind/100 m^2 to 10 ind/100 m^2) in comparison to reference values (2002/04). At that point, shallow areas had experienced great habitat changes, with an increase both in the mud portion of the substrate (driven by the deep seagrass decomposition process), and in the vegetated surface (due to nutrient input), thus modifying the mesohabitat selected by the species. Before the first fish kill (autumn 2019), the abundance of both species markedly increased, likely as a response to an anoxia-mediated refugee search by deep-habitat metapopulations. A slow recovery was detected after the 2020 recruitment period for both species, reaching record values for *P. marmoratus* (75 ind/100 m^2) in summer 2021, possibly related again to the search for more oxygenated shallow areas, since shortly thereafter a new dystrophic crisis occurred (summer 2021). Since then, the abundance of both species has steeply decreased to less than 2 ind/100 m^2, highlighting a critical threat to the long-term population viability of these conservation-concern species. As supported here, long-term monitoring programs provide insightful data on the response of fish species to acute human-related disturbance events, offering necessary information to guide the development of management and conservation actions.

Keywords: shallow areas; *Pomatoschistus marmoratus*; *Syngnathus abaster*; fish kill

Author Contributions: Conceptualization, A.G.-G., M.T. and F.J.O.-P.; methodology, A.Z.-L., M.T. and F.J.O.-P.; software, A.G.-G.; validation, A.Z.-L., A.G.-G. and P.L.-M.d.l.P.; formal analysis, A.Z.-L. and A.G.-G.; investigation, A.G.-G., A.Z.-L. and J.M.Z.-M.; resources, A.Z.-L. and A.G.-G.; data curation, A.Z.-L.; writing—original draft preparation, A.G.-G.; writing—review and editing, J.M.Z.-M., M.T., A.Z.-L. and F.J.O.-P.; visualization, J.M.Z.-M.; supervision, M.T. and F.J.O.-P.; project administration, M.T.; funding acquisition, M.T. and F.J.O.-P. All authors have read and agreed to the published version of the manuscript.

Funding: Part of this research was supported by the Environmental Service and Mar Menor and Service of the Autonomous Government of Murcia (Spain). Proyect (04812210001) "Study and spatio-temporal monitoring of priority faunal populations (ichthyofauna and the invasive species Callinectes sapidus): current situation in shallow and intertidal areas of the Mar Menor."

Institutional Review Board Statement: Not applicable.

Informed Consent Statement: Not applicable.

Data Availability Statement: The data presented in this study are available within the downloadable reports on this website https://canalmarmenor.carm.es/ciencia/estudios-de-investigacion/ (accessed on 31 March 2022).

Conflicts of Interest: The authors declare no conflict of interest.

biology and life sciences forum

Abstract

Colonization and Succession of Fish Assemblages in a New River Section: A Size Structure Approach [†]

Rafael Miranda [1,2,*,‡], **Mireia Bartrons** [3], **Sandra Brucet** [3,4] **and Lluís Benejam** [3]

[1] BIOMA Instituto de Biodiversidad y Medioambiente, Universidad de Navarra, 31008 Pamplona, Spain
[2] Biodiversity Data Analytics and Environmental Quality Group, Department of Environmental Biology, Universidad de Navarra, 31008 Pamplona, Spain
[3] Aquatic Ecology Group, University of Vic-Central University of Catalonia, 08500 Vic, Spain; mireia.bartrons@uvic.cat (M.B.); sandra.brucet@uvic.cat (S.B.); lluis.benejam@uvic.cat (L.B.)
[4] Catalan Institution for Research and Advanced Studies (ICREA), 08010 Barcelona, Spain
[*] Correspondence: rmiranda@unav.es; Tel.: +34-948-425600
[†] Presented at the IX Iberian Congress of Ichthyology, Porto, Portugal, 20–23 June 2022.
[‡] Presenting author (oral communication).

Citation: Miranda, R.; Bartrons, M.; Brucet, S.; Benejam, L. Colonization and Succession of Fish Assemblages in a New River Section: A Size Structure Approach. *Biol. Life Sci. Forum* **2022**, *13*, 42. https://doi.org/10.3390/blsf2022013042

Academic Editor: Alberto Teodorico Correia

Published: 7 June 2022

Publisher's Note: MDPI stays neutral with regard to jurisdictional claims in published maps and institutional affiliations.

Abstract: One of the severe impacts of rivers is the construction of new structures on the riverbed bringing about diversions and changes in the flow course. These new sections must then be colonized by the species present in the ecosystems. These colonizations are critical processes that regulate the communities' persistence and ecological resilience. Understanding these colonization and succession processes is essential for biodiversity conservation. However, studies that analyze these processes are rare for freshwater environments. This study explored the colonization and succession patterns of fish assemblages in a new channelized section of the Larraun River (Navarra, Spain) from its construction to the present (twenty-five year period). We also checked the usefulness of size-related variables as indicators of the changes in these processes for fish populations. Two sites (control and new section) were sampled by electrofishing from 1996 to 2020, physicochemical parameters were measured, and habitat was characterized. Our results show that constraining environmental features shape freshwater fish species' biological characteristics and distribution in the colonization processes. Initially, the mesohabitat of the new stream segment was simple and dominated by runs. Throughout the years, the complexity increased and mesohabitats became more diverse, with an increase in pools and higher width, depth and diversity of depths of the new section. The water temperature decreased due to the increased shade produced for the expansion of the riparian forest. Moreover, size-related variables of the fish community, such as size diversity, mean and maximum length, increased in the new section throughout the succession process, achieving values comparable to the control site. Regarding the slopes of the fish size spectra (i.e., the linear rate of decline of fish abundance as body size increases), it took twenty years for the new section to obtain values comparable to the control site, indicating a slow transition from a community dominated by small fish to a community where all sizes were well represented. This study suggests that habitat complexity (diversity of substrates, depths, water velocity, etc.) determines the structure of fish populations and that size-related variables can be effective ecological indicators when assessing the evolution of fish colonization and succession in temperate European rivers.

Keywords: ecological processes; temporal series; body size; habitat changes; biodiversity; human impacts; environmental management

Author Contributions: Conceptualization, R.M. and L.B.; formal analysis, M.B.; data curation, R.M.; writing—original draft preparation, R.M. and L.B.; writing—review and editing, M.B., S.B. and L.B.; funding acquisition, S.B. and L.B. All authors have read and agreed to the published version of the manuscript.

Funding: This research was funded by Ministerio de Ciencia, Innovación y Universidades (Government of Spain), grant number RTI2018-095363-B-I00.

Institutional Review Board Statement: Not applicable.

Informed Consent Statement: Not applicable.

Data Availability Statement: Not applicable.

Conflicts of Interest: The authors declare no conflict of interest.

Abstract

Probabilistic School Classification of Multiple Species in Acoustic Echograms Based on Machine Learning [†]

Aitor Lekanda *,[‡], Guillermo Boyra and Maite Louzao

Fundación AZTI, 48395 Sukarrieta, Spain; gboyra@azti.es (G.B.); mlouzao@azti.es (M.L.)
* Correspondence: alekanda@azti.es
† Presented at the IX Iberian Congress of Ichthyology, Porto, Portugal, 20–23 June 2022.
‡ Presenting author (Oral communication).

Abstract: Multifrequency trawl-acoustic surveys are used worldwide for continuous monitoring of pelagic ecosystems. Acoustic backscattering energy partitioning in different species is typically done by visual scrutiny of the echograms with the aid of trawl species composition, which may be subjective and time-consuming. Alternatively, machine learning techniques may provide well-stablished, objective, and reproducible methods for automatic school classification in acoustic echograms. The pelagic ecosystem is a diverse one, where many species co-occur in space and time, being mixed catches very common during scientific surveys. However, most of the school classification models are built using single species composition trawls due to difficulties to assign a class to each school in multispecific trawls. The present study has the aim of developing and comparing different probabilistic multivariate models to identify pelagic species in mixed scenarios based on trawl catch proportions. In addition to the standard predictors, a novel variable, collective mean TS per nautical mile measured on the periphery of the schools, has shown to play an important role in species discrimination. The methods were applied on data from seven consecutive years of an acoustic survey in the Bay of Biscay. Preliminary results yielded classification performances near 90% in classifying 5 different pelagic species.

Keywords: machine learning; classification; multivariate; pelagic species; mixed scenario

Citation: Lekanda, A.; Boyra, G.; Louzao, M. Probabilistic School Classification of Multiple Species in Acoustic Echograms Based on Machine Learning. *Biol. Life Sci. Forum* **2022**, *13*, 44. https://doi.org/10.3390/blsf2022013044

Academic Editor: Alberto Teodorico Correia

Published: 7 June 2022

Publisher's Note: MDPI stays neutral with regard to jurisdictional claims in published maps and institutional affiliations.

Author Contributions: Conceptualization, A.L., G.B. and M.L.; methodology, A.L. and G.B.; formal analysis, A.L. and G.B.; writing—original draft preparation, A.L.: writing—review and editing, A.L., G.B. and M.L. All authors have read and agreed to the published version of the manuscript.

Funding: This research was funded by the Pre-doctoral grant from the Department of Education of Basque Government, grant number PRE_2020_1_0129.

Institutional Review Board Statement: Not applicable.

Informed Consent Statement: Informed consent was obtained from all subjects involved in the study.

Data Availability Statement: The data underlying this article will be shared on reasonable request to the corresponding author.

Conflicts of Interest: The authors declare no conflict of interest.

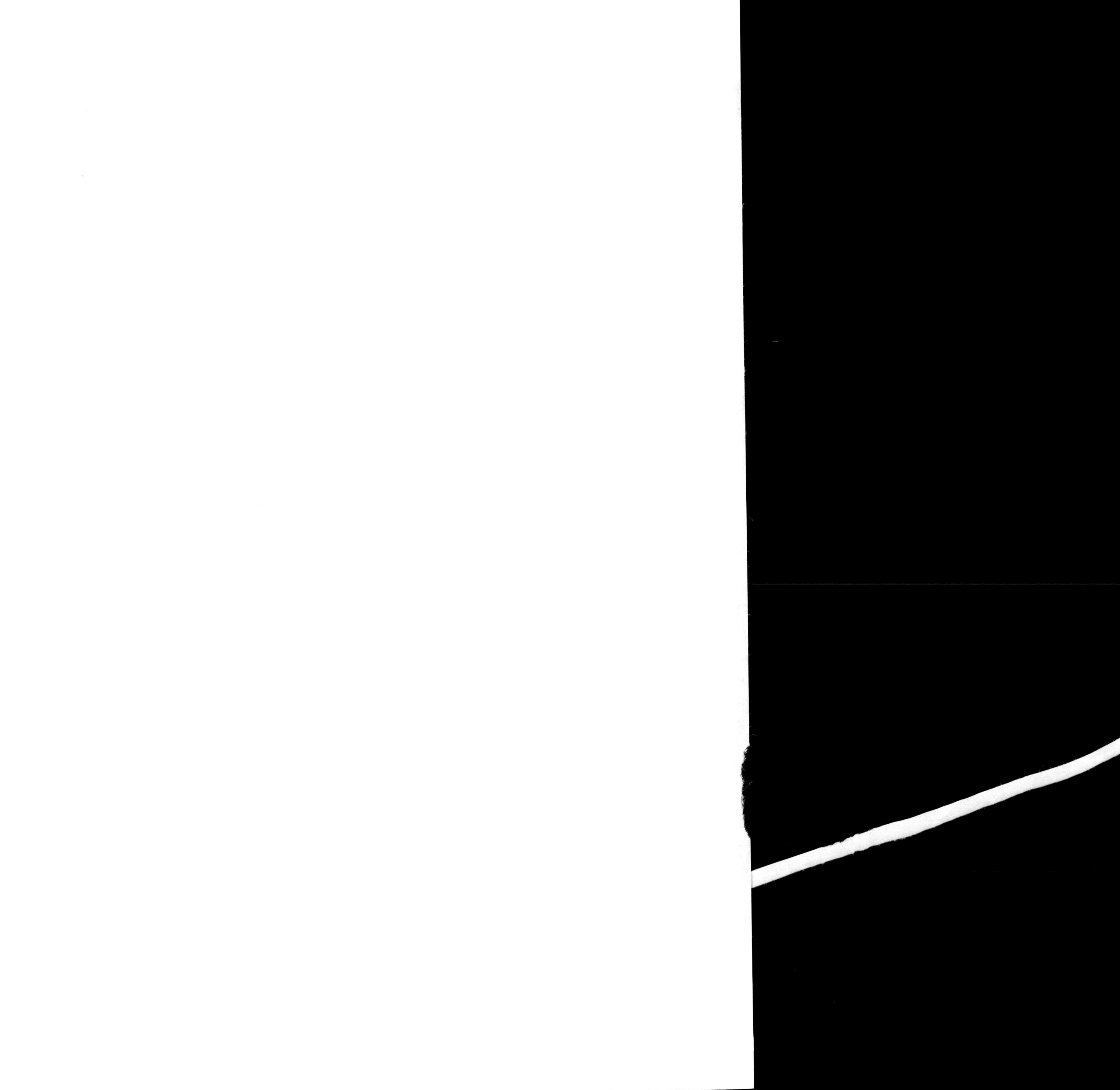

biology and life sciences forum

MDPI

Abstract

Mortality of Sardine and Horse Mackerel Eggs by Parasitism, Impact on the Survival of Early Life Stages [†]

Daniela Carriço [1,*,‡], Elisabete Henriques [2], Isabel Meneses [2], José Lino Costa [3], Ana Veríssimo [4], Catarina Vendrell [2], Cristina Nunes [2] and Maria Manuel Angélico [2,*]

[1] IPMA—Instituto Português do Mar e da Atmosfera, Faculdade de Ciências, Universidade de Lisboa, 1749-016 Lisboa, Portugal
[2] IPMA—Instituto Português do Mar e da Atmosfera, 1749-077 Lisboa, Portugal; ehenriques@ipma.pt (E.H.); imeneses@ipma.pt (I.M.); catarinavendrell@gmail.com (C.V.); cnunes@ipma.pt (C.N.)
[3] MARE—Centro de Ciências do Mar e do Ambiente, Faculdade de Ciências, Universidade de Lisboa, 1600-548 Lisboa, Portugal; jlcosta@fc.ul.pt
[4] CIBIO—Centro de Investigação em Biodiversidade e Recursos Genéticos, 4485-661 Vairão, Portugal; verissimoac@gmail.com
* Correspondence: danielacarrico22@gmail.com (D.C.); mmangelico@ipma.pt (M.M.A.)
† Presented at the IX Iberian Congress of Ichthyology, Porto, Portugal, 20–23 June 2022.
‡ Presenting author (Poster Presentation).

Citation: Carriço, D.; Henriques, E.; Meneses, I.; Costa, J.L.; Veríssimo, A.; Vendrell, C.; Nunes, C.; Angélico, M.M. Mortality of Sardine and Horse Mackerel Eggs by Parasitism, Impact on the Survival of Early Life Stages. *Biol. Life Sci. Forum* **2022**, *13*, 46. https://doi.org/10.3390/blsf2022013046

Academic Editor: Alberto Teodorico Correia

Published: 2 June 2022

Publisher's Note: MDPI stays neutral with regard to jurisdictional claims in published maps and institutional affiliations.

Abstract: Sardine (*Sardina pilchardus*) and horse mackerel (*Trachurus trachurus*) are two key species in marine food webs and have a high commercial value in the Iberian Peninsula. The populations of these species are known to have exhibited marked fluctuations in abundance over the years, however the causes for such variability are still not totally understood. One of the identified sources of mortality in the early life phase, is the parasitism by *Ichthyodinium chabelardi*, an endoparasite of pelagic fish eggs, which leads to the host death. A total of 4000 ichthyoplankton samples, and over 30,000 eggs, from 10 surveys conducted by IPMA, in the Iberian Atlantic area, were analyzed to assess the percentage of eggs infected by the parasite (prevalence). The results showed interannual variability in the infection prevalence, with values between 0.80% and 4.32% for sardine eggs and between 0.82% and 10.97% for horse mackerel eggs. The number of samples with infected eggs varied between 2.53% and 15.32% in the case of sardine eggs and between 2.00% and 12.59% for horse mackerel eggs. GLM analyses highlighted location variables, as the main responsible for the spatial distribution patterns observed. Comparisons of infection prevalence and total egg mortality values obtained by the daily egg production method estimations clarify the relevance of *I. chabelardi* infection for sardine egg mortality.

Keywords: *Ichthyodinium chabelardi*; fish eggs parasite; *Sardina pilchardus*; *Trachurus trachurus*; infection prevalence

Author Contributions: Conceptualization, I.M. and E.H.; methodology, I.M. and A.V.; validation, M.M.A., J.L.C. and C.N.; formal analysis, M.M.A.; investigation, D.C., I.M., C.V. and M.M.A.; resources, C.V. and M.M.A.; data curation, I.M.; writing—original draft preparation, D.C.; writing—review and editing, M.M.A. and J.L.C.; visualization, M.M.A.; supervision, M.M.A.; project administration, J.L.C.; funding acquisition, M.M.A. All authors have read and agreed to the published version of the manuscript.

Funding: SARDINHA2020—This research was funded by Abordagem Ecossistémica para a gestão da pesca da sardinha (MAR2020). (Ecosystem approach towards a sustainable sardine fishery exploitation). Mar2020-MAR-01.04.02-FEAMP-0009 and PNAB—Programa Nacional de Amostragem Biológica, Recolha de dados no âmbito da PCP. (EU National Plans for Data Collection; EU-DCF: Portuguese Marine Surveying Programme), EU Data Collection Framework (EU-DCF, FEAMP)-P03M02.

Institutional Review Board Statement: Not applicable.

Data Availability Statement: The data that supported these results can be found in IPMA database.

Conflicts of Interest: The authors declare no conflict of interest.

biology and life sciences forum

Abstract

First Recorded Case of Leucism in the Velvet Belly Lantern Shark *Etmopterus spinax* (Squaliformes: Etmopteridae) [†]

Juan Carlos Arronte [1,*,‡], Ana Antolínez [1], Rafael Bañón [2], José Rodríguez [1], Juan José Ortíz [3] and Juan Manuel Martínez [1]

[1] Instituto Español de Oceanografía (IEO-CSIC), C.O. Santander, 39004 Santander, Spain; ana.antolinez@ieo.csic.es (A.A.); jose.rodriguez@ieo.csic.es (J.R.); juanma.martinez@ieo.csic.es (J.M.M.)
[2] Consellería do Mar, Xunta de Galicia, 15701 Santiago de Compostela, Spain; anoplogaster@yahoo.es
[3] Facultad de Biología, Universidade de Santiago de Compostela, 15705 Santiago de Compostela, Spain; juanjoortiz42@gmail.com
* Correspondence: jcarlos.arronte@ieo.csic.es; Tel.: +34-942-29-17-16
† Presented at the IX Iberian Congress of Ichthyology, Porto, Portugal, 20–23 June 2022.
‡ Presenting author (Poster presentation).

Abstract: Albinism and leucism are genetically inherited conditions in which the pigment protein, melanin, is either absent or non-functional. The albino phenotype is characterized by the total lack of melanin in both the skin and iris, whereas leucism is the complete or partial loss of skin pigmentation but with a normal iris pigmentation. Colour abnormalities are known to occur in all vertebrate groups. Specifically, in Chondrichthyes, they have been reported in a wide variety of chimaera, ray, and shark species. Here, we report the occurrence of a case of leucism in the velvet belly lantern shark *Etmopterus spinax* (L. 1758), a small-sized (up to 45 cm total length) deep-water squalid shark inhabiting the eastern side of the Atlantic Ocean, from Iceland and Norway to Gabon and the Mediterranean Sea. The normal coloration of *E. spinax* is brown on the dorsal surface, the underside of the snout, and abdomen, with abruptly black and thin black marks above and behind the pelvic fins and along the caudal fin. One specimen of 11 cm TL showing abnormal coloration was caught in the Cantabrian Sea (Northwestern Atlantic) in 2021. The individual was whitish on the body and fins, greyish on the underside of the snout, abdomen, and margin of the caudal fin, and the eyes showed a normal retinal pigmentation. Hence, it was considered a leucistic shark. This is the first record of leucism in the species and the second of hypopigmentation for the genus. In addition, a thorough review of records of albinism and leucism in Chondrichthyans was undertaken. Albinism and leucism have been reported in 61 species (37 sharks, 23 rays, and one chimaera) corresponding to 31 families, accounting for about 4.9% of the known species.

Keywords: *Etmopterus spinax*; Shark; Albinism; Leucism

Citation: Arronte, J.C.; Antolínez, A.; Bañón, R.; Rodríguez, J.; Ortíz, J.J.; Martínez, J.M. First Recorded Case of Leucism in the Velvet Belly Lantern Shark *Etmopterus spinax* (Squaliformes: Etmopteridae). *Biol. Life Sci. Forum* **2022**, *13*, 48. https://doi.org/10.3390/blsf2022013048

Academic Editor: Alberto Teodorico Correia

Published: 7 June 2022

Publisher's Note: MDPI stays neutral with regard to jurisdictional claims in published maps and institutional affiliations.

Author Contributions: Conceptualization, J.C.A., A.A. and R.B.; methodology, J.C.A., A.A. and R.B.; formal analysis, J.C.A. and A.A.; investigation, A.A., R.B. and J.C.A.; writing—original draft preparation, J.C.A., A.A., R.B. and J.R.; writing—review and editing, J.C.A., A.A., R.B., J.R., J.J.O. and J.M.M. All authors have read and agreed to the published version of the manuscript.

Funding: This research received no external funding.

Institutional Review Board Statement: Not applicable.

Informed Consent Statement: Not applicable.

Data Availability Statement: No data availability statement necessary as there are no data.

Conflicts of Interest: The authors declare no conflict of interest.

Abstract

Restoration of Coastal Ecological Processes versus Fish Conservation: To Be or Not to Be … [†]

Nuno Caiola [1,2,*,‡] **and Carles Ibáñez** [1,2]

1   Climate Resilience Center (CRC), 08005 Barcelona, Spain; carles.ibanez@eurecat.org
2   Department of Climate Change, Eurecat Technology Center of Catalonia, 43870 Amposta, Spain
*   Correspondence: nuno.caiola@eurecat.org
†   Presented at the IX Iberian Congress of Ichthyology, Porto, Portugal, 20–23 June 2022.
‡   Presenting author (Oral communication).

Abstract: The H2020 project REST-COAST will build a framework for large-scale coastal restoration, encompassing the development of adequate monitoring plans and accurate assessment methods to evaluate the performance of restoration actions in terms of ecosystem-service delivery and bio-diversity enhancement. Hands-on restoration actions will be implemented in nine pilot sites representative of the European coastal archetype's variability. In many of these cases, the restoration actions imply the enhancement of the connectivity between coastal and marine habitats. In such cases, although the ecological processes are restored, the loss of coastal habitat isolation may affect fish species. The Ebro delta is one of the main REST-COAST pilot sites, and in a previous restoration project, abandoned commercial aquaculture ponds holding an abundant population of the endangered Spanish toothcarp (*Aphanius iberus*) were transformed into natural coastal wetlands. Moreover, the connectivity of the restored habitats with the sea was improved. The fish community was monitored following a before–after restoration design. Here, we use the Ebro delta as a REST-COAST case study to evaluate the effect of enhanced habitat connectivity on the fish community, with emphasis on the Spanish toothcarp. We analyzed the relationship between species richness, diversity and abundance, and the environmental variables affected by the restoration works. While species richness and diversity increased, Spanish toothcarp abundance decreased. This opens the debate on whether species conservation should be prioritized over ecosystem process restoration, even if it means maintaining a degraded habitat. Moreover, the adequacy of using fish as indicators of coastal ecosystem restoration performance is discussed. It is expected that the results will assist in establishing suitable criteria for coastal ecosystem restoration.

Keywords: coastal restoration; coastal wetlands; habitat connectivity; fish community; endangered Spanish toothcarp

Citation: Caiola, N.; Ibáñez, C. Restoration of Coastal Ecological Processes versus Fish Conservation: To Be or Not to Be … *Biol. Life Sci. Forum* **2022**, *13*, 51. https://doi.org/10.3390/blsf2022013051

Academic Editor: Alberto Teodorico Correia

Published: 2 June 2022

Funding: Funding was provided by the European Union's Horizon 2020 research and innovation programme under grant agreement number 101037097 (project REST-COAST).

Institutional Review Board Statement: Not applicable.

Informed Consent Statement: Not applicable.

Data Availability Statement: Not applicable.

Conflicts of Interest: The authors declare no conflict of interest.

 *biology and life sciences forum*

Abstract

Updating of the Carta Piscícola Española and Assessment of the State of the Conservation Status of Threatened Spanish Fishes [†]

Nora Escribano [1,2], Imanol Miqueleiz [1,2], Julen Torrens [1,2] and Rafael Miranda [2,3,*,‡]

1 Iberian Society of Ichthyology (SIBIC), 31008 Pamplona, Spain; nescribano@unav.es (N.E.); imiqueleiz@alumni.unav.es (I.M.); julentorrens@gmail.com (J.T.)
2 Biodiversity Data Analytics and Environmental Quality Group, Department of Environmental Biology, Universidad de Navarra, 31008 Pamplona, Spain
3 BIOMA Instituto de Biodiversidad y Medioambiente, Universidad de Navarra, 31008 Pamplona, Spain
* Correspondence: rmiranda@unav.es
† Presented at the IX Iberian Congress of Ichthyology, Porto, Portugal, 20–23 June 2022.
‡ Presenting author (oral communication).

Citation: Escribano, N.; Miqueleiz, I.; Torrens, J.; Miranda, R. Updating of the Carta Piscícola Española and Assessment of the State of the Conservation Status of Threatened Spanish Fishes. *Biol. Life Sci. Forum* **2022**, *13*, 52. https://doi.org/10.3390/blsf2022013052

Academic Editor: Alberto Teodorico Correia

Published: 7 June 2022

Publisher's Note: MDPI stays neutral with regard to jurisdictional claims in published maps and institutional affiliations.

Abstract: The Carta Piscícola Española (CPE) is currently the most extensive inventory of freshwater fish fauna in Spain. It is also a tool for knowledge management and outreach. Over the last three years, more than 4800 users from 73 countries have visited the CPE more than nine million times, consolidating this web as the primary source for the public interested in Spanish freshwater fishes. The CPE was first launched in 2014 and currently gives access to more than 40,000 records of 84 fish species in Spain. Conceived as a long-term project, the CPE needs to grow to keep on providing the same services while developing new ones. To this end, the SIBIC started the project "Update of the national inventory and evaluation of the conservation status of Iberian fish" in 2021. The aim of the project was threefold: (1) assess the conservation status of freshwater fish in the Iberian Peninsula, (2) update the CPE with new records, and (3) build Species Distribution Models (SDMs) of the fish species of the Iberian Peninsula. In this communication, we summarize the goals met in the project. Firstly, we have assessed the state of conservation of eleven freshwater fish species with the aim to propose their inclusion in the Spanish Catalogue of Threatened Species (CEEA). To date, we have indexed more than 40,000 new records of 89 fish species in the CPE, mainly coming from well-structured fish monitoring networks with solid protocols. Finally, we have built SDMs for 54 freshwater fish species, that will be included as an accessible resource on the webpage of the CPE. The CPE is an open project that benefits from the contribution of different sources while recognizing this contribution by a traceable citation. We invite the community of ichthyologists to support the CPE and enrich it by sharing fish records and spreading the word.

Keywords: biodiversity records; freshwater fishes; Species Distribution Models; extinction risk assessment; national inventory

---

Author Contributions: Conceptualization, R.M., N.E. and I.M.; methodology, J.T., N.E. and I.M.; formal analysis, J.T., N.E. and I.M.; writing—original draft preparation, R.M.; writing—review and editing, R.M., N.E. and I.M.; funding acquisition, R.M. All authors have read and agreed to the published version of the manuscript.

Funding: Fundación Biodiversidad, Ministerio para la Transición Ecológica y el Reto Demográfico, Government of Spain.

Institutional Review Board Statement: Not applicable.

Informed Consent Statement: Not applicable.

Data Availability Statement: Not applicable.

Conflicts of Interest: The authors declare no conflict of interest.

Abstract

Bioavailability and Ingestion of Microplastics by Fish Larvae in the Douro Estuary [†]

Sabrina M. Rodrigues [1,2,*,‡], **Liliana Sousa** [1], **Diogo Silva** [1,2], **Jacinto Cunha** [2,3], **Vânia Freitas** [2], **C. Marisa R. Almeida** [2] and **Sandra Ramos** [2]

1   ICBAS—Institute of Biomedical Sciences Abel Salazar, Porto University, 4050-313 Porto, Portugal; liliana.a.o.moreira@gmail.com (L.S.); silva.diogom@gmail.com (D.S.)
2   CIIMAR—Interdisciplinary Centre of Marine and Environmental Research, University of Porto, 4450-208 Matosinhos, Portugal; jfrcunha@gmail.com (J.C.); vpfreitas@ciimar.up.pt (V.F.); calmeida@ciimar.up.pt (C.M.R.A.); ssramos@ciimar.up.pt (S.R.)
3   CITAB—Centre for the Research and Technology of Agro-Environmental and Biological Sciences, University of Trás-os-Montes and Alto Douro, 5000-801 Vila Real, Portugal
*   Correspondence: smagalhaes@ciimar.up.pt
†   Presented at the IX Iberian Congress of Ichthyology, Porto, Portugal, 20–23 June 2022.
‡   Presenting author (Poster presentation).

Citation: Rodrigues, S.M.; Sousa, L.; Silva, D.; Cunha, J.; Freitas, V.; Almeida, C.M.R.; Ramos, S. Bioavailability and Ingestion of Microplastics by Fish Larvae in the Douro Estuary. *Biol. Life Sci. Forum* **2022**, *13*, 54. https://doi.org/10.3390/blsf2022013054

Academic Editor: Alberto Teodorico Correia

Published: 7 June 2022

Publisher's Note: MDPI stays neutral with regard to jurisdictional claims in published maps and institutional affiliations.

Abstract: In their early life stages, fish are highly susceptible to a wide range of biological and anthropogenic factors (e.g., habitat degradation or pollution) that can influence their growth and survival. Due to their size, microplastics (plastic particles with less than 5 mm) pose an additional threat to fish larvae since their size range coincides with their prey size. The ingestion of microplastics by fish larvae can cause gut blockage and limit food intake, and ultimately affect their growth, reproduction, and survival. This study aimed to evaluate the bioavailability of microplastics and quantify microplastic ingestion by fish larvae in an urban estuary. To this end, seasonal samplings surveys were performed in 2017 along the Douro estuary (NW Portugal). Sub-surface planktonic trawls were conducted along the estuarine horizontal gradient to collect fish larvae and microplastics. Samples were sorted, and fish larvae were identified and kept for further quantification of microplastics ingested. Microplastic bioavailability was determined using a previously optimized protocol. A total of 573 fish larvae were collected, with an average density of 14.63 fish larvae 100 m^{-3} and mostly composed of few but highly abundant taxa, such as Pomatoschistus spp. and Clupeidae n.i. A total of 609 microplastics were found in water samples, with an average density of 15.52 microplastics 100 m^{-3}—namely, fibers, particles, and films. In Summer, fish larvae presented the highest values of abundance, contrary to the other three seasons when microplastic density surpassed larval fish density. Preliminary tests were conducted to identify the best protocol for the digestion of fish larvae to quantify microplastic ingestion. Additionally, in accordance with those results, fish larvae are currently being digested using H_2O_2 for a period of 7 h at 65 °C, to evaluate microplastic ingestion by fish larvae and to compare these results with the microplastics collected in the water.

Keywords: plastic particles; ichthyoplankton; urban estuary; ingestion

Funding: This research was partially funded by Ocean3R (NORTE-01-0145-FEDER-000064) and ATLANTIDA (NORTE-01-0145-FEDER-000040) projects, supported by the Norte Portugal Regional Operational Programme (NORTE 2020) under the PORTUGAL 2020 Partnership Agreement and through the European Regional Development Fund (ERDF). Additionally, FCT is awarded by the Strategic Funding UIDB/04423/2020 and UIDP/04423/2020 through national funds provided by FCT and ERDF, and a PhD fellowship to S.M.R. (SFRH/BD/145736/2019), D.S. (2020.06088.BD) and J.C. (PD/BD/150359/2019) and a research contract to SRamos (DL57/2016/CP1344/CT0020).

Institutional Review Board Statement: Not applicable.

Informed Consent Statement: Not applicable.

Data Availability Statement: Not applicable.

Conflicts of Interest: The authors declare no conflict of interest.

64

biology and life sciences forum

Abstract

Biology and Ecology of Two Anadromous Species of the Genus *Alosa* (*A. alosa* and *A. fallax*) in the Galician Coastal Marine Environment Based on Bycatch Individuals: Proposals for the Improvement of Their Protection and Management [†]

David José Nachón [1,*,‡], Alejandro Pico [1], Rufino Vieira-Lanero [1], Sandra Barca [1], María del Carmen Cobo [2] and Fernando Cobo [1]

[1] Departmento de Zooloxía, Xenética e Antropoloxía Física, Facultade de Bioloxía, Universidade de Santiago de Compostela, 15782 Santiago de Compostela, Spain; alejandro.pico.calvo@rai.usc.es (A.P.); rufino.vieira@usc.es (R.V.-L.); sandra.barca@usc.es (S.B.); fernando.cobo@usc.es (F.C.)

[2] Department of Biological Sciences, University of Alabama, Tuscaloosa, AL 35487, USA; mariadelcarmen.cobo@usc.es

[*] Correspondence: davidjose.nachon@usc.es

[†] Presented at the IX Iberian Congress of Ichthyology, Porto, Portugal, 20–23 June 2022.

[‡] Presenting author (Oral communication).

Citation: Nachón, D.J.; Pico, A.; Vieira-Lanero, R.; Barca, S.; Cobo, M.d.C.; Cobo, F. Biology and Ecology of Two Anadromous Species of the Genus *Alosa* (*A. alosa* and *A. fallax*) in the Galician Coastal Marine Environment Based on Bycatch Individuals: Proposals for the Improvement of Their Protection and Management. *Biol. Life Sci. Forum* **2022**, *13*, 55. https://doi.org/10.3390/blsf2022013055

Academic Editor: Alberto Teodorico Correia

Published: 7 June 2022

Publisher's Note: MDPI stays neutral with regard to jurisdictional claims in published maps and institutional affiliations.

Abstract: The Allis shad, *Alosa alosa* (Linnaeus, 1758) and the Twaite shad, *Alosa fallax* (Lacépède, 1803) are two anadromous and congeneric Clupeidae species, i.e., closely related, even hybridising with each other, which makes it very difficult to differentiate between them. The marine phase represents the major part of the life cycle of these species, yet there are still large gaps in the knowledge of the biology, ecology and distribution during this phase. To overcome this lack of data, this study analysed the biometric characteristics, as well as different tissues and organs, of specimens of the genus *Alosa*, caught as bycatch by inshore fisheries on the continental shelf of the coastal region of Galicia, NW Iberian Peninsula. Specimens of the genus *Alosa* ($N = 345$) were acquired on daily first sale notes between January and March 2021 at the fish markets of A Guarda, Malpica and A Coruña, three of the most important landing sites for bycatches of these species in Galicia, whose fleet operates mainly on the continental shelf. Analysis of gill rakers revealed a slightly higher percentage of *A. fallax* than *A. alosa*, as well as the presence of probable F1 hybrids. At the demographic level, although there were significant differences, with *A. alosa* having the highest mean age, the results showed a typical age class distribution of the adult contingent, at the age of sexual maturity, for the three taxa. Condition index results clearly reflected the size differences between the parental species and the intermediate character of the hybrids, with *A. alosa* having the highest value for this index, the hybrids intermediate values and *A. fallax* the lowest values. Trophic spectrum was common to all three taxa, including fish, plankton, benthos and cephalopods; however, *A. fallax* showed more pronounced ichthyophagous behaviour than *A. alosa* and the hybrids. Both species and hybrids share habitat, simultaneously occupying both shallow coastal areas and deeper, more distant areas. Part of the occupied habitat is in Natura 2000 areas. Two new areas are proposed to increase the protection of these species, while fulfilling the EU's obligations towards them.

Keywords: anadromous species; bycatch; European shads; congeneric species; Natura 2000 network; hybridisation; onset of spawning migration; feeding behaviour

Author Contributions: Conceptualization, D.J.N., R.V.-L. and F.C.; methodology, D.J.N. and A.P.; software, D.J.N. and A.P.; validation, D.J.N., A.P., R.V.-L. and F.C.; formal analysis, A.P. and D.J.N.; investigation, D.J.N., A.P., R.V.-L. and F.C.; resources D.J.N., A.P., R.V.-L. and F.C.; data curation, A.P.

and D.J.N.; writing—original draft preparation, D.J.N.; writing—review and editing, D.J.N., A.P., R.V.-L., S.B., M.d.C.C. and F.C.; visualization, D.J.N. and A.P.; supervision, F.C. and R.V.-L.; project administration, F.C., R.V.-L. and D.J.N.; funding acquisition, F.C., R.V.-L. and D.J.N. All authors have read and agreed to the published version of the manuscript.

Funding: This work has been carried out within the framework of the project "Evaluación de las 'capturas incidentales' de *Alosa alosa* y *Alosa fallax* por la flota costera de Galicia: análisis del problema, sensibilización y proposición de medidas de gestión y protección (1MARDEALOSAS)", which has the collaboration of the Biodiversity Foundation of the Ministry for Ecological Transition and the Demographic Challenge, through the Pleamar Programme, co-financed by the FEMP.

Institutional Review Board Statement: Protocols used in this study conform to the ethical laws of the country and have been reviewed by the ethics committee of the University of Santiago de Compostela and the regional government (Xunta de Galicia).

Informed Consent Statement: Not applicable.

Data Availability Statement: Data from this research are available from the corresponding authors upon reasonable request.

Conflicts of Interest: The authors declare no conflict of interest.

Abstract

Analysis of Bycatches of Two Related Anadromous Shad Species in Fisheries along the Galician Atlantic Coast (NW Iberian Peninsula, Southwest Europe) [†]

David José Nachón [1,*,‡], Alejandro Pico [1], Rufino Vieira-Lanero [1], Sandra Barca [1], María del Carmen Cobo [2] and Fernando Cobo [1]

[1] Departmento de Zooloxía, Xenética e Antropoloxía Física, Facultade de Bioloxía, Universidade de Santiago de Compostela, 15782 Santiago de Compostela, Spain; alejandro.pico.calvo@rai.usc.es (A.P.); rufino.vieira@usc.es (R.V.-L.); sandra.barca@usc.es (S.B.); fernando.cobo@usc.es (F.C.)

[2] Department of Biological Sciences, University of Alabama, Tuscaloosa, AL 35487, USA; mariadelcarmen.cobo@usc.es

* Correspondence: davidjose.nachon@usc.es

† Presented at the IX Iberian Congress of Ichthyology, Porto, Portugal, 20–23 June 2022.

‡ Presenting author (Oral communication).

Citation: Nachón, D.J.; Pico, A.; Vieira-Lanero, R.; Barca, S.; Cobo, M.d.C.; Cobo, F. Analysis of Bycatches of Two Related Anadromous Shad Species in Fisheries along the Galician Atlantic Coast (NW Iberian Peninsula, Southwest Europe). *Biol. Life Sci. Forum* **2022**, *13*, 57. https://doi.org/10.3390/blsf2022013057

Academic Editor: Alberto Teodorico Correia

Published: 7 June 2022

Publisher's Note: MDPI stays neutral with regard to jurisdictional claims in published maps and institutional affiliations.

Abstract: Allis shad, *Alosa alosa* (Linnaeus, 1758) and Twaite shad, *Alosa fallax* (Lacépède, 1803) are two anadromous Clupeidae species that spend most of their life cycle in the marine environment, where they have a coastal distribution, entering the rivers only during the spawning migration. There is no commercial shad fishery at sea; however, due to their gregarious, shoal-forming behaviour close to the shore, they are frequently caught as "bycatch" along with the "intentional catches". In fact, both species have been arriving at Galician fish markets (NW Iberian Peninsula) for several decades. Hence, it was necessary to conduct a study in order to gain a deeper understanding of the extent of this phenomenon. To this end, we analysed the statistics on first-sale fishing in the Galician Fishing Platform. In addition, specimens of the genus *Alosa* ($N = 345$, sold in first sales not as *A. alosa*) were acquired at the A Guarda, Malpica and Coruña fish markets, from January to March 2021, in order to determine their specific identity. Finally, additional information regarding the capture of the specimens (precise location of capture, bathymetry, fishing method, target species, etc.) was requested from fishermen. There are records of catches of specimens of the genus *Alosa* in 14 Galician fish markets, from 1997 to 2020. Catches of *A. alosa* were regular throughout the period 1997–2020, amounting to a total of 23,956 kg. Regarding *A. fallax*, it is surprising to note the virtual absence of records throughout the period studied and the high pulses in certain years, with truly exceptional catches (5950 kilos in 1998 or 7320 kilos in 2018). Examination of the number of gill rakers, showed that 59% ($N = 203$) of the acquired specimens were not *A. alosa*. Of these 203 misidentified specimens, 81% ($N = 164$) were *A. fallax* and 19% ($N = 39$) were hybrids. Most of the shads were caught in the midst of flatfish, hake or sea bass fisheries, with driftnets being the main fishing gear, and the fishing vessels involved were of different sizes (from large vessels to traditional boats).

Keywords: bycatch; anadromous species; European shads; hybridization; artisanal fisheries; fisheries statistics; administrative areas; NW Iberian Peninsula

Author Contributions: Conceptualization, D.J.N., R.V.-L. and F.C.; methodology, D.J.N. and A.P.; software, D.J.N. and A.P.; validation, D.J.N., A.P., R.V.-L. and F.C.; formal analysis, A.P. and D.J.N.; investigation, D.J.N., A.P., R.V.-L. and F.C.; resources D.J.N., A.P., R.V.-L. and F.C.; data curation, A.P. and D.J.N.; writing—original draft preparation, D.J.N.; writing—review and editing, D.J.N., A.P., R.V.-L., S.B., M.d.C.C. and F.C.; visualization, D.J.N. and A.P.; supervision, F.C. and R.V.-L.; project administration, F.C., R.V.-L. and D.J.N.; funding acquisition, F.C., R.V.-L. and D.J.N. All authors have read and agreed to the published version of the manuscript.

Funding: This work has been carried out within the framework of the project "Evaluación de las 'capturas incidentales' de *Alosa alosa* y *Alosa fallax* por la flota costera de Galicia: análisis del problema, sensibilización y proposición de medidas de gestión y protección (1MARDEALOSAS)", which has the collaboration of the Biodiversity Foundation of the Ministry for Ecological Transition and the Demographic Challenge, through the Pleamar Programme, co-financed by the FEMP.

Institutional Review Board Statement: Protocols used in this study conform to the ethical laws of the country and have been reviewed by the ethics committee of the University of Santiago de Compostela and the regional government (Xunta de Galicia).

Informed Consent Statement: Not applicable.

Data Availability Statement: Data from this research are available from the corresponding authors upon reasonable request.

Conflicts of Interest: The authors declare no conflict of interest.

biology and life sciences forum

Abstract

Effects of a Pyrethroid Pesticide on the Behaviour of Native and Non-Native Cypriniformes Fish [†]

Paulo Branco [1,2,*,‡], Inês Vieira [1], Margarida Oliveira [1], Tamara Leite [1,2] and José Maria Santos [1,2]

[1]   Forest Research Centre, School of Agriculture, University of Lisbon, 1349-017 Lisbon, Portugal; ines.margarida2701@gmail.com (I.V.); mpoliveira2000@gmail.com (M.O.); tamaraleite@e-isa.ulisboa.pt (T.L.); jmsantos@isa.ulisboa.pt (J.M.S.)

[2]   Associated Laboratory TERRA, 1349-017 Lisbon, Portugal

*   Correspondence: pjbranco@isa.ulisboa.pt

†   Presented at the IX Iberian Congress of Ichthyology, Porto, Portugal, 20–23 June 2022.

‡   Presenting author (Oral communication).

Abstract: River ecosystems are exposed to a multitude of stressors, of which pesticide run-off is increasingly becoming a global environmental concern due to adverse effects on aquatic organisms. Of those, pyrethroids are now the fourth major group of insecticides in use worldwide; however, their sublethal effects on freshwater fish are still largely unknown. The present study aimed to assess the effects of an acute 2 h sublethal exposure to different levels of the pyrethroid pesticide esfenvalerate on the behaviour of two Cypriniformes species: the native Iberian barbel (*Luciobarbus bocagei*) and the non-native invasive bleak (*Alburnus alburnus*). Fish were previously exposed to three esfenvalerate concentrations (control, 1.2 µg/L (mild), and 2.0 µg/L (severe)), before being placed into a three-artificial-flume channel mesocosm for behavioural trials on (i) routine activity, (ii) shoal cohesion, and (iii) boldness. Significant differences in fish behaviour were detected for the barbel, as individuals were found to spend a higher proportion of time holding their position (i.e., resting) from the control to the severe esfenvalerate concentration. Behavioral changes were also detected for boldness, measured by the proportion of fish attempts to negotiate the upstream ramp, which were significantly higher in the control and in the severe concentration compared to the mild one. For the invasive bleak, there were no significant differences in any of the behavioural parameters upon previous exposure to an increasing esfenvalerate concentration. The present study demonstrated that even a short-term exposure to the pyrethroid esfenvalerate was sufficient to alter the behaviour of a native fish species, while no affecting the non-native, which may confer them greater competitive advantages in the context of global changes.

Keywords: pyrethroids; freshwater fish species; boldness; schooling; conservation

Citation: Branco, P.; Vieira, I.; Oliveira, M.; Leite, T.; Santos, J.M. Effects of a Pyrethroid Pesticide on the Behaviour of Native and Non-Native Cypriniformes Fish. *Biol. Life Sci. Forum* **2022**, *13*, 63. https://doi.org/10.3390/blsf2022013063

Academic Editor: Alberto Teodorico Correia

Published: 8 June 2022

Publisher's Note: MDPI stays neutral with regard to jurisdictional claims in published maps and institutional affiliations.

Author Contributions: Conceptualization: P.B. and J.M.S., investigation: I.V., M.O., T.L., P.B. and J.M.S., resources: J.M.S.; Coordination: P.B. and J.M.S.; Data Analisys: I.V., M.O. and T.L. All authors have read and agreed to the published version of the manuscript.

Funding: Tamara Leite was supported by a Ph.D. grant from the FLUVIO–River Restoration and Management pro-gramme funded by Fundação para a Ciência e a Tecnologia I. P. (FCT),Portugal (UI/BD/15052/2021). Paulo Branco was financed by national funds via FCT, under "Norma Transitória—DL57/2016/CP1382/CT0020". The study was partially funded by the project Dammed Fish (PTDC/CTA-AMB/4086/2021). Forest Research Centre (CEF) is a research unit funded by Fundação para a Ciência e a Tecnologia I.P. (FCT), Portugal (UIDB/00239/2020).

Institutional Review Board Statement: The study was conduct in line with national and international guidelines of animal welfare.

Data Availability Statement: Data is available upon request to the authors.

Conflicts of Interest: The authors declare no conflict of interest.

biology and life sciences forum

Abstract

Tuna Larvae (Scombridae) off Eastern Australia: When and Where Are They Spawned? [†]

Clare Cao [1], Anthony G. Miskiewicz [1,2,*,‡], Sharon A. Appleyard [3], Paloma Matis [1,4], Hayden Schilling [1,4] and Iain Suthers [1,4]

[1] School of Biological, Earth & Environmental Sciences, University of New South Wales, Sydney, NSW 2052, Australia; clare.cao3@gmail.com (C.C.); paloma.matis@sims.org.au (P.M.); hayden.schilling@sims.org.au (H.S.); i.suthers@unsw.edu.au (I.S.)

[2] Ichthyology Department, Australian Museum Research Institute, Australian Museum, Sydney, NSW 2010, Australia

[3] CSIRO Australian National Fish Collection, National Research Collections Australia, Hobart, TAS 7000, Australia; sharon.appleyard@csiro.au

[4] Sydney Institute of Marine Science, Chowder Bay Road, Mosman, NSW 2088, Australia

* Correspondence: tonymisk@yahoo.com

† Presented at the IX Iberian Congress of Ichthyology, Porto, Portugal, 20–23 June 2022.

‡ Presenting author (oral communication).

Citation: Cao, C.; Miskiewicz, A.G.; Appleyard, S.A.; Matis, P.; Schilling, H.; Suthers, I. Tuna Larvae (Scombridae) off Eastern Australia: When and Where Are They Spawned? *Biol. Life Sci. Forum* **2022**, 13, 66. https://doi.org/10.3390/blsf2022013066

Academic Editor: Alberto Teodorico Correia

Published: 8 June 2022

Publisher's Note: MDPI stays neutral with regard to jurisdictional claims in published maps and institutional affiliations.

Abstract: Tunas, mackerels, and bonitos (Scombridae) are commercially valuable fishes and contribute to the functioning of pelagic marine ecosystems worldwide, either as large predatory fishes or forage fishes. Despite this, the seasonality of larvae for most scombrids off eastern Australia is unknown. Using monthly plankton samples collected from 2014 to 2020 off Brisbane (27° S) and Sydney (34° S) and scombrid larvae in samples from several historical voyages at various times between 1983 and 2003 that were sampled between Brisbane and Sydney, we describe the spatial and temporal distribution of scombrid larvae occurring off eastern Australia. Based on morphology and mitochondrial DNA cytochrome c oxidase subunit I (COI) barcoding, we identified *Acanthocybium solandri* (wahoo), *Allothunnus fallai* (slender tuna), *Auxis rochei* (bullet tuna), *Auxis thazard* (frigate mackerel), *Euthynnus affinis* (mackerel tuna), *Katsuwonus pelamis* (skipjack tuna), *Sarda australis* (Australian bonito), *Thunnus albacares* (yellowfin tuna) and *Thunnus tonggol* (longtail tuna). *Auxis rochei* was the most abundant species, with predominately preflexion larvae present from October (mid spring) to February (late summer) off the coast of North Stradbroke Island (27° S). The water temperature significantly influenced the larval distributions of *A. rochei* (20–24 °C) and *E. affinis* (24–26 °C), while *E. affinis* larval abundances were positively associated with eddy kinetic energy. This highlights the importance of western boundary currents and their eddies in facilitating the spawning of scombrids.

Keywords: Scombridae; larvae; eastern Australia; cytochrome c oxidase subunit I (COI) barcoding

Author Contributions: Conceptualization, C.C., A.G.M., P.M. and I.S; methodology, C.C., A.G.M., S.A.A., P.M. and I.S; validation, C.C., A.G.M. and S.A.A.; formal analysis, all authors; investigation, C.C., A.G.M., P.M., H.S. and I.S.; resources, S.A.A.; data curation and writing—original, draft preparation, C.C. and A.G.M.; review and editing, all authors. All authors have read and agreed to the published version of the manuscript.

Funding: This research was funded by the Sydney Institute of Marine Science and co-funding from the University of New South Wales, Commonwealth Scientific Industrial Research Organisation and Australia's Integrated Marine Observing System (IMOS). Data was sourced from IMOS–IMOS is enabled by the National Collaborative Research Infrastructure Strategy (NCRIS).

Institutional Review Board Statement: Larval fish collections at the IMOS national reference stations were collected under ethics permit UNSW ACEC 19/96B.

Informed Consent Statement: Not applicable.

Data Availability Statement: Not applicable.

Conflicts of Interest: The authors declare no conflict of interest.

biology and life sciences forum

Abstract

Projecte Escanyagats: A Long-Term Strategy for the Conservation of Stickleback (*Gasterosteus aculeatus*) in Catalonia [†]

Quim Pou-Rovira [1,2,*,‡], Iago Pérez-Novo [2], Carla Juvinyà [2], Jesús Ríos [2], Eudald Vicens [2], Pau Ortega [2], Guillem Llenas [1], Andreu Porcar [1] and Eloi Cruset [1]

[1] Sorelló, Estudis al Medi Aquatic SL, 17003 Girona, Spain; guillem.llenas@sorello.net (G.L.); andreu.porcar@sorello.net (A.P.); eloi.cruset@sorello.net (E.C.)
[2] Associació la Sorellona, 17004 Girona, Spain; iago.perez@sorellona.org (I.P.-N.); carla.juvinya@sorellona.org (C.J.); jesus.rios@sorellona.org (J.R.); eudald.vicens@sorellona.org (E.V.); pau.ortega@sorellona.org (P.O.)
* Correspondence: quim.pou@sorello.net
† Presented at the IX Iberian Congress of Ichthyology, Porto, Portugal, 20–23 June 2022.
‡ Presenting author (Oral communication).

Citation: Pou-Rovira, Q.; Pérez-Novo, I.; Juvinyà, C.; Ríos, J.; Vicens, E.; Ortega, P.; Llenas, G.; Porcar, A.; Cruset, E. *Projecte Escanyagats*: A Long-Term Strategy for the Conservation of Stickleback (*Gasterosteus aculeatus*) in Catalonia. *Biol. Life Sci. Forum* **2022**, *13*, 67. https://doi.org/10.3390/blsf2022013067

Academic Editor: Alberto Teodorico Correia

Published: 8 June 2022

Publisher's Note: MDPI stays neutral with regard to jurisdictional claims in published maps and institutional affiliations.

Abstract: Stickleback (*Gasterosteus aculeatus*) is an endangered species in Catalonia. It has disappeared from most of its historical range, including at least five whole basins, and the remaining populations continue to decline. Our monitoring, carried out on most Catalan populations in the north-east of the country, indicates that over the last 15 years there has been a reduction of between 50% and 75% everywhere. Even worse, many of the remaining populations are in heavily modified Mediterranean river stretches, subjected to multiple anthropogenic pressures or highly vulnerable to increasingly drastic droughts due to climate change. In this context, in 2018 we started a project that aims to implement a long-term, low-cost strategy for the conservation of this species in Catalonia. We have opted for a combination of broad-spectrum alternative measures: (1) Strategic restoration of small stretches of river of high interest to the species, mainly intended to ensure the maintenance of flooded pools during extreme droughts. (2) Reintroduction in river stretches with historical presence, firstly in watersheeds with extant populations. (3) Gradually, application of assisted migration to water bodies without historical presence, but with good ecological conditions for a probable successful establishment. (4) Creation of new ex-situ populations in small artificial water bodies, from urban ornamental ponds to small irrigating reservoirs, reaching agreements with the owners, and often carrying out essential previous actions, such as the exotic fish removal. (5) Continuation and extension of regular sampling, in order to implement a permanent monitoring of its trends, and expand the specific knowledge on some key aspects of its ecology. (6) Implementation of a specific environmental awareness and education campaign, as well as environmental volunteering, for the involvement of society in the conservation of this small fish and its habitats. The known genetic differentiation of the remaining populations is being taken into account in the planning of these actions. We will present the results achieved so far, and the plans for the coming years.

Keywords: stickleback; *Gasterosteus aculeatus*; fish conservation; assisted migration; ex situ conservation; reintroduction; river restoration; *Projecte Escanyagats*

Author Contributions: Conceptualization, Q.P.-R., I.P.-N., C.J.; methodology, all authors; formal analysis, Q.P.-R., I.P.-N.; data curation, Q.P.-R., I.P.-N., C.J, J.R., E.V., P.O., G.L., A.P., E.C.; writing—review and editing, Q.P.-R., I.P.-N., C.J. All authors have read and agreed to the published version of the manuscript.

Funding: Several consecutive small grants coming from several public agents (Diputació de Girona, Fundació Zoo de Barcelona & Generalitat de Catalunya).

Institutional Review Board Statement: Ethical review and approval were waived for this study due to the use of fish standard sampling methods, under the authorization on national authorities on nature conservation and animal welfare.

Informed Consent Statement: Not applicable.

Data Availability Statement: Not applicable.

Conflicts of Interest: The authors declare no conflict of interest.

biology and life sciences forum

Abstract

Are River Springs Devoid of Fish?—The Case of the Maciço Calcário Estremenho [†]

Sofia Isabel Pardal [1,*,‡], Cristina S. Lima [1], Sofia Quaresma [2] and Carla Sousa-Santos [1]

[1] MARE—Centro de Ciências do Mar e do Ambiente, ISPA—Instituto Universitário de Ciências Psicológicas, Sociais e da Vida, 1149-041 Lisboa, Portugal; clima@ispa.pt (C.S.L.); carla.santos@ispa.pt (C.S.-S.)

[2] ICNF—Instituto da Conservação da Natureza e das Florestas, PNSAC—Parque Natural das Serras de Aires e Candeeiros, 2040-215 Rio Maior, Portugal; sofia.quaresma@icnf.pt

* Correspondence: spardal19@gmail.com

† Presented at the IX Iberian Congress of Ichthyology, Porto, Portugal, 20–23 June 2022.

‡ Presenting author (Poster presentation).

Abstract: The evolution of strictly freshwater fishes in the Iberian Peninsula is correlated with the paleogeomorphological evolution of Iberian basins and mountain ranges. Located in Portugal's central region, the Maciço Calcário Estremenho (MCE) is a unique limestone geological formation with river springs that separates two different biogeographic regions for freshwater ichthyofauna: at its western slope, small coastal streams with ancient connections to the Mondego Riverland, and at its eastern slope, sub-basins of the Tagus River. Springs from these rivers are located only a few kilometers apart and are typically seen as devoid of fish due to their intermittent regime, shallow water column, high slope, and/or existence of insurmountable barriers that virtually prevent upstream fish migration. Although the species richness is known for the region, little is known about which species have colonized the headwaters. The objectives of this study were to (1) assess which fish species are present at the springs and (2) characterize the genetic diversity at the inter- and intrapopulation levels of two target species (*Achondrostoma oligolepis* and *Squalius pyrenaicus*) existing on opposite slopes of the MCE, using one mitochondrial and one nuclear gene. Genetic analyses showed differences between populations from distinct rivers but also at the intrapopulation level, with unique haplotypes being found in some locations within a river basin. This study demonstrated that springs are not devoid of fish in the MCE and may instead be important conservation areas for native species, benefiting from the existing Natural Park established in the region. Despite their geographical proximity, the isolation of populations of the same species in distinct rivers resulted in significant interpopulation genetic differences. Moreover, intrapopulation genetic differences were also found, highlighting two important topics: the sampling scheme can influence the obtained genetic data and the genetic diversity may differ between upstream and downstream stretches of the same river. Sampling several locations within the same river leads, therefore, to more reliable results, and future landscape genetic studies using fast-paced markers are necessary to assess genetic connectivity and to depict how the genetic diversity of strictly freshwater fish species is distributed along a river course.

Keywords: genetic; Portugal; biogeographic; freshwater ichthyofauna

Citation: Pardal, S.I.; Lima, C.S.; Quaresma, S.; Sousa-Santos, C. Are River Springs Devoid of Fish?—The Case of the Maciço Calcário Estremenho. *Biol. Life Sci. Forum* **2022**, *13*, 68. https://doi.org/10.3390/blsf2022013068

Academic Editor: Alberto Teodorico Correia

Published: 8 June 2022

Publisher's Note: MDPI stays neutral with regard to jurisdictional claims in published maps and institutional affiliations.

Author Contributions: Conceptualization, C.S.-S. and S.Q.; methodology, C.S.-S., C.S.L. and S.I.P.; validation, C.S.-S.; formal analysis, C.S.-S. and S.I.P.; investigation, C.S.-S., C.S.L. and S.I.P.; resources, C.S.-S. and C.S.L.; writing—original draft preparation, C.S.-S. and S.I.P.; writing—review and editing, C.S.-S., C.S.L. and S.Q.; visualization, C.S.-S., C.S.L. and S.I.P.; supervision and project administration, C.S.-S. All authors have read and agreed to the published version of the manuscript.

Funding: This study had the support of national funds through Fundação para a Ciência e Tecnologia (FCT), under the strategic project MARE-UIDB/04292/2020 granted to MARE and the project LA/P/0069/2020 granted to the Associate Laboratory ARNET.

Institutional Review Board Statement: All sampling complied with Portuguese legislation and permits for field work were given by the National Institute for the Conservation of Nature and Forests (ICNF), Portugal (license nb. 515/2022/CAPT).

Informed Consent Statement: Not applicable.

Data Availability Statement: DNA sequences will be available in Genbank.

Conflicts of Interest: The authors declare no conflict of interest.

Abstract

Optimization of an Analytical Protocol for the Extraction of Microplastics from Seafood Samples with Different Levels of Fat [†]

Diogo M. Silva [1,2,*,‡], Marisa Almeida [2], Francisco Guardiola [3], Sabrina M. Rodrigues [1,2] and Sandra Ramos [2]

[1] ICBAS—Institute of Biomedical Sciences Abel Salazar, Porto University, Rua de Jorge Viterbo Ferreira n° 228, 4050-313 Porto, Portugal; smagalhaes@ciimar.up.pt
[2] CIIMAR—Interdisciplinary Centre of Marine and Environmental Research, University of Porto, Terminal de Cruzeiros do Porto de Leixões, Avenida General Norton de Matos, S/N, 4450-208 Matosinhos, Portugal; calmeida@ciimar.up.pt (M.A.); ssramos@ciimar.up.pt (S.R.)
[3] Faculty of Biology, C. Campus Universitario, University of Murcia, 5, 30100 Murcia, Spain; faguardiola@um.es
* Correspondence: silva.diogom@gmail.com; Tel.: +351-917580697
† Presented at the IX Iberian Congress of Ichthyology, Porto, Portugal, 20–23 June 2022.
‡ Presenting author (poster).

Citation: Silva, D.M.; Almeida, M.; Guardiola, F.; Rodrigues, S.M.; Ramos, S. Optimization of an Analytical Protocol for the Extraction of Microplastics from Seafood Samples with Different Levels of Fat. *Biol. Life Sci. Forum* **2022**, *13*, 75. https://doi.org/10.3390/blsf2022013075

Academic Editor: Alberto Teodorico Correia

Published: 9 June 2022

Publisher's Note: MDPI stays neutral with regard to jurisdictional claims in published maps and institutional affiliations.

Abstract: The global production of plastics increased from 1.5 million tons in 1950 to 370 million tons in 2020. Microplastics (MPs) are plastic particles smaller than 5 mm, with a concerning ubiquitous occurrence in marine environments, posing risks to wildlife and humans. Seafood is an important source of proteins (including all the essential amino acids), and other essential nutrients with numerous health benefits. Human exposure to MPs through the consumption of contaminated seafood, together with the potential of these particles to cause health risks, is motivating a better understanding of the security of our diets. In recent years, several methodologies have been used to investigate MP occurrence in seafood, although the existence of complex sample matrices, namely, a high level of fats, can pose severe difficulties and compromise the efficiency of MP quantification. To solve this issue, the present study aimed to develop a detailed protocol suitable to process seafood samples with different levels of fats (fish and mollusks, from fresh and canned sources). For the sample digestion, several tests were performed using two solutions (10% KOH, 30% H_2O_2) with different volumes, temperatures (40 °C, 65 °C) and durations (24, 48, 72 h) of incubation. To remove the fat remaining after digestion, three detergents (two laboratory surfactants and a commercial washing-up liquid) and 96% ethanol were tested. Manual recovery of the fat layer was also tested. For filtration, two filter membranes were compared (glass microfiber and nitrate cellulose filters, 0.45 μm pore size). The efficiency of the different experiments was determined through the observation and estimation, in percentage, of the organic matter digestion and post-digestion fat removal. The methodology optimized in this study combined a sample digestion with 30% H_2O_2 incubated at 65 °C, for 24 to 48 h, with a manual separation of the post-digestion fats with immediate observation in a stereomicroscope. After, this methodology was applied to different types of polymers (e.g., polyethylene, polypropylene, polyethylene-terephthalate, polystyrene), to investigate if these procedures altered the integrity of MPs. The results show that this methodology will allow us to efficiently process complex seafood samples with different fat levels, without compromising MPs' integrity.

Keywords: microplastics; seafood; canned seafood; fats; extraction protocol; contamination

Funding: This research was partially funded by Ocean3R (NORTE-01-0145-FEDER-000064) and ATLANTIDA (ref. NORTE-01-0145-FEDER-000040) projects, supported by the Norte Portugal Regional Operational Programme (NORTE 2020) under the PORTUGAL 2020 Partnership Agreement and through the European Regional Development Fund (ERDF). Also, FCT is awarded by the Strategic Funding UIDB/04423/2020 and UIDP/04423/2020 through national funds provided by FCT and ERDF, a PhD fellowship to DS (2020.06088.BD) and a research contract to SRamos (DL57/2016/CP1344/CT0020).

Institutional Review Board Statement: Not applicable.

Informed Consent Statement: Not applicable.

Conflicts of Interest: The authors declare no conflict of interest.

Abstract

The Role of Insular African Mangroves as Nursery Areas for the Early Life Stages of Fish [†]

Diogo Dias [1,*,‡], Filipe Ribeiro [1], Ana C. Brito [1,2], Filipa Afonso [1], Francisco Azevedo e Silva [1], João Medeiros [1], Joshua Heumüller [1], Paula Chainho [1,3], Ricardo Lima [3,4,5], Tomás Simões [1] and Pedro M. Félix [1,3]

[1] MARE—Marine and Environmental Sciences Centre, Faculdade de Ciências da Universidade de Lisboa, 1749-016 Campo Grande, Portugal; fmribeiro@fc.ul.pt (F.R.); acbrito@fc.ul.pt (A.C.B.); fmafonso@fc.ul.pt (F.A.); fhsilva@fc.ul.pt (F.A.e.S.); jpmedeiros@fc.ul.pt (J.M.); jaheumuller@fc.ul.pt (J.H.); pmchainho@fc.ul.pt (P.C.); tfsimoes@fc.ul.pt (T.S.); pmfelix@fc.ul.pt (P.M.F.)
[2] Departamento de Biologia Vegetal, Faculdade de Ciências da Universidade de Lisboa, 1749-016 Campo Grande, Portugal
[3] Departamento de Biologia Animal, Faculdade de Ciências da Universidade de Lisboa, 1749-016 Campo Grande, Portugal; rfaustinol@gmail.com
[4] Ce3C—Centre for Ecology, Evolution and Environmental Changes, Faculdade de Ciências da Universidade de Lisboa, 1749-016 Campo Grande, Portugal
[5] Associação Monte Pico, Monte Café, Mé Zóchi CP 1119, São Tomé and Príncipe
[*] Correspondence: dmlvmd@gmail.com; Tel.: +351-965092863
[†] Presented at the IX Iberian Congress of Ichthyology, Porto, Portugal, 20–23 June 2022.
[‡] Presenting author (oral presentation).

Citation: Dias, D.; Ribeiro, F.; Brito, A.C.; Afonso, F.; Azevedo e Silva, F.; Medeiros, J.; Heumüller, J.; Chainho, P.; Lima, R.; Simões, T.; et al. The Role of Insular African Mangroves as Nursery Areas for the Early Life Stages of Fish. *Biol. Life Sci. Forum* **2022**, *13*, 3080. https://doi.org/10.3390/blsf2022013080

Academic Editor: Alberto Teodorico Correia

Published: 13 June 2022

Publisher's Note: MDPI stays neutral with regard to jurisdictional claims in published maps and institutional affiliations.

Abstract: Mangroves have been recognized worldwide as crucial nursery areas for fish larvae and juveniles. Although they are critical for managing coastal fish stocks, information about larval fish communities in African island mangroves is scarce and these potential nursery areas in São Tomé Island have remained understudied. Fish larvae were collected over four weeks from October to November 2020 using light traps, passive plankton tows and seine nets in a multi-habitat approach. To overcome species identification constraints, both morphology and molecular analysis were considered. A total of 4 010 larvae were caught across all methods belonging to 16 families or 26 species. A few species dominated the ichthyoplankton community and the most abundant families were Cichlidae—especially the invasive *Oreochromis mossambicus* (47%)—and Gobiidae (43%), constituted by 7 *taxa*. The remaining 14 families only accounted for about 10% of total larvae captured. Three new species were recorded for the first time in the island mangroves and another three species were documented for the first time in the São Tomé Island. Taxa composition and richness varied considerably between sampling techniques. The highest taxa richness and diversity were recorded in the Malanza mangrove (25 species) while Praia das Conchas (9 species) was not able to sustain similar levels of biodiversity. Differences on fish larvae composition were found within the studied mangroves, depicting a strong influence of habitat type and a relative position within each system. These community composition patterns were marginally influenced by local environmental conditions such as temperature and dissolved oxygen. Overall, a total of eleven taxa have commercial interest and their presence as juveniles and larvae in São Tomé mangroves reinforces the need for conservation of these ecosystems and shows direct implications for the sustainability of the local fisheries.

Keywords: fish larvae; juvenile; São Tomé; West Africa; cytochrome c oxidase I (COI)

Author Contributions: Conceptualization, F.R., P.M.F. and D.D.; methodology, F.R., P.M.F. and D.D.; investigation, F.R., F.A., F.A.e.S., J.M., J.H., T.S., P.M.F. and D.D.; writing—original draft preparation, D.D.; writing—review and editing, D.D., F.R., A.C.B., F.A., F.A.e.S., J.M., J.H., P.C., R.L., T.S. and P.M.F.; visualization, D.D.; project administration, P.M.F.; funding acquisition, F.R., P.M.F. and R.L. All authors have read and agreed to the published version of the manuscript.

Funding: This study was funded by the Critical Ecosystem Partnership Fund (CEPF) "Participatory Management of Malanza and Praia das Conchas Mangroves in São Tomé" (GFWA-2018-LG-02); the Foundation for Science and Technology (FCT Ministério da Ciência, Tecnologia e Ensino Superior) supported this work through the strategic plan of the Marine and Environmental Sciences Centre (MARE) (UID/Multi/04326/2019) and through the Centre for Ecology, Evolution and Environmental Changes's (cE3c) Unit funding (UIDB/00329/2020).

Institutional Review Board Statement: Not applicable.

Informed Consent Statement: Not applicable.

Data Availability Statement: The data presented in this study are available on request from the corresponding author. The data are not publicly available due to privacy restrictions.

Conflicts of Interest: The authors declare no conflict of interest.

Abstract

On the Use of Stereo-Video System to Assess Microhabitat Preferences of the Spanish Toothcarp and Mosquitofish in Coastal Salt Marshes †

Lluís Zamora ‡

GRECO, Institute of Aquatic Ecology, University of Girona, Av. Maria Aurèlia Capmany 69, 17003 Girona, Spain; lluis.zamora@udg.edu; Tel.: +34-972418700
† Presented at the IX Iberian Congress of Ichthyology, Porto, Portugal, 20–23 June 2022.
‡ Presenting author (Oral communication).

Abstract: Stereo-video systems (hereafter SVS) have been widely applied to study fish ecology in marine coastal ecosystems and more recently in freshwater, especially in headwater streams, due to their dependence on water clarity. Here, we assess the use of these non-destructive methods to study microhabitat use, size structure, and the abundance of endangered Spanish toothcarp (*Apricaphanius iberus*) and the invasive mosquitofish (*Gambusia holbrooki*) in coastal salt marshes. Stereo-video measurements were obtained in situ by means of static pairs of GoPro HERO7 cameras in different shallow coastal lagoons of northeastern Spain. The analysis of stereo-video recordings were processed using the open-source videogrammetry software VidSync 1.661 in order to identify the species, sex, and total length of each fish as well as their relative position in the water column. A total of ninety 17.5 min long stereo-video clips containing more than 7300 fish positions were processed for this study. Fish assemblage and population size structure gathered with this method were compared with catches at the same places using fyke nets. The accuracy and precision of fish-length estimation using SVS was also tested in the lab. SVS revealed differential water-column use, with Spanish toothcarp occurring in a lower-water column. Larger mosquitofish tended to use the upper part of the water column, whereas no clear ontogenetic shift was observed for the Spanish toothcarp. Fyke nets and SVS yielded a similar species composition and considerably correlated with abundances for two species, particularly for mosquitofish, across the six coastal ponds. The size structure varied significantly with the two techniques, with fyke nets apparently being more size-selective as the smallest mosquitofish were underrepresented in fyke nets compared with SVS. Our results suggest that SVS is a non-destructive method that does not require capturing and handling the fish, and they also suggest that it is an ideal technique for studying endangered species, with enormous potential to improve the knowledge of microhabitat use and the behavior of fish species in natural conditions.

Keywords: *Apricaphanius iberus*; coastal lagoons; *Gambusia holbrooki*; microhabitat use; underwater videogrammetry; water column

Citation: Zamora, L. On the Use of Stereo-Video System to Assess Microhabitat Preferences of the Spanish Toothcarp and Mosquitofish in Coastal Salt Marshes. *Biol. Life Sci. Forum* **2022**, *13*, 83. https://doi.org/10.3390/blsf2022013083

Academic Editor: Alberto Teodorico Correia

Published: 13 June 2022

Funding: This research was funded by the Spanish Ministry of Science and Innovation ref. PID2019-103936GB-C21.

Institutional Review Board Statement: This study was authorized by the department of territory and sustainability of the Catalan Government, licenses 2021PNATMBTAUT007 and 2021PNATAAEAUT007 respectively.

Informed Consent Statement: Not applicable.

Data Availability Statement: Not applicable.

Conflicts of Interest: The author declares no conflict of interest.

biology and life sciences forum

Abstract

EcoPeak4Fish: A Multidisciplinary Project Targeting the Protection of Fish Populations Affected by Hydropeaking [†]

Isabel Boavida [1], José Maria Santos [2], Maria João Costa [1], Renan Leite [1,2,*,‡], Anthony Merianne [1], Maria Manuela Portela [1], Francisco Godinho [3], Pedro Leitão [3], Rui Mota [4], Jeffrey Tuhtan [5] and António Pinheiro [1]

[1] Civil Engineering for Research and Innovation for Sustainability, Instituto Superior Técnico, University of Lisbon, 1049-003 Lisbon, Portugal; isabelboavida@tecnico.ulisboa.pt (I.B.); mariajcosta@tecnico.ulisboa.pt (M.J.C.); anthony.merianne@grenoble-inp.org (A.M.); maria.manuela.portela@tecnico.ulisboa.pt (M.M.P.); antonio.pinheiro@tecnico.ulisboa.pt (A.P.)

[2] Forest Research Centre, School of Agriculture, University of Lisbon, 1649-004 Lisbon, Portugal; jmsantos@isa.ulisboa.pt

[3] Hidroerg-Projectos Energéticos Lda., 1300-327 Lisbon, Portugal; francisco.godinho@hidroerg.pt (F.G.); pedro.leitao@hidroerg.pt (P.L.)

[4] Nova School of Business, Universidade Nova de Lisboa, 1099-085 Lisbon, Portugal; rui.mota@novasbe.pt

[5] Department of Computer Systems, Tallinn University of Technology, 12616 Tallinn, Estonia; jeffrey.tuhtan@ttu.ee

* Correspondence: renanleite@edu.ulisboa.pt

† Presented at the IX Iberian Congress of Ichthyology, Porto, Portugal, 20–23 June 2022.

‡ Presenting author (oral communication).

Citation: Boavida, I.; Santos, J.M.; Costa, M.J.; Leite, R.; Merianne, A.; Portela, M.M.; Godinho, F.; Leitão, P.; Mota, R.; Tuhtan, J.; et al. EcoPeak4Fish: A Multidisciplinary Project Targeting the Protection of Fish Populations Affected by Hydropeaking. *Biol. Life Sci. Forum* **2022**, *13*, 85. https://doi.org/10.3390/blsf2022013085

Academic Editor: Alberto Teodorico Correia

Published: 14 June 2022

Publisher's Note: MDPI stays neutral with regard to jurisdictional claims in published maps and institutional affiliations.

Abstract: Hydroelectricity demand is still growing due to its reduced carbon impact and strong dispatchability. Concurrently, the necessity to support self-sustainable fish populations in a cost-effective way and restore water-related ecosystems is urgent. The impacts of rapid and artificial flow fluctuations caused by peak electricity demand, i.e., hydropeaking, on fish fauna are still largely unknown, particularly for cyprinid species. Flow-refuges (e.g., lateral deflectors) are believed to help fish with rapidly changing flows and high currents. Nonetheless, most studies addressing hydropeaking and its impacts focus on salmonids. Recent studies have assessed the utility of flow-refuges for cyprinids in controlled experimental conditions. However, fish responses to hydropeaking and the use of flow-refuges in peaking rivers remain unknown. Rethinking the hydropower operation to improve fish habitat during key lifecycle stages (e.g., spawning) can be an option. However, these measures may affect the hydropower production profit. Including habitat use in models of hydropower plant (HPP) optimal management is usually achieved by adding restrictions regarding minimum flows. The feedback between the available habitat and profit has not been explicitly modelled. Including a description of how the available habitat changes with water flow can help estimate tradeoffs between profit maximization and habitat preservation and to inform the development of flow restrictions. EcoPeak4Fish intends to answer these questions in a multidisciplinary approach with the 4E's: Ecology, Engineering and Economics in the profit of the Ecosystem protection. This project aims to assess the effects of hydropeaking in cyprinid species, propose a flow-refuge prototype and assess its cost-effectiveness, and develop a framework to adapt the HPP operation scheme to maximize profits and environmental benefits for a sustainable use of hydropower energy. The project intends to answer the following questions: How do fish react under hydropeaking conditions? Are flow-refuges an effective measure to mitigate impacts and contribute to the self-sustainability of fish populations? How can one find the HPP operation scheme that maximizes profit and power production while maximizing the suitable habitat for fish populations? This knowledge will be fundamental to implement new and redefine operational schemes, to recommend design criteria for flow-refuges, and to mobilize policy-makers to define legal instruments for hydropeaking.

Keywords: hydropeaking; hydropower plants; cyprinids; flow refuges; sustainability

Author Contributions: Conceptualization, I.B. and J.M.S.; Methodology, I.B., J.M.S. and M.J.C.; Validation, I.B., M.J.C. and J.M.S.; Formal analysis, I.B., R.L., M.J.C. and A.M.; Investigation, I.B., M.J.C., R.L. and A.M.; Resources, I.B., M.J.C., J.M.S. and A.P.; Data curation, I.B., M.J.C. and J.M.S.; Writing-original draft preparation, R.L. and A.M.; Writing-review and editing, I.B., M.J.C., J.M.S., R.L., A.M., M.M.P., F.G., P.L., R.M. and J.T.; Visualization, I.B., J.M.S. and M.J.C.; Supervision, I.B., J.M.S. and M.J.C.; Project Administration, I.B. and M.J.C.; Funding acquisition, I.B. All authors have read and agreed to the published version of the manuscript.

Funding: This research was funded by FCT "Fundação para a Ciencia e a Tecnologia", grant number: PTDC/EAM-AMB/4531/2020. The authors are grateful for the Foundation for Science and Technology's support through funding UIDB/04625/2020 from the research unit CERIS.

Institutional Review Board Statement: The Institute for Nature Conservation and Forests (ICNF) provided the necessary fishing and handling permits, and General Directorate of Food and Veterinary Medicine (DGAV) provided the authorization to perform experimental research with live animals. All research was conducted in accordance with national and international guidelines for animal welfare.

Informed Consent Statement: Not applicable.

Data Availability Statement: Data is available upon request from the authors.

Conflicts of Interest: The authors declare no conflict of interest.

biology and life sciences forum

Abstract

A Matter of Approach: Analysis of the Flow Refuge Preferences of Iberian Barbels during Pulsed Flows in Flume Conditions [†]

Renan Leite [1,2,*], Maria João Costa [1,*], Anthony Merianne [1,*,‡], Daniel Mameri [2,*], José Maria Santos [2,*], Antonio Nascimento Pinheiro [1,*] and Isabel Boavida [1,*]

[1] Civil Engineering for Research and Innovation for Sustainability, Instituto Superior Técnico, Campus of Alameda, University of Lisbon, 1049-003 Lisbon, Portugal

[2] Forest Research Centre (CEF) and Associate Laboratory TERRA, School of Agriculture, University of Lisbon, Tapada da Ajuda, 1349-017 Lisbon, Portugal

* Correspondence: renanleite@edu.ulisboa.pt (R.L.); mariajcosta@tecnico.ulisboa.pt (M.J.C.); anthony.merianne@grenoble-inp.org (A.M.); dmameri@isa.ulisboa.pt (D.M.); jmsantos@isa.ulisboa.pt (J.M.S.); antonio.pinheiro@tecnico.ulisboa.pt (A.N.P.); isabelboavida@tecnico.ulisboa.pt (I.B.)

† Presented at the IX Iberian Congress of Ichthyology, Porto, 20–23 June 2022.

‡ Presenting author (oral communication).

Citation: Leite, R.; Costa, M.J.; Merianne, A.; Mameri, D.; Santos, J.M.; Pinheiro, A.N.; Boavida, I. A Matter of Approach: Analysis of the Flow Refuge Preferences of Iberian Barbels during Pulsed Flows in Flume Conditions. *Biol. Life Sci. Forum* **2022**, *13*, 86. https://doi.org/10.3390/blsf2022013086

Academic Editor: Alberto Teodorico Correia

Published: 14 June 2022

Publisher's Note: MDPI stays neutral with regard to jurisdictional claims in published maps and institutional affiliations.

Abstract: To fight against global warming, we have to change our ways of consuming energy. Due to its low carbon impact and strong dispatchability, hydroelectric production will be one of the bases of this transition. However, peak electricity demand produces rapid and artificial flow fluctuations in tailwaters, i.e., hydropeaking, which has negative effects on fish biota. Thus, developing effective mitigation measures against hydropeaking is an urgent matter. The present study aims to limit the impact of this flow fluctuation on an Iberian cyprinid fish: the Iberian barbel (*Luciobarbus bocagei*). We experimentally tested different angles of flow refuge entrances (45° and 70°) in an indoor flume (6.5 m × 0.7 m × 0.8 m) to determine if this would affect the behavior of the fish. For each angle configuration, two refuges were installed and distanced 2.30 m from each other on the same side of the flume. Three possible resting locations were defined: downstream, inside, or upstream of each structure. Both angles were tested at 7 L/s (base flow), simulating the normal conditions of the river, and 60 L/s (peak flow), simulating a hydropeaking event. Each replicate comprised a group of five fish. For each, the frequency and residence time were quantified. The preliminary results indicated that the fish movement patterns changed when peak flow occurred. The downstream refuge was more frequently used in both configurations during peak flow. Additionally, the inside parts of the refuges were more frequently used, instead of the parts immediately downstream or upstream, and the time spent inside the refuge at peak flow was higher when compared to base flow. Additionally, hydraulic experiments were carried out at each configuration to determine the velocity field using ADV (Vectrino) technology. For the base flow, mean water depth and average velocity were 8 cm and 12 cm/s, respectively, increasing to 24 cm and 39 cm/s during peak flow. Measurements showed that velocity was equal to 74 cm/s in the narrowed area due to the refuge location, and velocity was null inside and directly downstream of the refuge. The results from this study will allow the development of guidelines for designing flow refuges for cyprinid fish, and hence mitigate the impact of hydropeaking.

Keywords: hydropeaking; hydropower plants; cyprinids; flow refuges; velocity

Author Contributions: Conceptualization, I.B. and M.J.C.; Methodology, I.B. and M.J.C.; Software, A.M. and R.L.; Validation, I.B., M.J.C., J.M.S. and A.N.P.; Formal analysis, I.B., R.L., M.J.C. and A.M.; Investigation, I.B., M.J.C., R.L., D.M. and A.M.; Resources, I.B., M.J.C., J.M.S. and A.N.P.; Data curation, I.B. and M.J.C.; Writing-original draft preparation, R.L. and A.M.; Writing-review and editing, I.B., M.J.C., J.M.S., R.L., D.M. and A.M.; Visualization, I.B. and M.J.C.; Supervision, I.B. and M.J.C.; Project Administration, I.B. and M.J.C.; Funding acquisition, I.B. All authors have read and agreed to the published version of the manuscript.

Funding: This research was funded by FCT "Fundação para a Ciencia e a Tecnologia", grant number: PTDC/EAM-AMB/4531/2020. The authors are grateful for the Foundation for Science and Technology's support through funding UIDB/04625/2020 from the research unit CERIS.

Institutional Review Board Statement: The Institute for Nature Conservation and Forests (ICNF) provided the necessary fishing and handling permits, and General Directorate of Food and Veterinary Medicine (DGAV) provided the authorization to perform experimental research with live animal. All research was conducted in accordance with national and international guidelines for animal welfare.

Informed Consent Statement: Not applicable.

Data Availability Statement: Data is available upon request from the authors.

Conflicts of Interest: The authors declare no conflict of interest.

biology and life sciences forum

Abstract

Meso- and Micro-Habitat Preferences of European River and Brook Lamprey in a Mediterranean River Basin [†]

Inês C. Oliveira [1,*,‡], Bernardo R. Quintella [2], Catarina S. Mateus [1], Carlos M. Alexandre [1] and Pedro R. Almeida [1,3]

[1] ARNET—Aquatic Research Network, MARE—Marine and Environmental Sciences Centre, University of Évora, 7004-516 Évora, Portugal; cspm@uevora.pt (C.S.M.); cmea@uevora.pt (C.M.A.); pmra@uevora.pt (P.R.A.)

[2] MARE—Marine and Environmental Sciences Centre, Department of Animal Biology, Faculty of Sciences, University of Lisbon, 1749-016 Lisbon, Portugal; bsquintella@fc.ul.pt

[3] Department of Biology, School of Science and Technology, College Luis António Verney, University of Évora, 7000-671 Évora, Portugal

* Correspondence: icso@uevora.pt

† Presented at the IX Iberian Congress of Ichthyology, Porto, Portugal, 20–23 June 2022.

‡ Presenting author (Oral presentation).

Abstract: The European river (*Lampetra fluviatilis* Linnaeus, 1758) and brook (*Lampetra planeri* Bloch, 1784) lampreys are considered 'paired species', i.e., they are closely related and morphologically very similar but have distinct modes of adult life (anadromous vs. resident). In the Iberian Peninsula, the southern limit of both species' distribution, they face different pressures (e.g., barriers, pollution, hydrological stress) that reduce the available habitat, which is exacerbated by the actual climate change context. The main objective of the present study was to evaluate meso- and microhabitat preferences of *Lampetra* sp. On the mesohabitat scale, the environmental variables that influence these species' distributions on the watershed scale were identified. On the microhabitat level, besides identifying the fine-scale variables that influence the presence and abundance of *Lampetra* sp., possible changes in habitat preferences throughout the larval stage (i.e., distinct size/age classes) were also assessed. Mesohabitat results suggest that the relative abundance of *Lampetra* sp. is related to variables such as pH and riparian vegetation. Regarding the microhabitat, the relative abundance of the size classes' distribution seems to be associated with variables such as substrate granulometry. The results in terms of habitat preferences on a Mediterranean basin are discussed in the context of a climate change scenario (e.g., decrease in habitat quality and availability) and management and conservation perspective.

Keywords: *Lampetra fluviatilis*; *Lampetra planeri*; lamprey ecology; environmental variables; habitat selection; size structure

Citation: Oliveira, I.C.; Quintella, B.R.; Mateus, C.S.; Alexandre, C.M.; Almeida, P.R. Meso- and Micro-Habitat Preferences of European River and Brook Lamprey in a Mediterranean River Basin. *Biol. Life Sci. Forum* **2022**, *13*, 93. https://doi.org/10.3390/blsf2022013093

Academic Editor: Alberto Teodorico Correia

Published: 15 June 2022

Publisher's Note: MDPI stays neutral with regard to jurisdictional claims in published maps and institutional affiliations.

Author Contributions: Conceptualization, B.R.Q., C.S.M., C.M.A. and P.R.A.; software, formal analysis, visualization, I.C.O.; data curation, writing—original draft preparation, I.C.O. and B.R.Q.; methodology, investigation, resources, writing—review and editing, I.C.O., B.R.Q., C.S.M., C.M.A. and P.R.A.; supervision and project administration, P.R.A.; funding acquisition, P.R.A., B.R.Q., C.S.M. and C.M.A. All authors have read and agreed to the published version of the manuscript.

Funding: This research was funded by EDP—Energias de Portugal, S.A. with the award Fundo EDP para a Biodiversidade 2008. Support was also provided by the Portuguese Foundation for Science and Technology (FCT) through the strategy plan for MARE (Marine and Environmental Sciences Centre), via project UIDB/04292/2020, and under the project LA/P/0069/2020 granted to the Associate Laboratory ARNET. FCT also supported this study through the individual contract attributed to Carlos M. Alexandre within the project CEECIND/02265/2018 and individual contract attributed to Bernardo Quintella 2020.02413.CEECIND.

Institutional Review Board Statement: The study was conducted according to the guidelines of the ICNF-Instituto da Conservação da Natureza e das Florestas.

Informed Consent Statement: Not applicable.

Data Availability Statement: Not applicable.

Conflicts of Interest: The authors declare no conflict of interest.

biology and life sciences forum

Abstract

Distribution of an Endemic Endangered Cyprinid *Anaecypris hispanica* in Extremadura Region (Southwestern Spain) [†]

Paloma Moreno [1,*,‡], Juan José Pérez [2], Carlos Rangel [3], Rafael Roso [3], César Fallola [2] and José Martín-Gallardo [4]

1   Vegas del Guadiana Aquaculture Centre, Gestión Pública de Extremadura, 06195 Badajoz, Spain
2   Vegas del Guadiana Aquaculture Centre, Junta de Extremadura, 06195 Badajoz, Spain; juanjose.perezg@juntaex.es (J.J.P.); cesar.fallola@juntaex.es (C.F.)
3   Ictios Consulting, 06400 Badajoz, Spain; ictiosconsulting@gmail.com (C.R.); rafaelroso28@gmail.com (R.R.)
4   Plant Biology, Ecology and Earth Science Department, Science Faculty, University of Extremadura, 06006 Badajoz, Spain; jomarga@unex.es
*   Correspondence: paloma.moreno@gpex.es; Tel.: +34-924012950
†   Presented at the IX Iberian Congress of Ichthyology, Porto, Portugal, 20–23 June 2022.
‡   Presenting author (Oral communication).

Citation: Moreno, P.; Pérez, J.J.; Rangel, C.; Roso, R.; Fallola, C.; Martín-Gallardo, J. Distribution of an Endemic Endangered Cyprinid *Anaecypris hispanica* in Extremadura Region (Southwestern Spain). *Biol. Life Sci. Forum* 2022, 13, 94. https://doi.org/10.3390/blsf2022013094

Academic Editor: Alberto Teodorico Correia

Published: 15 June 2022

Publisher's Note: MDPI stays neutral with regard to jurisdictional claims in published maps and institutional affiliations.

Abstract: The Iberian minnowcarp *Anaecypris hispanica* is one of the most endangered Iberian cyprinids, endemic to Guadiana and Bembezar river basins. Autochthonous fish populations are decreasing mainly due to habitat degradation, water quality decrease, and allochthonous species proliferation, especially in Guadiana's main rivers, so a monitoring program was started in temporary rivers in 2010. This temporary river monitoring program became more important in the following years, as several *Anaecypris* locations, some supposed to be extinguished, were found. Monitoring was carried out from 2010 to 2021 combining electrofishing and hand nets. Hand nets were extremely efficient in summer, when the species concentrates in summer ponds during the drought period. Anaecypris was found in twenty-three rivers, twenty one in Guadiana river basin and two in Bembézar river basin. Three of these rivers—Alcazaba, Lobo, and Pedruégano—are new locations for science. Although *Anaecypris* has a large distribution area, with localizations scattered all over Extremadura, populations are highly fragmented and, in some cases, reduced to one summer pond or summer refugee. The best populations are in Guadámez, Guadalemar, and Sotillo river basins, with higher fish densities and good habitat conditions. Other rivers have also high *A. hispanica* densities but are in danger because of pollution, habitat degradation, drought, water abstraction in summer ponds, and the continuous spread of allochthonous fishes. During the study period, two river populations have probably disappeared, Arroyoculebras due to pollution and habitat degradation and Alcorneo due to severe summer drought. This information has been used as a basis for the *Anaecypris hispanica* conservation plan in Extremadura. Conservation measures such as establishing appropriate conservation areas, under Natura 2000 network, summer ponds strict protection, habitat restoration, and allochthonous fish control, not only in the rivers with *A. hipanica* but in its whole drainage area, as well as captive breeding of the most sensible populations, are conservation measures that should be taken in the short term to reduce population extinction risks.

Keywords: *Anaecypris*; conservation; cyprinid; endangered; Iberian Peninsula

Author Contributions: Conceptualization and methodology, P.M. and J.J.P.; investigation, resources and data curation, P.M., J.J.P., C.R. and R.R.; validation, formal analysis, writing—original draft preparation, review and editing, P.M.; supervision, J.M.-G. and C.F.; project administration, C.F. All authors have read and agreed to the published version of the manuscript.

Funding: This research received no external funding.

Institutional Review Board Statement: Ethical review and approval were waived for this study due to fish monitoring was carried out according to the protocols specified by the Junta de Extremadura (Regional Government of Extremadura).

Informed Consent Statement: Not applicable.

Data Availability Statement: Ichthyological database. Junta de Extremadura (Regional Government of Extremadura).

Conflicts of Interest: The authors declare no conflict of interest.

biology and life sciences forum

Abstract

Guadiana Nase (*Pseudochondrostoma willkommii*) Reproduction in Still Water [†]

Paloma Moreno [1,*,‡], **Juan Carlos Ramírez** [1], **Guadalupe de la Cruz** [1] **and José Martín-Gallardo** [2]

[1]   Vegas del Guadiana Aquaculture Centre, Gestión Pública de Extremadura, 06011 Badajoz, Spain; juancarlos.ramirez@gpex.es (J.C.R.); guadalupe.delacruz@gpex.es (G.d.l.C.)

[2]   Plant Biology, Ecology and Earth Science Depertment, Science Faculty, University of Extremadura, 06006 Badajoz, Spain; jomarga@unex.es

*   Correspondence: paloma.moreno@gpex.es; Tel.: +34-924012950

†   Presented at the IX Iberian Congress of Ichthyology, Porto, Portugal, 20–23 June 2022.

‡   Presenting author (oral communication).

Abstract: *Pseudochondrostoma willkommii*, called Guadiana nase in the Iberian Peninsula, is an endemic cyprinid that lives in the middle stretches of rivers and is also common in still water. According to the literature, spawning occurs in April, after upstream migration while looking for shallow waters with a current and coarse substratum. There are no previous studies about captive breeding of this species, even though it has been a common species in the Guadiana River basin for years. Now, Guadiana nase populations are declining due to allochthonous fish introductions, river fragmentation, and pollution. The main objective of this study was the natural reproduction of *Pseudochondrostoma willkommii* in captivity. Guadiana nase captive breeding was started in 2017 at the Vegas del Guadiana fish farm; fish were captured in the wild by electrofishing and kept in spawning ponds with a natural photoperiod and temperature regime, with spawning substrates and without previous hormonal treatment. Guadiana nase spawned in captivity beginning in the first year. Spawning took place in still water, using the coarse substratum areas; no current was provided in the pond. It started in March and continued in April; at least two batches of larvae were recorded. The number of fingerlings was quite variable in different years; final juvenile fish densities ranged from 1.78 to 90.39 fish/m^3 and were not correlated with the initial number of spawners. Better results were obtained the sooner we introduced spawners to the pond, which enabled proper acclimatization and more complex habitat conditions. Spawners gathered in groups, except in the ponds with the lowest densities, where they were never in groups and showed agonistic behavior. This agonistic behavior did not affect reproduction success.

Keywords: cyprinid; reproduction; nase; captive breeding

Citation: Moreno, P.; Ramírez, J.C.; de la Cruz, G.; Martín-Gallardo, J. Guadiana Nase (*Pseudochondrostoma willkommii*) Reproduction in Still Water. *Biol. Life Sci. Forum* **2022**, *13*, 97. https://doi.org/10.3390/blsf2022013097

Academic Editor: Alberto Teodorico Correia

Published: 15 June 2022

Publisher's Note: MDPI stays neutral with regard to jurisdictional claims in published maps and institutional affiliations.

Author Contributions: Conceptualization, P.M.; methodology, P.M. and J.C.R.; formal analysis, P.M.; investigation, P.M.; resources, P.M., J.C.R. and G.d.l.C.; data curation, P.M. and G.d.l.C.; writing—original draft preparation, P.M.; writing—review and editing, P.M. and J.M.-G.; supervision, J.M.-G. All authors have read and agreed to the published version of the manuscript.

Funding: This research was funded by FEADER grant number 202012AGE035.

Institutional Review Board Statement: Ethical review and approval were waived for this study, due to no manipulation of normal reproductive behaviour, natural breeding and routine fish farms practices have been carried out.

Informed Consent Statement: Not applicable.

Data Availability Statement: Ichthyological database. Junta de Extremadura (Regional Government of Extremadura).

Conflicts of Interest: The authors declare no conflict of interest.

biology and life sciences forum

Abstract

Fishing Discards of Rays and Skates *Rajidae* in Galicia Waters [†]

Julio Valeiras [1,*], Esther Abad [1], Eva Velasco [1], Mateo Barreiro [2], José Carlos Fernández-Franco [2], Nair Vilas [1,‡] and María Grazia Pennino [1]

[1] Centro Nacional Instituto Español de Oceanografía-CSIC, CO-Vigo, Subida a Radio Faro 50-52, 36390 Vigo, Spain; esther.abad@ieo.csic.es (E.A.); eva.velasco@ieo.csic.es (E.V.); nair.vilas@ieo.csic.es (N.V.); grazia.pennino@ieo.csic.es (M.G.P.)

[2] OPPF-4 Organización de Productores de Pesca Fresca del Puerto de Vigo, Edificio Ramiro Gordejuela, Puerto Pesquero s/n, 36202 Vigo, Spain; mateobarreirosim@gmail.com (M.B.); josecarlos@arvi.org (J.C.F.-F.)

[*] Correspondence: julio.valeiras@ieo.csic.es

[†] Presented at the IX Iberian Congress of Ichthyology, Porto, Portugal, 20–23 June 2022.

[‡] Presenting author (Poster).

Citation: Valeiras, J.; Abad, E.; Velasco, E.; Barreiro, M.; Fernández-Franco, J.C.; Vilas, N.; Pennino, M.G. Fishing Discards of Rays and Skates *Rajidae* in Galicia Waters. *Biol. Life Sci. Forum* **2022**, *13*, 98. https://doi.org/10.3390/blsf2022013098

Academic Editor: Alberto Teodorico Correia

Published: 15 June 2022

Publisher's Note: MDPI stays neutral with regard to jurisdictional claims in published maps and institutional affiliations.

Abstract: Several skate and ray species are widely distributed in European Atlantic waters but many aspects still remain unknown: stock structure, species dynamics, migration movements, and spawning areas. Rays are vulnerable to overfishing and are bycatch species in the bottom fisheries in European waters of the International Council for the Exploration of the Sea (ICES). At least eight species of skate and rays inhabit north and northwestern Iberian waters. In offshore waters of the continental shelf, the most abundant is the thornback ray, *Raja clavata*, followed by *R. montagui* and *Leucoraja naevus*. In shallower waters, the most abundant species are *R. undulata*, *R. microocellata*, and *R. brachyura*. Some elasmobranchs are considered to have high survival rates, including the skates. In European waters, several studies have estimated the survival of discarded rays caught by different gear types. These species are usually discarded due to their small sizes or lack of fishing quota. Several commercial species are under an exemption for the landing obligation due to their high survivability when discarded in southwestern European waters. The estimates of the specific composition of landing skates are arduous to obtain due to the difficulties of certifying the identification of landed rays. This study presents the survival rates of discarded skates and rays caught by commercial trawlers and gillnetters operating in north Atlantic Spanish waters. Our results indicate that approximately 66.8% and 100% of sampled rays caught by bottom trawlers and trammel nets, respectively, survive fishing and handling operations on board. Detailed quantitative and biological data of species on catches and discards in Galician fisheries are also presented. Following the ICES recommendations, a tagging program has been carried out to improve knowledge of the status and spatial movements of species. Understanding the patterns of discarding and survivability rates could be used to reduce the fishing impact on skate and ray stocks.

Keywords: rays; skates; *Rajidae*; discards; survival; bycatch; landing obligation

---

Author Contributions: J.V. conceived the original idea, performed the analysis and interpretation of data, and wrote the manuscript. E.V. perfomed data curation and discards analysis., E.A., M.G.P., J.C.F.-F., M.B. and N.V., participate in work at sea. All authors have read and agreed to the published version of the manuscript.

Funding: This research was made within the DESCARSEL Project financed by European Maritime and Fisheries Funds (EMFF).

Institutional Review Board Statement: Not applicable.

Informed Consent Statement: Not applicable.

Data Availability Statement: All relevant data are within the published manuscript.

Conflicts of Interest: The authors declare no conflict of interest.

biology and life sciences forum

Abstract

The Fish Quality Index (FQI) Application in Extremadura (Spain) †

Manuela Rodríguez-Romero *,‡, Ángel Morales Hermoso, César E. Simón-Talero and Miguel A. Cotallo de Cáceres

Sección de Pesca, Acuicultura y Coordinación de la Dirección General de Política Forestal, Junta de Extremadura, Avda. de la Universidad, 10003 Cáceres, Spain; angel.morales@gpex.es (Á.M.H.); cesar.esteban@gpex.es (C.E.S.-T.); miguelangel.cotallo@juntaex.es (M.A.C.d.C.)
* Correspondence: manuela.rodriguez@juntaex.es
† Presented at the IX Iberian Congress of Ichthyology, Porto, Portugal, 20–23 June 2022.
‡ Presenting author (poster).

Abstract: Both Royal Decree 817/2015 which establishes in Spain the criteria for monitoring and evaluating the surface water status and environmental quality standards, and the Order ARM/2656/2008, by which the hydrological planning instruction is approved, contemplate among their indicators of the biological quality elements in rivers and lakes, the proportion of individuals of native species (ichthyological fauna) in the different water bodies. In Extremadura region, seven fish species have been declared invasive in all cases and another six species according to their habitat. As for the native fish species, there are ten threatened and two officially extinct. Different degrees of conservation are assigned to the rest of the continental native fish species, generally in regression. To evaluate these changes in fish communities, the Extremadura Fisheries Council approved a Fish Quality Index (FQI) in 2019, based on the presence/absence of exotics, without forgetting the importance that marine species contributed to the river environment before the implantation of the large dams and consequent loss of their migrations. This index completes the evaluation of the river environment in combination with the Biological Quality of waters (macroinvertebrates and diatoms) and its Geomorphological Quality (banks, fluvial continuity, minimum flows and generators, among others). Following the guidelines of the European and Spanish regulations applicable to hydrological planning, the FQI establishes five categories: (1) High quality for those water bodies with marine species; (2) Good, when there are native continental fishes and no exotic fish; (3) Moderate, when the river, reservoir, lagoon or pond can still lean towards one extreme or the other of the index due to the scarce or null presence of local or non-native fish; and (4) Poor and (5) Bad, for bodies of water with exotic species, according to their severity and abundance. The classification of water bodies within these categories of the FQI is based on the results obtained in more than 500 sampling stations over a period of more than 20 years. Then, it shows with high reliability the recent changes suffered in the environment and responds in an agile way before the decision-making needs in the management and investment of funds for the conservation of the river environment and for those who inhabit it.

Keywords: fish quality index; native fish species; presence or absence of alien species; management; conservation

Citation: Rodríguez-Romero, M.; Morales Hermoso, Á.; Simón-Talero, C.E.; Cotallo de Cáceres, M.A. The Fish Quality Index (FQI) Application in Extremadura (Spain). *Biol. Life Sci. Forum* **2022**, *13*, 99. https://doi.org/10.3390/blsf2022013099

Academic Editor: Alberto Teodorico Correia

Published: 16 June 2022

Publisher's Note: MDPI stays neutral with regard to jurisdictional claims in published maps and institutional affiliations.

Author Contributions: Conceptualization, designed and performed, Á.M.H., C.E.S.-T. and M.A.C.d.C.; writing and supervision, M.R.-R., Á.M.H., C.E.S.-T. and M.A.C.d.C. All authors have read and agreed to the published version of the manuscript.

Funding: This research received no external funding.

Institutional Review Board Statement: Not applicable.

Data Availability Statement: http://pescayrios.juntaextremadura.es/pescayrios/web/guest/seguimiento-de-los-rios-extremenos (accessed on 24 August 2021).

Conflicts of Interest: The authors declare no conflict of interest.

Abstract

The Future of RivTool [†]

Gonçalo Duarte [1,2,*,‡], **Pedro Segurado** [1,2], **Maria Teresa Ferreira** [1,2] and **Paulo Branco** [1,2]

[1] Forest Research Centre, School of Agriculture, University of Lisbon, 1349-017 Lisbon, Portugal; psegurado@isa.ulisboa.pt (P.S.); terferreira@isa.ulisboa.pt (M.T.F.); pjbranco@isa.ulisboa.pt (P.B.)
[2] Associate Laboratory TERRA, 1349-017 Lisbon, Portugal
* Correspondence: goncalofduarte@isa.ulisboa.pt
[†] Presented at the IX Iberian Congress of Ichthyology, Porto, Portugal, 20–23 June 2022.
[‡] Presenting author (poster communication).

Abstract: Effective research, conservation and management of freshwater ecosystems must take into consideration the environment, ecosystem functioning and human activities at multiple river scales. The river Network toolkit (RivTool) is a user-friendly and freely available software of universal applicability that enables the integration of these multiple inputs for large scale river network analysis. It is a table-driven application with several freely available datasets for European and South American basins. This software is currently implemented in all five continents, having been downloaded in nearly 70 countries, and represents a platform with high future scientific and management potential. As such, three new plugins are being conceptualized and developed to widen the contribution of this software to the freshwater community: the first is RivFish, a tool to integrate the resources of the rGBIF package with the river network framework of RivTool; the second RivConnect, an add-on for quantitative network connectivity analysis based on graph-theory, i.e., to enable the calculation of fragmentation metrics and river connectivity indexes; and a third tool, RivOpt, a multifunctional decision support system aiming at selecting an optimal portfolio of barrier removal and/or mitigation actions towards balancing competing environmental and socioeconomic objectives. Adding these new tools to RivTool will widen its scope of action and overall usefulness. The future of RivTool will provide, to the freshwater scientific and management community, a set of tools that will increase the ability to interpret and manage river systems, ultimately contributing to attain the European biodiversity goals.

Keywords: river networks; river management; river conservation; RivFish; RivConnect; RivOpt

Citation: Duarte, G.; Segurado, P.; Ferreira, M.T.; Branco, P. The Future of RivTool. *Biol. Life Sci. Forum* **2022**, *13*, 100. https://doi.org/10.3390/blsf2022013100

Academic Editor: Alberto Teodorico Correia

Published: 16 June 2022

Author Contributions: Conceptualization, G.D., P.B. and P.S.; methodology, G.D. and P.B.; data curation, G.D.; formal analysis, G.D. and P.B.; writing—original draft preparation G.D., P.B. and P.S.; writing—review and editing, all authors; figure development, G.D. and P.B.; supervision, M.T.F. and P.S.; project administration, G.D. and P.B.; Funding acquisition, M.T.F. All authors have read and agreed to the published version of the manuscript.

Funding: GD was supported by national funds via FCT—Fundação para a Ciência e a Tecnologia, I.P. (PTDC/ASP-AGR/29771/2017 and UIDP/00239/2020). P.B. was supported by national funds through Fundação para a Ciência e Tecnologia I.P., Portugal (FCT) under 'Norma Transitória—DL57/2016/CP1382/CT0020'. This research was funded by the Forest Research Centre, a research unit funded by Fundação para a Ciência e a Tecnologia I.P. (FCT), Portugal (UIDB/00239/2020) and by the Laboratory for Sustainable Land Use and Ecosystem Services (LA/P/0092/2020).

Institutional Review Board Statement: Not applicable.

Informed Consent Statement: Not applicable.

Data Availability Statement: All data used is freely available.

Conflicts of Interest: The authors declare no conflict of interest.

MDPI

Abstract

Inferring Past Occurrences of Diadromous Fish Species—The iPODfish Framework [†]

Gonçalo Duarte [1,2,*,‡], Paulo Branco [1,2], Gertrud Haidvogl [3], Maria Teresa Ferreira [1,2], Didier Pont [3] and Pedro Segurado [1,2]

[1] Forest Research Centre, School of Agriculture, University of Lisbon, 1349-017 Lisbon, Portugal; pjbranco@isa.ulisboa.pt (P.B.); terferreira@isa.ulisboa.pt (M.T.F.); psegurado@isa.ulisboa.pt (P.S.)

[2] Associate Laboratory TERRA, 1349-017 Lisbon, Portugal

[3] Institute of Hydrobiology and Aquatic Ecosystem Management, University of Natural Resources and Life Sciences Vienna (BOKU), Gregor-Mendel-Straße 33, 1180 Vienna, Austria; gertrud.haidvogl@boku.ac.at (G.H.); didier.pont@boku.ac.at (D.P.)

[*] Correspondence: goncalofduarte@isa.ulisboa.pt

[†] Presented at the IX Iberian Congress of Ichthyology, Porto, Portugal, 20–23 June 2022.

[‡] Presenting author (oral communication).

Citation: Duarte, G.; Branco, P.; Haidvogl, G.; Ferreira, M.T.; Pont, D.; Segurado, P. Inferring Past Occurrences of Diadromous Fish Species—The iPODfish Framework. *Biol. Life Sci. Forum* **2022**, *13*, 101. https://doi.org/10.3390/blsf2022013101

Academic Editor: Alberto Teodorico Correia

Published: 16 June 2022

Publisher's Note: MDPI stays neutral with regard to jurisdictional claims in published maps and institutional affiliations.

Abstract: Historical information on diadromous fish species is commonly incomplete or truncated across the species distribution range and spatial scales. However, historical insights have proven to be very relevant for river management and conservation. The iPODfish is a new methodological framework that enables the inference of a more thorough representation of the historical occurrence of diadromous fish over their complete distribution range. The method is based on the interplay between freshwater network features, diadromous fish species ecology and known historical occurrence from which assumptions, rules and thresholds are derived. It has five steps: main river segments vs. tributary segments; segment specificities; relative distance threshold; strahler value threshold; and sub-basin strahler threshold, divided into 2 moments of application (tributaries after main river) and can be expressed by a tree-like representation. iPODfish can cope with data bias and deal with multiple sources of information with distinct resolution scales to generate historical pseudo-presence/absence records of diadromous fish at a fine-scale spatial unit, such as river segments and along river networks. It allows for the enlarging of diadromous fish historical distributions and could be applied in any river network throughout the globe because despite its inference nature, it remains a conservative approach that uses concepts and definitions derived from common features of diadromous ecology and freshwater networks. The outputs obtained may prove useful in biogeographical and/or macroecological studies using historical occurrences and targeting the conservation and management of diadromous fish species.

Keywords: data-informed methodology; freshwater ecology; species historical data; pseudo-absences; pseudo-presences; segment scale distribution

---

Author Contributions: Conceptualization, G.D., P.B., G.H. and P.S.; methodology, G.D., P.B. and P.S.; data curation, G.D. and G.H.; formal analysis, G.D., P.B. and P.S.; writing—original draft preparation G.D., P.B. and P.S.; writing—review and editing, all authors; figure development, G.D., P.B. and P.S.; supervision, G.H:, M.T.F., D.P. and P.S.; project administration, M.T.F.; Funding acquisition, M.T.F.; All authors have read and agreed to the published version of the manuscript.

Funding: GD was supported by national funds via FCT—Fundação para a Ciência e a Tecnologia, I.P. (PTDC/ASP-AGR/29771/2017 and UIDP/00239/2020). P.B. was supported by national funds through Fundação para a Ciência e Tecnologia I.P., Portugal (FCT) under 'Norma Transi-tória—DL57/2016/CP1382/CT0020'. This research was funded by the Forest Research Centre, a research unit funded by Fundação para a Ci^encia e a Tecnologia I.P. (FCT), Portugal (UIDB/00239/2020) and by the Laboratory for Sustainable Land Use and Ecosystem Services (LA/P/0092/2020).

Institutional Review Board Statement: Not applicable.

Informed Consent Statement: Not applicable.

Data Availability Statement: All data used is freely available or available upon request.

Conflicts of Interest: The authors declare no conflict of interest.

Abstract

Local Breeding Centres: Engaging Local Anglers in Native Trout Conservation [†]

Guadalupe de la Cruz [*,‡], **Seila López, Paloma Moreno and Daniel Sánchez**

Vegas del Guadiana Aquaculture Centre, Gestión Pública de Extremadura, 06011 Badajoz, Spain; seila.lopez@gpex.es (S.L.); paloma.moreno@gpex.es (P.M.); daniel.sanchezs@gpex.es (D.S.)
* Correspondence: guadalupe.delacruz@gpex.es; Tel.: +34-652036355
† Presented at the IX Iberian Congress of Ichthyology, Porto, Portugal, 20–23 June 2022.
‡ Presenting author (Poster presentation).

Abstract: Trout (*Salmo trutta*) in the Extremadura region live in the Sierra de Gredos foothills, which is the southern distribution limit for this species in the Iberian Peninsula. Trout populations are declining mostly due to water resource development schemes—mainly cherry orchards—increases in temperatures due to climate change, the fragmentation of rivers, and habitat alterations. The Fisheries and Aquaculture Service of the Extremadura Government has been working on the enhancement of river habitats and biota. To improve trout populations, several management actions have been taken, such as so-called "Local Breeding Centres" (LBC), which have provided collaboration between local anglers and fishery technicians since 2019. Around late October, expert local anglers make sure breeding males and females are captured and are temporarily kept in breeding centres supervised by fish specialists. Genetic samples are collected from every individual to guarantee conservation units. When trout reach annual sexual maturity around late November, technicians assist with manual spawning, and the eggs are fertilized. Immediately after, adults are returned to the original river point where they were caught, and the eggs are transported to "Centro de Salmónidos de Jerte" (an aquaculture centre) in order to be incubated in controlled conditions. Once the critic larvae term is over, juvenile fish are moved back into the river. In the spring of 2020, around 3600 eggs and 7000 young fish were repopulated. Every year, there are more angler societies showing interest in participating in this program to improve trout populations in rivers. Thanks to these efforts, local citizens have become involved in environmental action, and help the government with the preservation of habitat, flora and fauna. Although this may seem an irrelevant measure, cooperation between the government and citizens is crucial, as it helps to promote ecological awareness.

Keywords: trout; conservation; salmonids; endangered; Iberian Peninsula; angler

Citation: de la Cruz, G.; López, S.; Moreno, P.; Sánchez, D. Local Breeding Centres: Engaging Local Anglers in Native Trout Conservation. *Biol. Life Sci. Forum* **2022**, *13*, 103. https://doi.org/10.3390/blsf2022013103

Academic Editor: Alberto Teodorico Correia

Published: 16 June 2022

Publisher's Note: MDPI stays neutral with regard to jurisdictional claims in published maps and institutional affiliations.

Author Contributions: Conceptualization, P.M., D.S. and S.L.; methodology, P.M. and D.S.; formal analysis, P.M., G.d.l.C. and D.S.; investigation, P.M. and D.S.; resources, P.M. and S.L.; data curation, P.M. and G.d.l.C.; writing—original draft preparation, G.d.l.C. and S.L.; writing—review and editing, G.d.l.C.; supervision, P.M. All authors have read and agreed to the published version of the manuscript.

Funding: This research was funded by FEADER grant number 202012AGE035.

Institutional Review Board Statement: Ethical review and approval were waived for this study, due to no manipulation of normal reproductive behaviour, natural breeding and routine fish farms practices have been carried out.

Informed Consent Statement: Not applicable.

Data Availability Statement: Ichthyological database. Junta de Extremadura.

Conflicts of Interest: The authors declare no conflict of interest.

Abstract

Management and Conservation of Fish Populations in Mountain Streams: An Holistic Approach in the Framework of LIFE DIVAQUA Project [†]

Alejandra Goldenberg-Vilar [1], Mario Álvarez Cabria [1], Francisco J. Peñas [1], Alexia González Ferreras [1], Francisco Javier Sanz-Ronda [2,‡], Noel Quevedo [3] and José Barquín [1,*]

[1] IHCantabria—Instituto de Hidráulica Ambiental de la Universidad de Cantabria, Parque Científico y Tecnológico de Cantabria C/Isabel Torres, N° 15-C.P., 39011 Santander, Spain; agoldenberg@unican.es (A.G.-V.); mario.alvarez@unican.es (M.Á.C.); penasfj@unican.es (F.J.P.); gferrerasam@unican.es (A.G.F.)

[2] Area of Hydraulics and Hydrology, Department of Agroforestry Engineering, University of Valladolid, Avenida de Madrid 44, Campus La Yutera, 34004 Palencia, Spain; jsanz@uva.es

[3] Red Cambera, Apdo. Correos 4013, 39011 Santander, Spain; noel@redcambera.org

[*] Correspondence: jose.barquin@unican.es

[†] Presented at the IX Iberian Congress of Ichthyology, Porto, Portugal, 20–23 June 2022.

[‡] Presenting author (Poster presentation).

Citation: Goldenberg-Vilar, A.; Cabria, M.Á.; Peñas, F.J.; González Ferreras, A.; Sanz-Ronda, F.J.; Quevedo, N.; Barquín, J. Management and Conservation of Fish Populations in Mountain Streams: An Holistic Approach in the Framework of LIFE DIVAQUA Project. *Biol. Life Sci. Forum* **2022**, *13*, 106. https://doi.org/10.3390/blsf2022013106

Academic Editor: Alberto Teodorico Correia

Published: 16 June 2022

Publisher's Note: MDPI stays neutral with regard to jurisdictional claims in published maps and institutional affiliations.

Abstract: The recovery of threatened and endangered fish species is among the highest priorities for biodiversity conservation in national parks and fisheries management in nearby areas. Threats to fish populations are numerous and include habitat fragmentation and degradation, proliferation of invasive and pathogen species, and climate change. Moreover, mountain areas often share the most critical threats. However, there does not exist a common strategy that integrates conservation and management plans for fish populations in mountain areas. In this regard, LIFE DIVAQUA designed a conservation strategy that integrates new knowledge gained from scientific research and long-term monitoring data, and considers the main threats to fish populations in mountain areas: (1) A long term monitoring program has been already implemented for 10 years, revealing temporal trends of fish populations in mountain streams. (2) Modeling of fish population by the use of environmental DNA allowed analyzing fish distributions in areas with scarce data and evaluating habitat suitability maps. (3) Fishways construction and removal of river barriers substantially increased the distribution area of endangered species. (4) The analysis of climate change effects in water temperature and hydrology led to the implementation of environmental flows under a climate change scenario; (5) Monitoring fish diseases, their occurrence, and temporal changes (e.g., Aeromonas spp.) can be used as an early warning signal of ecosystem unbalance. A pilot study for the implementation of this conservation and management plan in the LIFE DIVAQUA project is showing promising results. However, the success of conservation and management strategies requires a broader approach. This includes the participation of a wide range of partners and stakeholders and utilizes independent scientific oversight, assessment, and project adjustments to ensure conservation goals are met.

Keywords: fish conservation; fish management; national parks; environmental DNA; long term monitoring

Author Contributions: Conceptualization, A.G.-V., F.J.S.-R., M.Á.C., F.J.P. and J.B.; methodology, A.G.-V., F.J.P., A.G.F., M.Á.C. and F.J.S.-R.; validation, M.Á.C., A.G.F. and A.G.-V.; formal analysis, M.Á.C., A.G.F., A.G.-V. and J.B.; investigation, F.J.S.-R., M.Á.C., A.G.F., A.G.-V. and J.B.; data curation, M.Á.C., A.G.-V. and A.G.F.; writing—original draft preparation, A.G.-V.; writing—review and editing, A.G.-V., M.Á.C., A.G.F., F.J.P., F.J.S.-R., N.Q. and J.B.; visualization, A.G.-V., F.J.S.-R. and N.Q.; supervision, project administration, funding acquisition, J.B. All authors have read and agreed to the published version of the manuscript.

Funding: This research was financed by the European Commission Life Program Nature and Biodiversity. Reference: LIFE18 NAT/ES/000121.

Institutional Review Board Statement: The study was conducted according to the European Union ethical guidelines (Directive 2010/63/UE) and Spanish Act RD 53/2013 and approved by the competent authorities (Regional Government on Natural Resources of Castilla y León, Principado de Asturias and Cantabria).

Informed Consent Statement: Not applicable.

Data Availability Statement: Not applicable.

Conflicts of Interest: The authors declare no conflict of interest.

Abstract

Living on the Edge: Management and Conservation of Atlantic Salmon at the Southern Limit of the Species Distribution [†]

Carlos M. Alexandre [1,*,‡], Sara Silva [1], Catarina S. Mateus [1], Maria J. Lança [2], Bernardo R. Quintella [3,4], Ana F. Belo [1], Andreia Domingues [1], Ana S. Rato [1], Roberto Oliveira [1], André Moreira [1], Joana Pereira [1], Inês Raposo [1], Pedro Sousa [1], Yorgos Stratoudakis [5], Pablo Caballero [6] and Pedro R. Almeida [1,7]

[1] MARE-Centro de Ciências do Mar e do Ambiente/ARNET-Rede de Investigação Aquática, Universidade de Évora, 7004-516 Evora, Portugal; ssrs@uevora.pt (S.S.); cspm@uevora.pt (C.S.M.); affadb@uevora.pt (A.F.B.); afdomingues@uevora.pt (A.D.); assrato@uevora.pt (A.S.R.); rlpo@uevora.pt (R.O.); andre.moreira@uevora.pt (A.M.); joana_gomespereira@hotmail.com (J.P.); senifox@gmail.com (I.R.); fc50839@alunos.fc.ul.pt (P.S.); pmra@uevora.pt (P.R.A.)
[2] MED-Instituto Mediterrânico para a Agricultura, Ambiente e Desenvolvimento & Departamento de Zootecnia, Escola de Ciências e Tecnologia, Universidade de Évora, 7004-516 Evora, Portugal; mjlanca@uevora.pt
[3] MARE-Centro de Ciências do Mar e do Ambiente, Faculdade de Ciências, Universidade de Lisboa, 1600-548 Lisboa, Portugal; bsquintella@fc.ul.pt
[4] Departamento de Biologia Animal, Faculdade de Ciências, Universidade de Lisboa, 1749-016 Lisboa, Portugal
[5] Departamento do Mar, IPMA-Instituto Português do Mar e da Atmosfera, 1749-077 Lisboa, Portugal; yorgos@ipma.pt
[6] Dirección Xeral de Patrimonio Natural, Consellería de Medio Ambiente, Xunta de Galicia, 15781 Santiago de Compostela, Spain; pblo.cbllero.javierre@xunta.gal
[7] Departamento de Biologia, Escola de Ciências e Tecnologias, Universidade de Évora, 7004-516 Evora, Portugal
*   Correspondence: cmea@uevora.pt; Tel.: +351-962858816
†   Presented at the IX Iberian Congress of Ichthyology, Porto, Portugal, 20–23 June 2022.
‡   Presenting author (Oral communication).

Citation: Alexandre, C.M.; Silva, S.; Mateus, C.S.; Lança, M.J.; Quintella, B.R.; Belo, A.F.; Domingues, A.; Rato, A.S.; Oliveira, R.; Moreira, A.; et al. Living on the Edge: Management and Conservation of Atlantic Salmon at the Southern Limit of the Species Distribution. *Biol. Life Sci. Forum* **2022**, *13*, 107. https://doi.org/10.3390/blsf2022013107

Academic Editor: Alberto Teodorico Correia

Published: 16 June 2022

Publisher's Note: MDPI stays neutral with regard to jurisdictional claims in published maps and institutional affiliations.

Abstract: Atlantic salmon (*Salmo salar* L.) are an emblematic anadromous fish species that inhabit marine and freshwater ecosystems in the Northern Hemisphere. In Portugal, the southern limit of its global distribution range, the species is classified as *Critically Endangered* (CR), and occurs only in the Minho and Lima rivers, which hold the most abundant populations, with occasional confirmed occurrences in the Cávado and Douro rivers. In this region, the species faces several highly detrimental threats (e.g., dams and other obstacles, unsuitable fishing legislation, and climate change), but some knowledge gaps about the biology, ecology, and structure of these southern populations often impair any attempt to define and implement effective management and conservation programs. We present the objectives, actions, and preliminary results of a set of projects and partnerships, recently implemented in Portugal, focused on increasing knowledge about local salmon populations to contribute to the development of suitable management guidelines for the target species. Within the project "SALMONLINK—Contribution of scientists and fishermen to the conservation and participatory management of Atlantic salmon populations in Portugal", several actions have been implemented for the past three years to improve knowledge about this species, including the assessment of salmon distribution and abundance, adult and juvenile migration patterns, and population structure. In complement to these measures, we are also implementing (in Portugal) the international projects "SMOLTrack III & IV—Quantifying smolt survival from source to sea: informing management strategies to optimize returns", which are specifically focused on smolt seaward migration and aim to obtain more information on this particularly vulnerable life-stage. Combined with a strong link with the local commercial and recreational fishing communities, who are providing data on salmon catches and contributing to an overview of the socioeconomic value of salmon in Portugal, these projects will contribute to increasing the knowledge of these populations, and at the same time, within the context of the constant transfer of knowledge between all the involved parties, advise the adaptation of the current fishing legislation to the conservation and management requirements of this highly endangered species.

Keywords: Atlantic salmon; fisheries management; sustainability; population ecology and structure

---

Author Contributions: Conceptualization, C.M.A. and P.R.A.; methodology, C.M.A., P.R.A., C.S.M., B.R.Q., Y.S., P.C.; formal analysis, C.M.A., S.S., C.S.M., I.R.; investigation, C.M.A., C.S.M., S.S., A.S.R., J.P., R.O., A.M., A.F.B., A.D., P.S.; resources, C.M.A. and P.R.A.; data curation, S.S., C.M.A., J.P., C.S.M.; writing—original draft preparation, C.M.A.; writing—review and editing, S.S., C.S.M., B.R.Q., M.J.L., Y.S., P.C., P.R.A.; supervision, C.M.A., C.S.M., P.R.A.; project administration, C.M.A. and P.R.A.; funding acquisition, C.M.A., M.J.L., P.R.A. All authors have read and agreed to the published version of the manuscript.

Funding: The present study was supported by European Funds (EMFF - European Maritime and Fisheries Fund), through the project SMOLTRACK (S12.817653) and, more specifically, by the Operational Program MAR2020 (project SALMONLINK: MAR-01.03.02-FEAMP-0048). Funding was also provided by the Portuguese Science Foundation through the strategy plan for MARE (Marine and Environmental Sciences Centre), via project UIDB/04292/2020, and under the project LA/P/0069/2020 granted to the Associate Laboratory ARNET. FCT also supported this study through the individual contracts attributed to Carlos M. Alexandre (CEECIND/02265/2018) and to Bernardo R. Quintella (2020.02413.CEECIND), and the PhD scholarships attributed to Sara Silva (2021.05558.BD), Ana S. Rato (2021.05339.BD) and Andreia Domingues (2021.05644.BD).

Institutional Review Board Statement: Not applicable.

Informed Consent Statement: Informed consent was obtained from all subjects involved in the study.

Data Availability Statement: Data is available from correspondence author, upon reasonable request.

Conflicts of Interest: The authors declare no conflict of interest.

Abstract

The Red List Threats to European Freshwater Fishes—Spatial Patterns and Knowledge Gaps †

Gonçalo Duarte [1,2,*,‡], Maria João Costa [1], Pedro Segurado [1,2], Afonso Teixeira [1], José Maria Santos [1,2], Maria Teresa Ferreira [1,2] and Paulo Branco [1,2]

[1] Forest Research Centre, School of Agriculture, University of Lisbon, 1349-017 Lisbon, Portugal; mjcscosta@hotmail.com (M.J.C.); psegurado@isa.ulisboa.pt (P.S.); afonsolest@gmail.com (A.T.); jmsantos@isa.ulisboa.pt (J.M.S.); terferreira@isa.ulisboa.pt (M.T.F.); pjbranco@isa.ulisboa.pt (P.B.)

[2] Associate Laboratory TERRA, 1349-017 Lisbon, Portugal

* Correspondence: goncalofduarte@isa.ulisboa.pt

† Presented at the IX Iberian Congress of Ichthyology, Porto, Portugal, 20–23 June 2022.

‡ Presenting author (oral communication).

Abstract: Freshwater fish species comprise 40% of all fish diversity and provide multiple ecosystem services. Recently, several populations of this faunal group have faced declines and range contractions due to several threats. The Red List of Threatened Species, established by the International Union for Conservation of Nature (IUCN), is the most comprehensive system for evaluating species' risk of extinction and holds information on the species' populations, distribution, ecology, and threats. This work aims to use this database to (1) identify the major threats to European freshwater fishes; (2) portray the geographical patterns of threat incidence with species richness, conservation status, and migratory phenology; and (3) identify the knowledge gaps in terms of valid published scientific literature supporting the threat data in the IUCN Red List database. The analysis includes 434 species, of which 41.2% are threatened, accounting for 837 threats, whereas only 11 register a valid and published scientific output. "Resident" was the migratory phenology with the highest percentage of threatened species (46.3%), and "Dams & water management/Use" was the most frequent (>50%) threat type. Across Europe, there is a high level of imperilment in freshwater fish species, with a particular incidence in southern regions and some coastal areas. Southern Europe, particularly the Iberian Peninsula, has a comparatively low species richness but a high proportion of threatened species with a high threat incidence. Overall, only 1.6% of the species and 1.3% of all threats identified are supported by valid published scientific literature. In sum, the present level of imperilment of European freshwater fish fauna is high, particularly in Iberia, and river network fragmentation will likely be the most challenging threat to future restoration efforts. The lack of valid scientific support for the IUCN Red List assessment affects its reliability and may hamper efforts of threat mitigation and species conservation.

Keywords: UICN; species conservation; river network; dams and water management; threat incidence

Citation: Duarte, G.; Costa, M.J.; Segurado, P.; Teixeira, A.; Santos, J.M.; Ferreira, M.T.; Branco, P. The Red List Threats to European Freshwater Fishes—Spatial Patterns and Knowledge Gaps. *Biol. Life Sci. Forum* **2022**, *13*, 109. https://doi.org/10.3390/blsf2022013109

Academic Editor: Alberto Teodo-rico Correia

Published: 17 June 2022

Publisher's Note: MDPI stays neutral with regard to jurisdictional claims in published maps and institutional affiliations.

Author Contributions: Conceptualization, P.B., J.M.S., P.S., G.D. and M.T.F.; methodology, P.B., M.J.C. and G.D.; data curation, M.J.C. and A.T.; writing—original draft preparation P.B., M.J.C. and G.D.; writing—review and editing, all authors; figure development, P.B, P.S. and G.D.; supervision, M.T.F., P.B. and G.D.; project administration, P.B. and G.D. All authors have read and agreed to the published version of the manuscript.

Funding: GD was supported by national funds via FCT – Fundação para a Ciência e a Tecnologia, I.P. (PTDC/ASP-AGR/29771/2017 and UIDP/00239/2020). P.B. was supported by national funds through Fundação para a Ciência e Tecnologia I.P., Portugal (FCT) under 'Norma Transitória— DL57/2016/CP1382/CT0020'. This research was funded by the Forest Research Centre, a research unit funded by Fundação para a Ciência e a Tecnologia I.P. (FCT), Portugal (UIDB/00239/2020) and by the Laboratory for Sustainable Land Use and Ecosystem Services (LA/P/0092/2020).

Data Availability Statement: All data used is freely available.

Conflicts of Interest: The authors declare no conflict of interest.

Abstract

Modeling Long-Term Changes in Atlantic Salmon Abundance in the Ulla River (Galicia—Spain) through Multivariate Time-Series Analysis [†]

Esteban L. Alvarez-Romero [1,*,‡], Marcos Barrio-Anta [2], Carlos Garcia de Leaniz [3], Pablo Caballero [4] and Paloma Moran [5]

1. DIRENA, Indurot-Institute of Natural Resources and Territorial Planning, Universidad de Oviedo, Escuela Politécnica de Mieres, 33600 Mieres, Spain
2. Departamento de Biología de Organismos y Sistemas, Area de Ingeniería Agroforestal, Universidad de Oviedo, Escuela Politécnica de Mieres, 33600 Mieres, Spain; barriomarcos@uniovi.es
3. BioSciences, CSAR-Centre for Sustainable Aquatic Research, Swansea University, Swansea SA2 8PP, UK; c.garciadeleaniz@swansea.ac.uk
4. Dirección Xeral de Patrimonio Natural, Consellería de Medio Ambiente, Xunta de Galicia, 36002 Santiago de Compostela, Spain; pablo.caballero.javierre@xunta.gal
5. Departamento de Bioquímica, Genética e Inmunología, Facultad de Biología, Universidad de Vigo, 33610 Vigo, Spain; paloma@uvigo.es
* Correspondence: author: laloalvarez@me.com; Tel.: +34-616966437
† Presented at the IX Iberian Congress of Ichthyology, Porto, Portugal, 20–23 June 2022.
‡ Presenting author (Oral communication).

Citation: Alvarez-Romero, E.L.; Barrio-Anta, M.; Garcia de Leaniz, C.; Caballero, P.; Moran, P. Modeling Long-Term Changes in Atlantic Salmon Abundance in the Ulla River (Galicia—Spain) through Multivariate Time-Series Analysis. *Biol. Life Sci. Forum* **2022**, *13*, 111. https://doi.org/10.3390/blsf2022013111

Academic Editor: Alberto Teodorico Correia

Published: 17 June 2022

Publisher's Note: MDPI stays neutral with regard to jurisdictional claims in published maps and institutional affiliations.

Abstract: Due to the decline in catches of Atlantic salmon at the end of the 1980s, the Natural Heritage Service of the Xunta de Galicia has been obtaining data on the returning adults and the downstream migration of juvenile at the Ximonde fish-trap station since the year 1992 to 2021, as well as on the recapture of marked specimens from artificial restocking. At the same time, and during the same period, catches made by the fishing industry and the recruitment obtained through population censuses of different areas of the basin have been analyzed. A first exploratory analysis of the data indicates that the distribution area and flow regimes did not vary significantly during the study period. Throughout the time series, the number of juveniles stocked in the river changed, and even in some years none were stocked. The multivariate analysis of the time series indicates that the observed variations in the abundance of the population can be predicted based on the type of management carried out during the period studied, and that the methodology used in the Ulla river can potentially be used in other rivers with similar characteristics, especially those located in the southern areas of its natural distribution, affected by dams and with a reduced spawning habitat.

Keywords: multivariate time series; Atlantic salmon; fisheries management; sustainability; population ecology and structure

Author Contributions: Conceptualization, methodology, formal analysis, investigation and writing—original draft preparation, E.L.A.-R.; validation and project administration, M.B.-A.; resources and visualization, P.C.; data curation, P.M.; supervision and writing—review and editing, C.G.d.L. All authors have read and agreed to the published version of the manuscript.

Funding: This research received no external funding.

Institutional Review Board Statement: Samples come from salmon caught in recreational fishing, which is regulated by Spanish laws and regulations, we do not slaughter any animals. We use samples from salmon that are already legally dead.

Informed Consent Statement: Not applicable.

Data Availability Statement: This data can be found at https://cmatv.xunta.gal (accessed on 31 March 2022).

Conflicts of Interest: The authors declare no conflict of interest.

biology and life sciences forum

MDPI

Abstract

Environmental Factors Promoting Upstream Movement of Yellow Eel (*Anguilla anguilla* L.) in the Mondego River [†]

Rui Monteiro [1,*,‡], Bernardo R. Quintella [1,2], Pedro R. Almeida [3,4], José Lino Costa [1,2], Esmeralda Pereira [3], Ana Filipa Belo [3], Teresa Portela [1], Carlos Batista [5], Ana Telhado [6], Verónica Pinto [6], Maria Felisbina Quadrado [6] and Isabel Domingos [1,2]

[1] MARE—Marine and Environmental Sciences Centre, Faculdade de Ciências da Universidade de Lisboa, 1749-016 Lisboa, Portugal; bsquintella@fc.ul.pt (B.R.Q.); jlcosta@fc.ul.pt (J.L.C.);tmportela@fc.ul.pt (T.P.); idomingos@fc.ul.pt (I.D.)

[2] Departamento de Biologia Animal, Faculdade de Ciências da Universidade de Lisboa, 1749-016 Lisboa, Portugal

[3] MARE—Marine and Environmental Sciences Centre, Universidade de Évora, 7004-516 Évora, Portugal; pmra@uevora.pt (P.R.A.); ecdp@uevora.pt (E.P.); affadb@uevora.pt (A.F.B.)

[4] Departamento de Biologia, Escola de Ciências e Tecnologia, Universidade de Évora, 7004-516 Évora, Portugal

[5] Departamento do Litoral e Proteção Costeira (DOS), Agência Portuguesa do Ambiente, I.P., 2610-124 Amadora, Portugal; carlos.batista@apambiente.pt

[6] Departamento de Recursos Hídricos, Agência Portuguesa do Ambiente, I.P., 2610-124 Amadora, Portugal; ana.telhado@apambiente.pt (A.T.); veronica.pinto@apambiente.pt (V.P.); maria.quadrado@apambiente.pt (M.F.Q.)

* Correspondence: rmmonteiro@fc.ul.pt

† Presented at the IX Iberian Congress of Ichthyology, Porto, Portugal, 20–23 June 2022.

‡ Presenting author (Oral communication).

Citation: Monteiro, R.; Quintella, B.R.; Almeida, P.R.; Costa, J.L.; Pereira, E.; Belo, A.F.; Portela, T.; Batista, C.; Telhado, A.; Pinto, V.; et al. Environmental Factors Promoting Upstream Movement of Yellow Eel (*Anguilla anguilla* L.) in the Mondego River. *Biol. Life Sci. Forum* **2022**, *13*, 112. https://doi.org/10.3390/blsf2022013112

Academic Editor: Alberto Teodorico Correia

Published: 17 June 2022

Publisher's Note: MDPI stays neutral with regard to jurisdictional claims in published maps and institutional affiliations.

Abstract: One of the main reasons for the collapse of the European eel (*Anguilla anguilla* L.) is habitat loss resulting from the severe obstruction of rivers throughout the species' range. Specific fish passes (i.e., eel ladders) can be installed in weirs and dams to mitigate this impact and to promote the upstream movement of eels. In the Mondego River, one of the most important basins for diadromous species in Portugal, an eel ladder was installed in 2015 at the Coimbra weir, the first obstacle to fish migration in this basin. This ladder is equipped with a monitoring trap at the upstream exit, where the eels that have successfully overcome the eel ladder are counted and measured before being released. The timing and environmental variables that promote the upstream movement of these individuals (counts at the eel trap were considered as a proxy for upstream movement activity) were assessed between January 2017 and August 2019. A total of 12,019 eels with a length ranging from 60 mm to 287 mm (median = 138 mm) used the eel ladder to move upstream the obstacle. The upstream movement occurred throughout the year, but a clear peak in activity was observed between May and June (~74%). Eels < 150 mm appeared mainly in early summer (May–June), with 64% of the total number of individuals counted in the eel trap belonging to this size class, but no differences in total body length were found between these years. Generalized Additive Models (GAMs) were used to determine the environmental factors that explained these upstream movements. Among the predictors considered (river flow, precipitation, water temperature and photoperiod), minimum water temperature had the strongest explanatory power. The results from this study are crucial for the management of the species, particularly at obstacles where no transposition equipment exists, and human intervention may be required to assist in their upstream progression.

Keywords: European eel; river basin colonization; eel ladder; GAMs; minimum water temperature

Author Contributions: All authors have contributed to this study in several stages of the work. Conceptualization, P.R.A., I.D., B.R.Q., J.L.C. and R.M.; methodology, P.R.A., I.D., B.R.Q., J.L.C., R.M.;

formal analysis, R.M.; writing—original draft preparation, R.M.; writing—review and editing, All authors; supervision, I.D., B.R.Q.; project administration, P.R.A., B.R.Q., I.D.; funding acquisition, P.R.A., B.R.Q., I.D. All authors have read and agreed to the published version of the manuscript.

Funding: This research was financially supported by the projects "Habitat restoration for diadromous fish in river Mondego" (PROMAR 31-03-02-FEP-5), funded by the Ministry of Agriculture and the Sea, and "Eel Pilot Project" funded by Instituto Português do Mar e da Atmosfera under the Data Collection Framework 2017–2019 (DCF), both co-funded by the European Fisheries Fund through PROMAR. Additionally, this work was financially supported by Fundação para a Ciência e a Tecnologia (FCT), through the PhD Grant awarded to R.M. (PD/BD/142778/2018) and the strategic project UID/MAR/04292/2019 granted to MARE.

Conflicts of Interest: The authors declare no conflict of interest.

biology and life sciences forum

Abstract

Fish Communities in the Lower Tagus Inland Wetlands: From Anthropogenic Pressures to Conservation Management [†]

Diogo Ribeiro [1,2,*,‡], Diogo Dias [2], Gil Santos [2], João Gago [1,2], Beatriz Serrano [2], Luis Almeida [3], Joana Martelo [2,3], Luis M. da Costa [2], Filipe Ribeiro [2], João Catalão [4] and Maria Filomena Magalhães [3]

[1] Escola Superior Agrária de Santarém, Quinta do Galinheiro–S. Pedro, 2001-904 Santarém, Portugal; joao.gago@esa.ipsantarem.pt

[2] Centro de Ciências do Mar e do Ambiente (MARE), Faculdade de Ciências da Universidade de Lisboa, Campo Grande, 1749-016 Lisboa, Portugal; dmlvmd@gmail.com (D.D.); gilsaraivasantos@gmail.com (G.S.); b.spserrano@hotmail.com (B.S.); joanamartelo@gmail.com (J.M.); dacosta.luis@gmail.com (L.M.d.C.); fmvribeiro@gmail.com (F.R.)

[3] Centro de Ecologia, Evolução e Alterações Ambientais (cE3c), Faculdade de Ciências da Universidade de Lisboa, Campo Grande, 1749-016 Lisboa, Portugal; lpralmeida225@gmail.com (L.A.); mfmagalhaes@fc.ul.pt (M.F.M.)

[4] Instituto Dom Luiz, Faculdade de Ciências da Universidade de Lisboa, Campo Grande, 1749-016 Lisboa, Portugal; jcfernandes@fc.ul.pt

* Correspondence: diogorrribeiro@gmail.com

† Presented at the IX Iberian Congress of Ichthyology, Porto, Portugal, 20–23 June 2022.

‡ Presenting author (Oral communication).

Citation: Ribeiro, D.; Dias, D.; Santos, G.; Gago, J.; Serrano, B.; Almeida, L.; Martelo, J.; da Costa, L.M.; Ribeiro, F.; Catalão, J.; et al. Fish Communities in the Lower Tagus Inland Wetlands: From Anthropogenic Pressures to Conservation Management. *Biol. Life Sci. Forum* **2022**, *13*, 114. https://doi.org/10.3390/blsf2022013114

Academic Editor: Alberto Teodorico Correia

Published: 17 June 2022

Publisher's Note: MDPI stays neutral with regard to jurisdictional claims in published maps and institutional affiliations.

Abstract: Inland wetlands are important biodiversity hotspots and amongst the most impacted ecosystems worldwide. Conservation management and restauration actions in wetlands are thus urgently needed to reverse trends in species loss and habitat degradation, particularly in regions harbouring already endangered endemic species. Inland wetlands may play an important role in supporting endemic endangered fishes in the Lower Tagus basin, where anthropogenic pressures have been increasing, but there is a lack of studies on fish communities, and few areas are identified as inland wetlands. Here, we aim to identify small inland wetlands in the Lower Tagus River and their potential role in supporting fish species, constituting the first study to identify and evaluate the most important fish communities. Inland wetlands were identified through the usage of remote sensing techniques and the calculation of a Normalized Difference Water Index (NDWI) with Sentinel-2 imagery for the Lower Tagus region. From a total of 486 locations identified, 31 were recognized as wetlands as having potential to host fish communities, with 11 being selected for sampling after in loco assessment. Fish sampling was conducted between 6 May and 11 June 2021. Furthermore, for each wetland, we evaluated anthropogenic stressors and land use changes between 2007 and 2018, using national land use data (i.e., Carta de Uso e Ocupação do Solo). A total of 7727 fishes from eight non-native and five native species were captured. Overall, fish communities were dominated by non-native species (97% catches), but both European eel (*Anguilla anguilla*) and Lisbon arched-mouth nase (*Iberochondrostoma olisiponense*), which are globally classified as critically endangered (CR), were found in at least two wetlands. Our results suggest that, over the last 10 years, intensive agriculture decreased (on average ≈ 3%) in the areas surrounding these wetlands, being replaced by extensive agriculture or natural uses. Despite non-native fish prevalence, some wetlands may act as refuge habitats for CR fish species. These results are important for guiding the restoration of inland wetlands and promoting conservation management actions to help reverse fish diversity loss.

Keywords: remote sensing; aquatic habitats; endangered fish species; invasive species; freshwater biodiversity

Author Contributions: Conceptualization, D.R. and F.R.; methodology, D.R., F.R., J.C. and M.F.M.; investigation, D.R., D.D., G.S., J.G., B.S., L.A., J.M., L.M.d.C., F.R., J.C. and M.F.M.; writing—original draft preparation, D.R., F.R. and M.F.M.; writing—review and editing, D.R., J.G., J.M., L.M.d.C., F.R., J.C. and M.F.M.; supervision, F.R. and M.F.M.; funding acquisition, F.R. All authors have read and agreed to the published version of the manuscript.

Funding: This study was conducted in the frame of the projects SONICINVADERS (FCT ref.PTDC/CTA-AMB/28782/2017).

Institutional Review Board Statement: Not applicable.

Informed Consent Statement: Not applicable.

Data Availability Statement: The data presented in this study are available on request from the corresponding author. The data are not publicly available due to privacy restriction.

Conflicts of Interest: The authors declare no conflict of interest.

Abstract

Evaluating Environmental DNA Efficiency in the Detection of Freshwater Species in a System with High Endemism [†]

Sofia Batista [1,*,‡], Manuel Curto [1], Ana Veríssimo [2], Filipe Ribeiro [1], Maria Judite Alves [3], Francisco Pina-Martins [4], Carlos David Santos [5], Sissel Jentoft [6] and Hugo Gante [7,8]

1. MARE—Marine and Environmental Sciences Center, Faculty of Sciences, University of Lisbon, 1600-548 Lisbon, Portugal; macurto@fc.ul.pt (M.C.); fmvribeiro@gmail.com (F.R.)
2. CIBIO—Research Center in Biodiversity and Genetic Resources, University of Porto, 4485-661 Vairão, Portugal; verissimoac@gmail.com
3. cE3c—Center for Ecology, Evolution and Environmental Changes & Museu Nacional de História Natural e da Ciência, Universidade de Lisboa, 1250-102 Lisbon, Portugal; mjalves@museus.ulisboa.pt
4. cE3c—Center for Ecology, Evolution and Environmental Changes & CHANGE—Global Change and Sustainability Institute, Faculdade de Ciências, Universidade de Lisboa, Campo Grande, 1749-016 Lisboa, Portugal; f.pinamartins@gmail.com
5. Centre for Environmental and Marine Studies, Faculty of Sciences, University of Lisbon, 1749-016 Lisboa, Portugal; cdsantos@fc.ul.pt
6. Centre for Ecological and Evolutionary Synthesis (CEES), Department of Biosciences, University of Oslo, NO-0371 Oslo, Norway; sissel.jentoft@ibv.uio.no
7. Department of Biology, KU Leuven, 3000 Leuven, Belgium; hugo.gante@kuleuven.be
8. Royal Museum for Central Africa, Section Vertebrates, 3080 Tervuren, Belgium
* Correspondence: sofia.cupertino7@gmail.com
† Presented at the IX Iberian Congress of Ichthyology, Porto, Portugal, 20–23 June 2022.
‡ Presenting author (Poster presentation).

Citation: Batista, S.; Curto, M.; Veríssimo, A.; Ribeiro, F.; Alves, M.J.; Pina-Martins, F.; Santos, C.D.; Jentoft, S.; Gante, H. Evaluating Environmental DNA Efficiency in the Detection of Freshwater Species in a System with High Endemism. *Biol. Life Sci. Forum* **2022**, *13*, 116. https://doi.org/10.3390/blsf2022013116

Academic Editor: Alberto Teodorico Correia

Published: 17 June 2022

Publisher's Note: MDPI stays neutral with regard to jurisdictional claims in published maps and institutional affiliations.

Abstract: Freshwater fish diversity is highly imperiled due to environmental change, habitat destruction, over-fishing and invasive species. Consequently, it is crucial to develop efficient and easy-to-standardize tools to describe these communities. Traditional fishing methods might not be able to capture all-size and well-hidden organisms, which can lead to an underestimation of biodiversity. In theory, it is possible to detect all species present in a certain locality by sequencing DNA traces left in the environment (eDNA). This approach is especially promising in freshwater systems, which may function as eDNA reservoirs of their respective drainage area. In this context, eDNA metabarcoding is the established method where amplicons for a specific locus are sequenced. However, primers affinity towards certain taxa can lead to false negatives and biased species abundances estimates. In this study, we tested eDNA ability and efficiency in detecting fish species in a threatened and highly endemic ecosystem: the Portuguese area occupied by the Tagus Basin. DNA present in the collected water samples from the Tagus River was amplified with 12S rRNA gene primers and sequenced with Illumina MiSeq technology. Primer choice was previously carried out by in silico PCR test which evaluates primers taxonomic coverage and species identification reliability. The obtained sequences were converted to 312 Amplicon Sequence Variance (ASV), filtered to decrease contamination errors, and compared to different reference databases. Obtained results were compared with conventional methods, such as electric fishing, considering that differences might shed light on which factors constrain species detection using eDNA metabarcoding. Preliminary results show congruences in fish species distributions between electrofishing and eDNA metabarcoding. However, eDNA metabarcoding detected exotic species that had not been previously detected in the Tagus basin, but which are known to exist in other Iberian basins: the South American cichlid—*ustraloheros facetus*—and a European leuciscid—*Squalius cephalus*. Nevertheless, traditional methods were more efficient in describing fish communities at some sites along the basin, showing the importance of using both approaches in a complementary way.

Keywords: eDNA metabarcoding; freshwater biodiversity; Tagus basin

Author Contributions: All authors participated in the conceptualization of the project. M.C., F.R., M.J.A. and H.G. conducted filed work; M.C., A.V. conducted the laboratory work under the supervision of H.G. and M.J.A.; and data analysis was done by S.B., M.C., C.D.S., S.J. and H.G. with extensive support of F.P.-M.; S.B. wrote the first draft of the manuscript and all authors contributed to improve it. All authors have read and agreed to the published version of the manuscript.

Funding: This work was funded by the Portuguese Foundation for Science and Technology (FCT) though the project PTDC/BIA-CBI/31644/2017.

Institutional Review Board Statement: Not applicable.

Informed Consent Statement: Not applicable.

Data Availability Statement: All sequence data were deposit in public databases.

Conflicts of Interest: The authors declare no conflict of interest.

Abstract

Identifying the Threatened Ecosystem Services Provided by Diadromous Species [†]

Estibaliz Díaz [1,*,‡], Arantza Murillas [1], Matthew Ashley [2], Angela Muench [3], Cristina Marta Pedroso [4], Patrick Lambert [5] and Geraldine Lassalle [5]

[1] AZTI, 48395 Sukarrieta, Spain; amurillas@azti.es
[2] School of Biological and Marine Science, University of Plymouth, Plymouth PL4 8AA, UK; matthew.ashley@plymouth.ac.uk
[3] Centre for Environment Fisheries and Aquaculture Science (Cefas), Suffolk NR33 OHT, UK; angela.muench@cefas.co.uk
[4] MARETEC, Instituto Superior Técnico, Universidade de Lisboa, 1049-001 Lisboa, Portugal; cristina.marta@tecnico.ulisboa.pt
[5] INRAE, EABX, 50 Avenue de Verdun Gazinet, F-33612 Cestas, France; patrick.mh.lambert@inrae.fr (P.L.); geraldine.lassalle@inrae.fr (G.L.)
* Correspondence: ediaz@azti.es
[†] Presented at the IX Iberian Congress of Ichthyology, Porto, Portugal, 20–23 June 2022.
[‡] Presenting author (Oral communication).

Citation: Díaz, E.; Murillas, A.; Ashley, M.; Muench, A.; Pedroso, C.M.; Lambert, P.; Lassalle, G. Identifying the Threatened Ecosystem Services Provided by Diadromous Species. *Biol. Life Sci. Forum* **2022**, *13*, 120. https://doi.org/10.3390/blsf2022013120

Academic Editor: Alberto Teodorico Correia

Published: 17 June 2022

Publisher's Note: MDPI stays neutral with regard to jurisdictional claims in published maps and institutional affiliations.

Abstract: Diadromous fish are declining across their Atlantic distribution, and the status of their future is very worrying with the additional threat posed by climate change right now. European stakeholders and policy makers know very well the benefits provided by these species, from uses such as selling fish. However, in addition to that, the diadromous fish populations provide other lesser known benefits to society, known as ecosystem services (ESs), that are now in danger. In this research, developed under the framework of the INTERREG Atlantic Area DiadES Project, ESs linked to diadromous fish are identified by reviewing existing evidence and considering ESs provided in a set of case studies across the AA (from the Gipuzkoa rivers in Spain and Loire and Mondego rivers in France and Portugal, to the Rivers Tamar, Frome and Taff in the UK). ESs identified to be related to diadromous fish populations include food provision (provisioning service), nutrient exchanges between coastal and inland habitats (regulating service) and recreational fishing and tourism linked to the societal interest for diadromous fish (cultural service). The contribution of diadromous species to supporting gastronomic festivals, brotherhoods, the knowledge systems (environmental education and research), the local identity, traditional know-how or even to the natural heritage around diadromous fish also relates to cultural ESs. Potential trade-offs are identified between services provided by diadromous fish populations and other services provided in AA rivers that support alternative benefits (i.e., flood control; electricity production; agriculture (pollution); sand extraction). By providing this list and a monetary assessment of ESs derived from diadromous fishes, DiadES wants to convey to stakeholders and policy makers the importance of these ESs, as they must consider them as part of the decision-making process. Enhancing the assessment of ESs related to diadromous fish species, including the full diversity of ESs the species contribute to (across provisioning, regulating and cultural ESs) and the health of the habitats that support them, is a major necessity to advance towards an ecosystem approach to diadromous fishes' management.

Keywords: river ecosystems; diadromous fish; Atlantic Area; ecosystem services; assessment framework; empirical knowledge

---

Author Contributions: Conceptualization, methodology, software, validation, formal analysis, investigation, resources and data curation: A.M. (Arantza Murillas), M.A., A.M. (Angela Muench), C.M.P. and E.D.; original draft preparation: E.D. and A.M. (Arantza Murillas); review and editing: M.A.,

---

A.M. (Angela Muench), C.M.P., P.L. and G.L.; project administration: P.L., G.L. and E.D.; funding acquisition: P.L., G.L., E.D. and A.M. (Arantza Murillas). All authors have read and agreed to the published version of the manuscript.

Funding: This research was funded by DiadES project (Interreg Atlantic Area Programme through the European Regional Development Fun).

Institutional Review Board Statement: Not applicable.

Informed Consent Statement: Not applicable.

Data Availability Statement: Not applicable.

Conflicts of Interest: The authors declare no conflict of interest.

Abstract

Resident Brown Trout (*Salmo trutta*) Populations in Portugal: Status, Threats, and Fishery Management Requirements [†]

Amílcar Teixeira [1,*,‡], João M. Oliveira [2], Filipe Sá [2], Nuno Pereira [2], António Faro [3], Pedro Segurado [3], Fernando Miranda [1], Fernando Teixeira [1], António Martinho [4] and Maria Filomena Magalhães [2]

[1] Centro de Investigação de Montanha (CIMO), Instituto Politécnico de Bragança, 5300-253 Bragança, Portugal; fernando.veloso.miranda@hotmail.com (F.M.); fteiga@ipb.pt (F.T.)
[2] Centre for Ecology, Evolution and Environmental Changes (cE3c), Faculty of Sciences, University of Lisbon, 1349-017 Lisboa, Portugal; joliveira@fciencias-id.pt (J.M.O.); fsilvasa5@gmail.com (F.S.); nunopereira96@hotmail.com (N.P.); mfmagalhaes@fc.ul.pt (M.F.M.)
[3] Forest Research Centre (CEF), School of Agriculture, University of Lisbon, 1349-017 Lisboa, Portugal; antoniotfaro@isa.ulisboa.pt (A.F.); psegurado@isa.ulisboa.pt (P.S.)
[4] Instituto de Conservação da Natureza e das Florestas, 5000-567 Vila-Real, Portugal; antonio.martinho@icnf.pt
[*] Correspondence: amilt@ipb.pt
[†] Presented at the IX Iberian Congress of Ichthyology, Porto, Portugal, 20–23 June 2022.
[‡] Presenting author (oral communication).

Citation: Teixeira, A.; Oliveira, J.M.; Sá, F.; Pereira, N.; Faro, A.; Segurado, P.; Miranda, F.; Teixeira, F.; Martinho, A.; Magalhães, M.F. Resident Brown Trout (*Salmo trutta*) Populations in Portugal: Status, Threats, and Fishery Management Requirements. *Biol. Life Sci. Forum* **2022**, *13*, 128. https://doi.org/10.3390/blsf2022013128

Academic Editor: Alberto Teodorico Correia

Published: 20 June 2022

Publisher's Note: MDPI stays neutral with regard to jurisdictional claims in published maps and institutional affiliations.

Abstract: Brown trout (*Salmo trutta*) has high ecological and socio-economic importance in many cold-water rivers of northern and central Portugal. However, no studies have addressed the ecology of this species on a large scale. To understand factors driving Brown trout populations in diverse Portuguese streams, we sampled 33 sites, during the summer season of 2020, in cold-water rivers of Minho, Lima, Neiva, Cávado, Ave, Douro, and Vouga basins. Brown trout populations were characterised by several populations and growth parameters, including abundance, density, biomass, age, and body condition. The relationships of these parameters with environmental variables, biotic factors, and fishery management regulations were analysed, and the reference parameters of the populations were defined. There was a good/excellent ecological integrity in most sites, assessed by several water-quality-related, hydromorphological, and biological metrics. Nevertheless, several threats were identified—namely, the riparian degradation (e.g., mortality of alder trees), the presence of exotic species, habitat fragmentation, overfishing, stocking, and more recently, extended dry periods. Portuguese populations showed higher growth rates but lower density, biomass, and physical condition. Intra- and interspecific competition did not seem to play relevant roles in the regulation of brown trout populations, and the common fishery management appears to have no clear positive impact on the natural sustainability of these wild populations. In this context, under the POSEUR 03-2215-FC-000096 project, several studies were developed for in situ conservation and ex situ reproduction of wild brown trout populations. Global genetic characterisation of brown trout populations was made, and eight wild stocks were selected and acclimated to the captivity for implementation of stocking programs. Furthermore, the rehabilitation of riparian corridors and the implementation of specific exploitation and management plans were also developed for the conservation of sympatric populations of brown trout and the critically endangered freshwater pearl mussel *Margaritifera margaritifera* in Portuguese mountain rivers.

Keywords: population parameters; salmonid streams; ecological integrity; conservation; management

Author Contributions: Conceptualization, A.T., J.M.O. and M.F.M.; methodology, F.S., N.P., A.F., F.M. and F.T.; formal analysis, P.S.; resources, A.M.; writing—original draft preparation, A.T., F.S. and J.M.O.; writing—review and editing, M.F.M.; All authors have read and agreed to the published version of the manuscript.

Funding: This research was funded by POSEUR (03-2215-FC-000096) and Portuguese Environmental Fund.

Institutional Review Board Statement: The study was conducted according to the guidelines of the Declaration of Helsinki, and approved by Instituto de Conservação da Natureza e das Florestas (Licença 172/2022/CAPT; Credencial Pesca nº 35/2022).

Informed Consent Statement: Not applicable.

Data Availability Statement: Data availability can be found in project website https://margaritifer amargaritifera.pt/ (accessed on 8 April 2022).

Conflicts of Interest: The authors declare no conflict of interest.

biology and life sciences forum

Abstract

Habitat Suitability of Threespine Stickleback (*Gasterosteus aculeatus* L.) in the Southern Limit of Its Global Distribution: Implications for Species Management and Conservation [†]

Andre Moreira [1,‡], Joana Boavida-Portugal [1], Pedro R. Almeida [1,2], Sara Silva [1] and Carlos M. Alexandre [1,*]

[1] MARE—Centro de Ciências do Mar e do Ambiente/ARNET–Aquatic Research NETwork, Universidade de Évora, 7004-516 Évora, Portugal; andre.moreira@uevora.pt (A.M.); jbp@uevora.pt (J.B.-P.); pmra@uevora.pt (P.R.A.); ssrs@uevora.pt (S.S.)

[2] Departamento de Biologia, Escola de Ciências e Tecnologia, Universidade de Évora, Largo dos Colegiais 2, 7004-516 Évora, Portugal

* Correspondence: cmea@uevora.pt; Tel.: +351-21-750-01-48; Fax: +351-21-750-00-09

† Presented at the IX Iberian Congress of Ichthyology, Porto, Portugal, 20–23 June 2022.

‡ Presenting author (oral communication).

Citation: Moreira, A.; Boavida-Portugal, J.; Almeida, P.R.; Silva, S.; Alexandre, C.M. Habitat Suitability of Threespine Stickleback (*Gasterosteus aculeatus* L.) in the Southern Limit of Its Global Distribution: Implications for Species Management and Conservation. *Biol. Life Sci. Forum* **2022**, *13*, 129. https://doi.org/10.3390/blsf2022013129

Academic Editor: Alberto Teodorico Correia

Published: 20 June 2022

Publisher's Note: MDPI stays neutral with regard to jurisdictional claims in published maps and institutional affiliations.

Abstract: The conservation of fish biodiversity requires reliable information on their distribution and habitat use, especially of endemic species that generally occur in restricted geographic areas and specific habitats. This is the case of threespine stickleback (*G. aculeatus* L.), that is a small freshwater fish listed as Endangered (EN) in Portugal, which represents the southern limit of the species global distribution. The monitoring and conservation of populations placed near to the species distribution limits is very important because in these places, small environmental changes can lead to the significant decline of local populations. However, due to the lack of knowledge about its regional distribution, ecology, and macrohabitat preferences, few measures have been proposed aiming the conservation of this species. This project aimed to identify which macro-scale environmental factors determine regional distribution of *G. aculeatus*, to predict their potential distribution and, therefore, define the most important areas for their protection and conservation. The occurrence data, from existing databases and specific sampling campaigns, together with 15 environmental macrohabitat predictors, were used to model the potential distribution of sticklebacks in Portugal, through an ensemble of species distributions models (SDM). Through the results of our ensemble model, we found that threespine stickleback may occur predominantly at lower stretches of river systems, where sandy substrate is dominant, and flow is higher. Sticklebacks are also more likely to occur in sites with high levels of rainfall in the driest month, thus avoiding locations with high potential for drying during summer, which tend to be common in the Iberian Peninsula. The species also tends to avoid steep slope areas, with high levels of annual precipitation. Based on our results, a probability map of occurrence was generated; from this, some river stretches were categorized into levels according to their importance for the conservation of the species. With the results obtained, it was also possible to identify some atypical populations, occurring in areas with low or null habitat suitability, which need to be further studied, because they must have developed physiological characteristics that allow them to subsist in places that are not conducive to their survival.

Keywords: Gasterosteidae; species distribution models; ensemble forecasting; species distribution and conservation; Iberian Peninsula

Author Contributions: Conceptulization, A.M., C.M.A., P.R.A.; methodology, A.M., C.M.A., J.B.-P.; software, A.M., J.B.-P.; formal analysis, A.M., C.M.A., J.B.-P.; investigation, A.M., C.M.A., J.B.-P., P.R.A., S.S.; data curation, A.M., C.M.A., J.B.-P., S.S.; writing—original draft preparation, A.M.; writing—review and editing, A.M., C.M.A., J.B.-P., P.R.A., S.S.; supervision, C.M.A., P.R.A.; project administration, C.M.A., P.R.A.; funding acquisition, C.M.A., P.R.A.; All authors have read and agreed to the published version of the manuscript.

Funding: The present study was supported by the project "Colmatação de lacunas de informação sobre a distribuição e abundância de peixes dulciaquícolas e migradores (diádromos) em Portugal Continental no âmbito da elaboração do Livro Vermelho", developed by the University of Évora, in partnership with UTAD-University of Trás-os-Montes and Alto Douro and the IPB-Polytechnic Institute of Bragança, and funded by FCiências.ID. Funding was also provided by the Portuguese Science Foundation through the strategy plan for MARE (Marine and Environmental Sciences Centre), via project UIDB/04292/2020, and under the project LA/P/0069/2020 granted to the Associate Laboratory ARNET. FCT also supported this study through the individual contracts attributed to Carlos M. Alexandre (CEECIND/02265/2018) and the PhD scholarship attributed to Sara Silva (2021.05558.BD).

Institutional Review Board Statement: Not applicable.

Informed Consent Statement: Not applicable.

Data Availability Statement: Data is available from correspondence author, upon reasonable request.

Conflicts of Interest: The authors declare no conflict of interest.

Abstract

Management Measures for the Conservation of Freshwater Pearl Mussel (*Margaritifera margaritifera*) and Brown Trout (*Salmo trutta*) in Portugal [†]

Amílcar Teixeira [1,*,‡], João M. Oliveira [2], Manuel Lopes-Lima [3], Joaquim Jesus [4], Joaquim Reis [5], Tânia Barros [6], António Martinho [7] and Ronaldo Sousa [8]

[1] CIMO-Centro de Investigação de Montanha, 5300-253 Bragança, Portugal
[2] cE3c-Centre for Ecology, Evolution and Environmental Changes, Faculdade de Ciências da Universidade de Lisboa, 1749-016 Lisboa, Portugal; joliveira@fciencias-id.pt
[3] CIBIO/InBIO-Centro de Inv. Biodiversidade e Recursos Genéticos, Universidade do Porto, 4485-661 Vairão, Portugal; mlopeslima@cibio.up.pt
[4] Freshwater, Sociedade Unipessoal Lda, Estrada do Cando 42, 5400-010 Chaves, Portugal; joaquimbarreira@gmail.com
[5] BIOTA Lda-Estudos e Divulgação em Ambiente, Lda., 2510-718 Gaeiras, Portugal; jreis@biota.pt
[6] CESAM-Centro de Estudos do Ambiente e do Mar, Universidade de Aveiro, 3810-193 Aveiro, Portugal; taniabarros@ua.pt
[7] ICNF-Instituto de Conservação da Natureza e das Florestas, 5000-567 Vila-Real, Portugal; antonio.martinho@icnf.pt
[8] CBMA-Centro de Biologia Molecular e Ambiental, Universidade do Minho, 4710-057 Braga, Portugal; ronaldo@bio.uminho.pt
[*] Correspondence: amilt@ipb.pt
[†] Presented at the IX Iberian Congress of Ichthyology, Porto, Portugal, 20–23 June 2022.
[‡] Presenting author (poster presentation).

Citation: Teixeira, A.; Oliveira, J.M.; Lopes-Lima, M.; Jesus, J.; Reis, J.; Barros, T.; Martinho, A.; Sousa, R. Management Measures for the Conservation of Freshwater Pearl Mussel (*Margaritifera margaritifera*) and Brown Trout (*Salmo trutta*) in Portugal. *Biol. Life Sci. Forum* **2022**, *13*, 134. https://doi.org/10.3390/blsf2022013134

Academic Editor: Alberto Teodorico Correia

Published: 20 June 2022

Publisher's Note: MDPI stays neutral with regard to jurisdictional claims in published maps and institutional affiliations.

Abstract: The project "Recovery and Protection of *Margaritifera margaritifera*" (POSEUR-03-2215-FC-000096) was developed to protect and restore freshwater pearl mussel *M. margaritifera* and brown trout *S. trutta* populations in Portugal. To conduct this, in the last 3 years, several in situ and ex situ conservation measures were applied, and several tasks were defined, namely the following: (1) assessment of the spatial distribution and conservation status of the target species; (2) evaluation of the biological and ecological quality of salmonid rivers; (3) analysis of vulnerability to human disturbances; (4) an increase in habitat suitability; (5) definition of management plans for salmonid watercourses; 6) captive reproduction of *M. margaritifera* and *S. trutta*; (7) monitoring of *M. margaritifera* and *S. trutta* stocking programs; (8) genetic characterization of wild populations of *S. trutta*. Results showed that *M. margaritifera* is currently distributed in seven rivers (mainly in Rabaçal and Tuela rivers, and residually in Mente, Paiva, Neiva, Tâmega and Beça rivers). In Portugal, only *S. trutta* functions as a suitable host for *M. margaritifera*. Both species are highly vulnerable to human disturbances and the main threats include habitat fragmentation and degradation (e.g., dams and alder tree disease), invasive alien species (e.g., American crayfishes, *Pacifastacus leniusculus* and *Procambarus clarkii*) and climate change (e.g., droughts and heatwaves). A riparian area of 2.91 ha was restored, and several management plans were specifically developed in different rivers to contribute to the recovery of both native populations. Moreover, the ex situ reproduction of *M. margaritifera* produced more than 100,000 juveniles, allowing the implementation of reintroduction actions. Complementarily, eight wild brown trout stocks were selected, based on the genetic diversity, and acclimated to captivity in ICNF fishfarms (Castrelos, Bragança and Torno, Amarante) for future stocking programs according to national legislation. Overall, this project is a first step to improve the conservation status of pearl mussel *M. margaritifera* and its fish host, brown trout *S. trutta*.

Keywords: endangered species; brown trout; threats; riparian rehabilitation; management plans

Author Contributions: The authors contributed in equal parts for the several components of the manuscript; All authors have read and agreed to the published version of the manuscript.

Funding: This research was funded by POSEUR (03-2215-FC-000096) and Portuguese Environmental Fund.

Institutional Review Board Statement: The study was conducted according to the guidelines of the Declaration of Helsinki, and approved by Instituto de Conservação da Natureza e das Florestas (Licença 172/2022/CAPT; Credencial Pesca n° 35/2022).

Informed Consent Statement: Not applicable.

Data Availability Statement: Data availability can be found in project website https://margaritifer amargaritifera.pt/ (accessed on 9 April 2022).

Conflicts of Interest: The authors declare no conflict of interest.

Abstract

Suitability Models at Mesohabitat Scale of Native Freshwater Fish and Mussels for Their Application in Environmental Flows Assessment in the NE of the Iberian Peninsula [†]

Anna Costarrosa [1,*,‡], Dídac Jorda-Capdevila [2], Andreu Porcar [3], Julio C. López-Doval [2], Quim Pou-Rovira [3], Albert Herrero [1], Giovanni Negro [4], Roberta Colucci [4], Beatrice Pinna [4] and Paolo Vezza [4]

[1] Engisic Solucions and Consulting, 17001 Catalonia, Spain; aherrero@ensigic.com
[2] BAC Engineering Consultancy Group, 08018 Barcelona, Spain; djorda@bacecg.com (D.J.-C.); jclopezdoval@gmail.com (J.C.L.-D.)
[3] Sorelló Estudis al Medi Aquàtic, 17007 Girona, Spain; andreu.porcar@sorello.net (A.P.); quim.pou@sorello.net (Q.P.-R.)
[4] Department of Environment, Land and Infrastructure Engineering (DIATI), Politecnico di Torino, 10129 Torino, Italy; giovanni.negro@polito.it (G.N.); s272551@studenti.polito.it (R.C.); beatrice.pinna@polito.it (B.P.); paolo.vezza@polito.it (P.V.)
* Correspondence: acostarrosa@engisic.com; Tel.: +34-630204901
† Presented at the IX Iberian Congress of Ichthyology, Porto, Portugal, 20–23 June 2022.
‡ Presenting author (Oral communication).

Citation: Costarrosa, A.; Jorda-Capdevila, D.; Porcar, A.; López-Doval, J.C.; Pou-Rovira, Q.; Herrero, A.; Negro, G.; Colucci, R.; Pinna, B.; Vezza, P. Suitability Models at Mesohabitat Scale of Native Freshwater Fish and Mussels for Their Application in Environmental Flows Assessment in the NE of the Iberian Peninsula. *Biol. Life Sci. Forum* **2022**, *13*, 138. https://doi.org/10.3390/blsf2022013138

Academic Editor: Alberto Teodorico Correia

Published: 20 June 2022

Publisher's Note: MDPI stays neutral with regard to jurisdictional claims in published maps and institutional affiliations.

Abstract: In Mediterranean streams and rivers in general, aquatic organisms use a specific habitat for rearing, growing, breeding, and wintering. Multiple studies have focused on this subject, but few for the specific purpose of developing suitability models that feed hydrobiological models for the analysis of flow regimes and the design of environmental flows. Therefore, this study analyzes the habitat preferences of five freshwater species of fish and mussels in the NE of the Iberian Peninsula for that purpose. We use simple decision trees and random forest (RF), a machine learning technique based on the aggregation of multiple decision trees, to develop suitability models that relate the habitat preferences of the five species—separately adults and juveniles—to different attributes of a physical habitat at the meso-scale. Selected attributes are the surface percentage of different levels of depth (0–15 cm, 15–30 cm, ... , >120 cm), velocity (0–15 cm/s, 15–30 cm/s, ... , >120 cm/s) and abiotic/biotic substrate (e.g., gigalithal, megalithal, detritus, phytal), and absence/presence of refuges (boulder, canopy shading, emerging vegetation, undercut banks, woody debris, roots). The models were developed in order to predict three ranks of habitat suitability: absence, presence and abundance, depending on the mentioned attributes of the mesohabitat analysed. Our study provides quantitative results concerning the correspondence between the presence and abundance of different species and habitat characteristics, confirming qualitative observations stated in previous studies. We proved now that the adult mussels of *Unio* genus require a minimum of 5% of sand or silt, low velocities, and undercut banks and roots; that *Barbus meridionalis* habitat changes considerably among seasons; that *Salaria fluviatilis* needs coarse substrates (megalithal, macrolithal and mesolithal) and velocities above 15 cm/s; and that the adult *Squalius laietanus* prefers glides and pools with depths above 60 cm and velocities below 45 cm/s, depending on the season; and that *Anguilla anguilla* prefers intermediate size substrates (macrolithal, mesolithal and microlithal). These results are essential for the modeling of environmental flows in rivers where these species are present. Thus, by analyzing how their physical habitat changes according to the flow regime, one can see whether the available habitat of fish and mussels increases or decreases and predict periods of danger for the species.

Keywords: environmental flows; freshwater bivalves; freshwater fish; Mediterranean streams; mesohabitat scale; random forest; rivers; suitability models

Author Contributions: Conceptualization, A.C. and D.J.-C.; methodology, A.C., D.J.-C., J.C.L.-D., Q.P.-R., R.C., B.P. and A.P.; software, G.N. and P.V.; validation, D.J.-C.; investigation, A.C., D.J.-C., J.C.L.-D. and Q.P.-R.; writing—original draft preparation, D.J.-C.; writing—review and editing, A.C., D.J.-C., J.C.L.-D., Q.P.-R. and A.H.; supervision, D.J.-C.; funding acquisition, D.J.-C. and A.H. All authors have read and agreed to the published version of the manuscript.

Funding: This research was funded by the Catalan Water Agency (ACA).

Institutional Review Board Statement: Not applicable.

Informed Consent Statement: Not applicable.

Data Availability Statement: Not applicable.

Conflicts of Interest: The authors declare no conflict of interest.

biology and life sciences forum

Abstract

Assessment of the Reproductive Status of Captive Populations of Endangered Leuciscid Species from the Iberian Peninsula: *A. hispanica, I. lusitanicum* and *A. occidentale* [†]

Ana Hernández [1], Fátima Gil [2], Carla Sousa-Santos [3], Elsa Cabrita [1], Pedro M. Guerreiro [1] and Victor Gallego [1,*,‡]

[1] Center of Marine Sciences (CCMAR), University of Algarve, 8005-139 Faro, Portugal; ana.hernandez149@alu.ulpgc.es (A.H.); ecabrita@ualg.pt (E.C.); pmgg@ualg.pt (P.M.G.)

[2] Aquário Vasco de Gama, 1495-718 Cruz Quebrada, Portugal; avg.aqua@marinha.pt

[3] MARE—Marine and Environmental Sciences Centre, ISPA, 3004-517 Coimbra, Portugal; carla_santos@ispa.pt

[*] Correspondence: vgalbiach@ualg.pt

[†] Presented at the IX Iberian Congress of Ichthyology, Porto, Portugal, 20–23 June 2022.

[‡] Presenting author (Poster presentation).

Abstract: Populations of freshwater fish species endemic to the Iberian Peninsula have been declining since the mid-20th century, and the captive breeding of highly endangered species is considered to be a useful tool to restock declining populations. A pioneer project of supportive breeding of critically endangered fish started in 2007 at the Aquário Vasco da Gama (AVG), and this work aims to show the reproductive status of the breeders which make up the current captive broodstoks. Populations of different endangered leuciscid species (*Anaecypris hispanica, Iberochondrostoma lusitanicum* and *Achondrostoma occidentale*) were sampled at AVG during the spring of 2022. Breeders were counted and sexed, and males were stripped to check for the presence of spermatozoa. The sperm volume was assessed visually, and spermatozoa motility was assessed by a CASA system. Sperm samples were classified into four classes based on the percentage of motile cells: C-I $\leq$ 25%, C-II = 25–50%, C-III = 50–75%; and C-IV > 75%. The captive population of *A. hispanica* consisted of 63 individuals and showed a 40% of spermiating males, with an average volume of 5–10 µL. The histogram of sperm quality reported that 15% males had sperm motility of C-II, 50% of males had sperm motility of C-III and, finally, 35% of males had sperm with the high-quality class (C-IV). The population of *I. lusitanicum* consisted of 599 individuals and showed 93% of spermiating males, with an average volume of 15–20 µL. The histogram of sperm quality reported that most part of the males had good sperm quality belonging to C-III and C-IV class (26% and 71%, respectively), while just 1 male showed bad quality sperm (C-II). The captive population of *A. occidentale* consisted of 193 individuals, showing a 62% of spermiating males with an average volume of 20–25 µL. The histogram showed that 6% males had sperm motility of C-I, 26% of males had sperm motility of C-II, the most part of the males (45%) showed a sperm quality of C-III and, finally, 23% of males had sperm with the high-quality class (C-IV). Since the project began in 2007, more than 12,000 fish of these three critically endangered species have been released to restock the populations from which the respective wild breeders were caught. All captive fish were released after a maximum of three consecutive generations in captivity, and new stocks were established with wild adults from the target populations, to avoid the negative effects of inbreeding and genetic drift on the original genetic pool.

Keywords: gamete quality; ex-situ conservation; sperm kinetics

Citation: Hernández, A.; Gil, F.; Sousa-Santos, C.; Cabrita, E.; Guerreiro, P.M.; Gallego, V. Assessment of the Reproductive Status of Captive Populations of Endangered Leuciscid Species from the Iberian Peninsula: *A. hispanica, I. lusitanicum* and *A. occidentale* . *Biol. Life Sci. Forum* **2022**, *13*, 139. https://doi.org/10.3390/blsf2022013139

Academic Editor: Alberto Teodorico Correia

Published: 20 June 2022

Publisher's Note: MDPI stays neutral with regard to jurisdictional claims in published maps and institutional affiliations.

Author Contributions: Conceptualization, V.G. and E.C.; methodology, A.H., V.G., F.G., C.S.-S. and P.M.G.; formal analysis, A.H.; investigation, V.G. and C.S.-S.; draft preparation, A.H. and V.G.; writing—review editing, V.G. and E.C.; supervision, V.G.; project administration, V.G.; funding acquisition, V.G. and E.C. All authors have read and agreed to the published version of the manuscript.

Funding: This study was funded by the call "Marie Skłodowska-Curie Actions—Individual Fellowships)", within the project entitled CRYO-FISH (Grant Agreement number: 101038049).

Institutional Review Board Statement: The study was conducted according to the guidelines of the CCMAR´s Ethiomittee-ORBEA.

Informed Consent Statement: Not applicable.

Data Availability Statement: Not applicable.

Conflicts of Interest: The authors declare no conflict of interest.

biology and life sciences forum

Abstract

Sperm Characterization of Endangered Leuciscids Endemic from the Iberian Peninsula: Gamete Storage as a Tool for Helping Ex-Situ Breeding Programs [†]

Ana Hernández [1], Carla Sousa-Santos [2], Fátima Gil [3], Pedro M. Guerreiro [1], Elsa Cabrita [1] and Victor Gallego [1,*,‡]

[1] Center of Marine Sciences (CCMAR), University of Algarve, 8005-139 Faro, Portugal; ana.hernandez149@alu.ulpgc.es (A.H.); pmgg@ualg.pt (P.M.G.); ecabrita@ualg.pt (E.C.)

[2] MARE—Marine and Environmental Sciences Centre, ISPA, 3004-517 Coimbra, Portugal; carla_santos@ispa.pt

[3] Aquário Vasco de Gama, 1495-718 Cruz Quebrada, Portugal; avg.aqua@marinha.pt

[*] Correspondence: vgalbiach@ualg.pt

[†] Presented at the IX Iberian Congress of Ichthyology, Porto, Portugal, 20–23 June 2022.

[‡] Presenting author (Oral communication).

Citation: Hernández, A.; Sousa-Santos, C.; Gil, F.; Guerreiro, P.M.; Cabrita, E.; Gallego, V. Sperm Characterization of Endangered Leuciscids Endemic from the Iberian Peninsula: Gamete Storage as a Tool for Helping Ex-Situ Breeding Programs. *Biol. Life Sci. Forum* **2022**, *13*, 140. https://doi.org/10.3390/blsf2022013140

Academic Editor: Alberto Teodorico Correia

Published: 20 June 2022

Publisher's Note: MDPI stays neutral with regard to jurisdictional claims in published maps and institutional affiliations.

Abstract: Populations of freshwater fish species endemic to the Iberian Peninsula have been declining since the mid-20th century, and several types of actions (from in situ to ex situ measurements) have been applied over the past decades for preserving these species. However, limited knowledge about their reproductive biology makes it necessary to investigate different aspects of the reproductive cycle for improving breeding programs. The main objectives of this work were to advance knowledge concerning sperm kinetics and spermatozoa morphology and to develop protocols for the short- and long- term storage of gametes. Populations of different endangered leuciscid species (*Anaecypris hispanica, Iberochondrostoma lusitanicum, Achondrostoma occidentale, and Squalius aradensis*) were sampled during the spring of 2022 both in captive populations kept at Aquário Vasco da Gama (AVG) and in wild populations from different Portuguese rivers. Sperm samples were collected and sperm motion parameters were assessed, for the first time, for these four species. Sperm kinetics differed between species in motility and velocity traits, also showing a different number of sperm subpopulations. The longevity of sperm (swimming period) was also different among species: the shortest period was obtained for the wild population of *S. aradensis* (values close to zero at 40 s), and the longest swimming period for the captive population of *I. lusitanicum* (values close to zero at 120 s). Furthermore, different storage trials were carried out diluting the sperm in a extender solution (75 mM NaC1, 70 mM KC1, 2 mM CaC12, 1 mM, MgSO4, l0 mM Hepes, pH 8) at a ratio 1:20 (sperm:extender). Sperm quality (>40% of motile cells) was kept for a maximum of four days of storage, depending on the species. In addition, new cryopreservation protocols (using DMSO, Methanol and/or egg yolk) were tested for cryobanking the sperm of these threatened species. Cryopreserved samples showed significantly lower motility when compared with fresh samples, and the best results were obtained for *I. lusitanicum*, reaching 20% of motile cells after thawing using 10% of DMSO supplemented with 10% of egg yolk. This study is the first of its kind to successfully achieve gamete cryopreservation of Iberian endemic and endangered freshwater fish species, developing new and useful tools to complement the management and conservation programs.

Keywords: gamete quality; cryopreservation; sperm subpopulations

---

Author Contributions: Conceptualization, V.G. and E.C.; methodology, A.H., V.G., F.G., C.S.-S. and P.M.G.; formal analysis, A.H.; investigation, V.G. and C.S.-S.; draft preparation, A.H. and V.G.; writing—review editing, V.G. and E.C.; supervision, V.G.; project administration, V.G.; funding acquisition, V.G. and E.C. All authors have read and agreed to the published version of the manuscript.

Funding: This study was funded by the call "Marie Skłodowska-Curie Actions—Individual Fellowships)", within the project entitled CRYO-FISH (Grant Agreement number: 101038049).

Institutional Review Board Statement: The study was conducted according to the guidelines of the CCMAR's Ethiomittee-ORBEA.

Informed Consent Statement: Not applicable.

Data Availability Statement: Not applicable.

Conflicts of Interest: The authors declare no conflict of interest.

Abstract

A First Guide to Freshwater and Diadromous Fish in Portugal [†]

Maria João Collares-Pereira [1,*], Maria Judite Alves [1,2], Filipe Ribeiro [3,‡], Isabel M. Domingos [3,4], Pedro Raposo de Almeida [5,6], Luís Moreira da Costa [2,3], Hugo F. Gante [7,8], Ana Filipa Filipe [9], Maria Ana Aboim [3], Patrícia Marta [10] and Maria Filomena Magalhães [1,4]

[1] cE3c—Centre for Ecology, Evolution and Environmental Changes, Faculty of Sciences, University of Lisbon, 1749-016 Lisbon, Portugal; mjalves@museus.ulisboa.pt (M.J.A.); mfmagalhaes@fc.ul.pt (M.F.M.)

[2] National Museum of Natural History and Science, University of Lisbon, 1250-102 Lisbon, Portugal; ldcosta@fc.ul.pt

[3] MARE—Marine and Environmental Sciences Centre, Faculty of Sciences, University of Lisbon, 1749-016 Lisbon, Portugal; fmribeiro@fc.ul.pt (F.R.); imdomingos@fc.ul.pt (I.M.D.); maaboim@fc.ul.pt (M.A.A.)

[4] Animal Biology Department, Faculty of Sciences, University of Lisbon, 1749-016 Lisbon, Portugal

[5] Department of Biology, School of Science and Technology, University of Évora, 7004-516 Évora, Portugal; pmra@uevora.pt

[6] MARE—Marine and Environmental Sciences Centre, University of Évora, 7004-516 Évora, Portugal

[7] Department of Biology, KU Leuven, 3000 Leuven, Belgium; hugo.gante@kuleuven.be

[8] Royal Museum for Central Africa, Section Vertebrates, 3080 Tervuren, Belgium

[9] CIBIO/InBIO—Research Centre in Biodiversity and Genetic Resources, University of Porto, 4485-661 Porto, Portugal; affilipe@gmail.com

[10] BIOTA—Biodiversity and Water Consultancy, Rua Carlos Ramos, Nº 9 A, 2620-529 Odivelas, Portugal; prodrigues@biota.pt

* Correspondence: mjpereira@fc.ul.pt

† Presented at the IX Iberian Congress of Ichthyology, Porto, Portugal, 20–23 June 2022.

‡ Presenting author (oral communication).

Citation: Collares-Pereira, M.J.; Alves, M.J.; Ribeiro, F.; Domingos, I.M.; Raposo de Almeida, P.; Moreira da Costa, L.; Gante, H.F.; Filipe, A.F.; Aboim, M.A.; Marta, P.; et al. A First Guide to Freshwater and Diadromous Fish in Portugal. *Biol. Life Sci. Forum* **2022**, *13*, 142. https://doi.org/10.3390/blsf2022013142

Academic Editor: Alberto Teodorico Correia

Published: 28 June 2022

Publisher's Note: MDPI stays neutral with regard to jurisdictional claims in published maps and institutional affiliations.

Abstract: Fresh waters are among the most imperiled ecosystems in the world. Not surprisingly, freshwater and diadromous fishes, which show remarkably high levels of endemism, are declining dramatically worldwide. In Portugal, more than 60% of the native fishes are already threatened and risk further population declines due to the intensification of multiple threats such as habitat loss and degradation, non-native species, and climate change. However, there is a lack of a comprehensive information about this vertebrate group in Portugal, and also in Spain. *A Guide to Freshwater and Migratory Fishes in Portugal* is the first book with dedicated information about the 62 fishes listed for freshwater ecosystems in the region, including data on their morphological characteristics, distribution, ecology, conservation, and economic value, and original scientific drawings by Claudia Baeta and Pedro Salgado. The guide is directed to the general public, scholars and fishermen, constitutes a groundbreaking mark for the Portuguese and Iberian ichthyology, and hopes to attract more research and public engagement to the conservation of the freshwater ecosystems.

Keywords: freshwater conservation; biological diversity; Iberian Peninsula; identification book; rivers

Author Contributions: All authors contributed for conceptualization, resources, data curation, writing, review and editing, and fundraising. All authors have read and agreed to the published version of the manuscript.

Funding: This study was supported by Fundação para Ciência e Tecnologia (FCT) throught the grants UID/BIA/00329/2013 and UID/BIA/00329/2019 (to Ce3C) and UIDB/04292/2020 (to MARE), and sponsored by BES–Biodiversity prize, FCiências. ID through the Fish Identification Course, Faculty of Sciences of the University of Lisbon, University of Évora, BIOTA–Biodiversity and Water Consultancy, and by the Lisbon Municipality–Lisbon the European Green Capital 2020.

Institutional Review Board Statement: Not applicable.

Informed Consent Statement: Not applicable.

Data Availability Statement: Not applicable.

Conflicts of Interest: The authors declare no conflict of interest.

Abstract

Migratory Patterns of *Centropomus parallelus* across Estuaries of the Abrolhos Bank Shelf as Determined by Otolith Elemental Analysis [†]

Felippe A. Daros [1,*,‡]**, Sabrina Luz** [1]**, Marcelo Soeth** [2]**, Mario Condini** [3] **and Maurício Hostim-Silva** [3,4]

[1] Curso de Engenharia de Pesca, Universidade Estadual Paulista "Júlio de Mequita Filho" UNESP Registro, Nelson Brihi Badur 430, Registro 11900-000, Brazil; sabrina.luz@unesp.br

[2] Programa de Pós-Graduação em Sistemas Costeiros e Oceânicos, Universidade Federal do Paraná, P.O. Box 61, Pontal do Paraná 83255-976, Brazil; marcelosoeth@yahoo.com.br

[3] Laboratório de Ecologia de Peixes Marinhos—LEPMAR, UniversidadeFederal do Espirito Santo UFES, São Mateus 29932-540, Brazil; mvcondini@gmail.com (M.C.); mauricio.silva@ufes.br (M.H.-S.)

[4] Departamento de Ciências Agrárias e Biológicas DCAB CEUNES, Progrma de Pós-Graduação em Oceanografia Ambiental (PPGOAM), UFES, Vitória 29075-910, Brazil

* Correspondence: felippe.daros@unesp.br; Tel.: +55-1338282900

† Presented at the IX Iberian Congress of Ichthyology, Porto, Portugal, 20–23 June 2022.

‡ Presenting author (Oral communication).

Citation: Daros, F.A.; Luz, S.; Soeth, M.; Condini, M.; Hostim-Silva, M. Migratory Patterns of *Centropomus parallelus* across Estuaries of the Abrolhos Bank Shelf as Determined by Otolith Elemental Analysis. *Biol. Life Sci. Forum* **2022**, *13*, 14. https://doi.org/10.3390/blsf2022013014

Academic Editor: Alberto Teodorico Correia

Published: 6 June 2022

Publisher's Note: MDPI stays neutral with regard to jurisdictional claims in published maps and institutional affiliations.

Abstract: The fat snook (*Centropomus parallelus* Poey, 1860) is an important recreational and commercial amphidromous fish species along Brazilian coastal environments. Herein, we used otolith elemental analysis to investigate migratory patterns of *C. parallelus* across estuaries from Abrolhos Bank shelf, eastern Brazil. Ninety-four adults were sampled between January and April 2019 in four estuaries, including the Rio Doce, Ipiranga, São Mateus (in Espírito Santo estate), and Caravelas (in Bahia estate). The individuals were weighed, measured, and sagittal otoliths were removed, cleaned, and prepared for further microchemical analyses. The migratory patterns were compared between estuaries through the elemental composition profile (Sr:Ca ratio), from core to edge of otoliths, considering that Sr:Ca ratios (mmol/mol) from 2–5, 5–8, and 8–11 mmol/mol were considered fresh, estuarine, and seawater residency, respectively. The Sr:Ca ratio profiles indicated seven distinct habitat-use patterns. Considering all individuals, 43% were estuarine residents (type 2), 27% were migrants between estuarine and saltwater (type 6), and 16% were migrants between estuarine and freshwater (type 5). Considering each estuary, in Rio Doce, Ipiranga, and São Mateus, 32%, 71%, and 45% of fish were classified as type 2, respectively. In the Caravelas estuary, 58% were classified as type 6. Types 1, 3, 4, and 7 presented 3%, 3%, 6%, and 2%, respectively Our data show that *C. parallelus* can occupy diverse salinity habitats and migrate among marine, estuarine, and freshwater areas, showing high environmental plasticity and adaptation throughout their lifetime. The conservation of this species requires the preservation of freshwater and marine environments, but mainly estuaries, which allows the connectivity between habitats.

Keywords: *Centropomidae*; migration; salinity gradient; environmental plasticity

Funding: Technical-Scientific Cooperation Agreement n° 30/2018 between Espitito Santo Foundatoin of Technology (FEST) and Renova Foundation.

Institutional Review Board Statement: Not applicable.

Informed Consent Statement: Not applicable.

Data Availability Statement: Not applicable.

Conflicts of Interest: The authors declare no conflict of interest.

biology and life sciences forum

Abstract

Estimating Fish Passage over Velocity Barriers for Non-Uniform Flow Conditions: A Case Study in Flat-V Gauging Weirs [†]

Francisco Javier Sanz-Ronda [1,*,‡], Juan Francisco Fuentes-Pérez [1], Francisco Javier Bravo-Córdoba [2], Ana García-Vega [2], Jorge Ruiz-Legazpi [2] and Andrés Martínez de Azagra [1]

[1] Area of Hydraulics and Hydrology, Department of Agroforestry Engineering, University of Valladolid, Avenida de Madrid 44, 34004 Palencia, Spain; juanfrancisco.fuentes@uva.es (J.F.F.-P.); a.m.d.a.p@uva.es (A.M.d.A.)

[2] Centro Tecnológico Agrario y Agroalimentario—ITAGRA.CT, Avenida de Madrid 44, 34004 Palencia, Spain; fjbravo@itagra.com (F.J.B.-C.); a.g.-v.ega@itagra.com (A.G.-V.); jorge.ruiz.legazpi@iaf.uva.es (J.R.-L.)

* Correspondence: jsanz@uva.es; Tel.: +34-979108358

† Presented at the IX Iberian Congress of Ichthyology, Porto, Portugal, 20–23 June 2022.

‡ Presenting author (Oral communication).

Citation: Sanz-Ronda, F.J.; Fuentes-Pérez, J.F.; Bravo-Córdoba, F.J.; García-Vega, A.; Ruiz-Legazpi, J.; Martínez de Azagra, A. Estimating Fish Passage over Velocity Barriers for Non-Uniform Flow Conditions: A Case Study in Flat-V Gauging Weirs. *Biol. Life Sci. Forum* **2022**, *13*, 20. https://doi.org/10.3390/blsf2022013020

Academic Editor: Alberto Teodorico Correia

Published: 6 June 2022

Publisher's Note: MDPI stays neutral with regard to jurisdictional claims in published maps and institutional affiliations.

Abstract: When the flow velocity over a river structure exceeds the swimming capacity of fish, it behaves as a velocity barrier. Depending on the hydrodynamic circumstances of the structure as well as the fish's swimming ability and motivation, the barrier can be permanent, partial, or intermittent. This is the case of flat-V gauging weirs, a common type of velocity barrier in Spanish rivers and in other European rivers. Flat-V weirs are broadly used as they provide precise information about river discharge for water resource management under different hydraulic scenarios, especially during low flow conditions. However, depending on their size, local river morphology, and the river flow scenario, they can produce excessive velocities and thus, reduce or hinder fish upstream movements. Due to their variable geometry, velocity barriers exhibit a non-uniform flow velocity field, which means that flow velocity varies along the barrier. Therefore, any predictive swimming model to assess the barrier effect on fish must consider the spatial variation to achieve a valuable forecast. This work aims to estimate fish passage over Flat-V weirs by linking their 3D hydraulic performance with the swimming capacity of fish. For this, a predictive model is developed using as target species the Iberian barbel (*Luciobarbus bocagei*), combining research on their swimming ability with 3D models of the structure. Results of the model show the river conditions and weir dimensions that permit the ascent of this species through the sloped wall of the weir. This information has direct implications for the design and assessment of velocity barriers as well as for the retrofitting of velocity barriers, making them compatible with the fish migration.

Keywords: gauging weirs; V-Crump; predictive models; potamodromous cyprinid; Iberian fish; swimming performance

Author Contributions: Conceptualization, F.J.S.-R.; methodology, F.J.S.-R.; software, J.F.F.-P. and F.J.B.-C.; validation, F.J.S.-R., F.J.B.-C., J.F.F.-P. and A.G.-V.; formal analysis, F.J.S.-R., F.J.B.-C. and J.F.F.-P.; investigation, F.J.S.-R., F.J.B.-C., J.F.F.-P. and A.G.-V.; data curation, F.J.B.-C. and J.F.F.-P.; writing—original draft preparation, F.J.S.-R., F.J.B.-C. and J.F.F.-P.; writing—review and editing, J.F.F.-P., F.J.S.-R., A.G.-V., F.J.B.-C., J.R.-L. and A.M.d.A.; visualization, J.F.F.-P.; supervision, F.J.S.-R.; project administration, F.J.S.-R.; funding acquisition, F.J.S.-R. All authors have read and agreed to the published version of the manuscript.

Funding: This research was financed by the Spanish Duero River Authority (Confederación Hidrográfica del Duero, CHD). J.F. Fuentes-Pérez contribution was funded by the European Research Council (ERC) under the European Union's Horizon 2020 research and innovation program (Smart fishways—grant agreement n° 101032024).

Institutional Review Board Statement: The study was conducted according to the European Union ethical guidelines (Directive 2010/63/UE) and Spanish Act RD 53/2013 and approved by the competent authorities (Regional Government on Natural Resources of Castilla y León and Water Management Authority of Duero river basin) and the Ethics Committee of University of Valladolid (protocol code 7904309; 3 August 2018).

Informed Consent Statement: Not applicable.

Conflicts of Interest: The authors declare no conflict of interest.

Abstract

Horizontal and Vertical Movements of Swordfish in the Atlantic Ocean and Mediterranean Sea [†]

Daniela Rosa [1,2,*,‡], Fulvio Garibaldi [3], Derke Snodgrass [4], Eric Orbesen [4], Catarina C. Santos [1,2], David Macias [5], Josetxu Ortiz de Urbina [5], Rodrigo Forselledo [6], Philip Miller [6], Andrés Domingo [6], Craig Brown [4] and Rui Coelho [1,2]

[1] IPMA—Instituto Português do Mar e da Atmosfera, Av. 5 de Outubro s/n, 8700-305 Olhão, Portugal; catarina.santos@ipma.pt (C.C.S.); rpcoelho@ipma.pt (R.C.)
[2] CCMAR—Centro de Ciências do Mar da Universidade do Algarve, Campus de Gambelas, 8005-139 Faro, Portugal
[3] Dipartimento di Scienze della Terra dell'Ambiente e della Vita, UNIGE—Università di Genova, C. so Europa, 26, 16132 Genoa, Italy; fulvio.garibaldi@unige.it
[4] NOAA-SFSC—United States NOAA Fisheries Service, Southeast Fisheries Science Center, 75 Virginia Beach Drive, Miami, FL 33149, USA; derke.snodgrass@noaa.gov (D.S.); eric.orbesen@noaa.gov (E.O.); craig.brown@noaa.gov (C.B.)
[5] IEO—Instituto Español de Oceanografía, C.O. Málaga. Pto. Pesquero s/n, 29640 Fuengirola, Spain; david.macias@ieo.es (D.M.); urbina@ieo.es (J.O.d.U.)
[6] DINARA—Dirección Nacional de Recursos Acuáticos, Laboratorio de Recursos Pelágicos (LaRPe), Constituyente 1497, CP 11200 Montevideo, Uruguay; rforselledo@gmail.com (R.F.); philip.miller@cicmar.org (P.M.); dimanchester@gmail.com (A.D.)
[*] Correspondence: daniela.rosa@ipma.pt
[†] Presented at the IX Iberian Congress of Ichthyology, Porto, Portugal, 20–23 June 2022.
[‡] Presenting author (Oral communication).

Citation: Rosa, D.; Garibaldi, F.; Snodgrass, D.; Orbesen, E.; Santos, C.C.; Macias, D.; Ortiz de Urbina, J.; Forselledo, R.; Miller, P.; Domingo, A.; et al. Horizontal and Vertical Movements of Swordfish in the Atlantic Ocean and Mediterranean Sea. *Biol. Life Sci. Forum* **2022**, *13*, 26. https://doi.org/10.3390/blsf2022013026

Academic Editor: Alberto Teodorico Correia

Published: 6 June 2022

Publisher's Note: MDPI stays neutral with regard to jurisdictional claims in published maps and institutional affiliations.

Abstract: The swordfish (*Xiphias gladius*) is an epi- and mesopelagic oceanic species with a wide geographical range within the tropical and temperate waters of all oceans, and is one of the most important target species in surface-longline fisheries. In order to study the vertical habitat-use and migration patterns of swordfish, and to help delimit the stock boundaries and mixing rate of swordfish between the Mediterranean Sea and the North and South Atlantic, satellite telemetry tagging is used. A total of 26 miniPAT tags have been deployed so far in the North (n = 13) and South Atlantic Oceans (n = 9) and the Mediterranean Sea (n = 4). Of the deployed tags, eight individuals suffered post-release mortality; one was fished after one day; three did not transmit; three tags had premature releases with less than 30 days; four had premature releases with more than 30 days; six tags reached full term; and one individuals' tag is still at large. The data from ten tags were analyzed for horizontal and vertical habitat use. The results presented herein are preliminary, as more tag deployments are planned. The results show that swordfish moved in several directions, travelling considerable distances in both the North and South Atlantic Ocean, while having shorter displacements in the Mediterranean Sea. Regarding vertical habitat use, swordfish spent most of the day-time in deeper waters, and were closer to the surface during the night-time. The deepest dive recorded was 1480 m. Regarding temperature, swordfish inhabited waters with temperatures ranging from 3.9 °C to 30.5 °C, mostly residing in waters between 10–12 °C during the day-time and in waters >20 °C during the night-time. The migration of swordfish in this study agrees with the current stock boundaries defined for this species in the Atlantic Ocean, and shows a high vertical overlap with pelagic longline fisheries that are set during the night-time.

Keywords: *Xiphias gladius*; telemetry; habitat use; movement patterns; Atlantic; Mediterranean

Funding: The tagging activities, funded by ICCAT, have been carried out with the financial support of the European Union through Grant Agreement "Strengthening the scientific basis for decision-making in ICCAT", as well as the ICCAT science funds provided within the regular annual budget. Financial support for NOAA tags was provided by the NOAA Fisheries Cooperative Research Program and the NOAA Fisheries International Science Program. D. Rosa and C. Santos are supported by an FCT Doctoral grant (Ref: SFRH/BD/136074/2018 and SFRH/BD/139187/2018, respectively) funded by National Funds and the European Social Fund.

Institutional Review Board Statement: Not applicable.

Informed Consent Statement: Not applicable.

Data Availability Statement: Not applicable.

Acknowledgments: The authors are grateful to all fishery observers and longline skippers from the nations involved in the tagging for this study.

Conflicts of Interest: The authors declare no conflict of interest.

Abstract

Insights in the Stock Mixing Dynamics of Atlantic Bluefin Tuna in the North Atlantic [†]

Natalia Díaz-Arce [1,*,‡], Igaratza Fraile [1], Noureddine Abid [2], Piero Addis [3], Simeon Deguara [4], Fambaye N. Sow [5], Alex Hanke [6], Firdes Saadet Karakulak [7], Pedro G. Lino [8], David Macias [9], Leif Nøttestad [10], Isik K. Oray [7], Enrique Rodriguez-Marin [11], Yohei Tsukahara [12], Jose Luis Varela [13], Haritz Arrizabalaga [1] and Naiara Rodriguez-Ezpeleta [1]

[1] Marine Research Division, AZTI, 48395 Sukarrieta, Spain; ifraile@azti.es (I.F.); harri@azti.es (H.A.); nrodriguez@azti.es (N.R.-E.)

[2] Institute National de Reserche Halieutique, Tanger 90000, Morocco; noureddine.abid65@gmail.com

[3] Department of Environmental and Life Sciences, University of Cagliari, 09124 Cagliari, Italy; addisp@unica.it

[4] AquaBio Tech Ltd., Central Complex, MST1761 Mosta, Malta; dsd@aquabt.com

[5] Centre de Recherches Océanographiques de Dakar-Thiaroye, Dakar B.P. 2241, Senegal; ngomfambaye2015@gmail.com

[6] St Andrews Biological Station, Fisheries and Oceans Canada, St. Andrews, NB E5B 0E4, Canada; alex.hanke@dfo-mpo.gc.ca

[7] Faculty of Aquatic Sciences, Istanbul University, Istanbul 34134, Turkey; karakul@istanbul.edu.tr (F.S.K.); isikoray@hotmail.com (I.K.O.)

[8] Instituto Português do Mar e da Atmosfera, 8700-305 Olhão, Portugal; plino@ipma.pt

[9] Instituto Español de Oceanografía, Centro Oceanográfico de Málaga, 29640 Málaga, Spain; david.macias@ieo.es

[10] Institute of Marine Research, NO-5817 Bergen, Norway; leif.noettestad@imr.no

[11] National Center Spanish Institute of Oceanography, Spanish Council for Scientific Research, Santander Oceanographic Center, 39004 Santander, Spain; enrique.rmarin@ieo.es

[12] Fisheries Resources Institute, Japan Fisheries Research and Education Agency, Kanagawa 220-6115, Japan; tsukahara_y@affrc.go.jp

[13] Campus de Excelencia International del Mar, Universidad de Cádiz, 11519 Cádiz, Spain; joseluis.varela@gm.uca.es

[*] Correspondence: ndiaz@azti.es

[†] Presented at the IX Iberian Congress of Ichthyology, Porto, Portugal, 20–23 June 2022.

[‡] Presenting author (Oral communication).

Citation: Díaz-Arce, N.; Fraile, I.; Abid, N.; Addis, P.; Deguara, S.; Sow, F.N.; Hanke, A.; Karakulak, F.S.; Lino, P.G.; Macias, D.; et al. Insights in the Stock Mixing Dynamics of Atlantic Bluefin Tuna in the North Atlantic. *Biol. Life Sci. Forum* **2022**, *13*, 30. https://doi.org/10.3390/blsf2022013030

Academic Editor: Alberto Teodorico Correia

Published: 6 June 2022

Publisher's Note: MDPI stays neutral with regard to jurisdictional claims in published maps and institutional affiliations.

Abstract: Effective fisheries management requires accurate stock identification, which can be challenging in mixed stock fisheries such as the Atlantic bluefin tuna (*Thunnus thynnus*). This species is currently managed considering two stocks known to spawn in the Mediterranean Sea and Gulf of Mexico, respectively. However, recent studies have shown that individuals from both spawning components can interbreed at a recently discovered spawning ground, located in the Slope Sea. A better understanding of the mixing patterns, as well as the proportion in which both stocks interbreed in the Slope Sea are valuable for a reliable Atlantic bluefin tuna stock assessment. With this aim, we assigned genetic origin of 2000 individuals captured at feeding aggregates across the North Atlantic using a 96 SNP panel and analyzed the genetic profile of 500 individuals including 200 potential Slope Sea spawners (i.e., spawning capable individuals captured in this area at the spawning season), using a 8000 SNP array. We confirmed that stock mixing occurs across different feeding aggregates in the North Atlantic, being stronger in the Northwest Atlantic, where the Mediterranean component was a majority at some locations within and near the Slope Sea spawning ground. The analysis of Slope Sea spawner candidate individuals showed nearly equal representation from both Mediterranean and Gulf of Mexico genetic origin individuals, suggesting similar contribution to the Slope Sea origin offspring. Our findings constitute an important progress towards the understanding of the Atlantic bluefin tuna stock mixing dynamics and the relevance of the recently discovered Slope Sea spawning ground for the conservation of the species.

Keywords: Atlantic bluefin tuna; stock mixing dynamics; genetic assignment

Funding: This research was financially supported by the ICCAT GBYP program, as well as many partners that provided samples, and the Basque Government funded GENPES project.

Institutional Review Board Statement: Not applicable.

Informed Consent Statement: Not applicable.

Data Availability Statement: Not applicable.

Conflicts of Interest: The authors declare no conflict of interest.

Abstract

Two-Way Migration of a Potamodromous Cyprinid in a Small Hydropower Plant with a Pool Type Fishway [†]

Francisco Javier Bravo-Córdoba [1,‡], **Ana García-Vega** [1], **Juan Francisco Fuentes-Pérez** [2], **Leandro Fernandes-Celestino** [3], **Sergio Makrakis** [4] **and Francisco Javier Sanz-Ronda** [2,*]

[1] Centro Tecnológico Agrario y Agroalimentario—ITAGRA.CT, 34004 Palencia, Spain; fjbravo@itagra.com (F.J.B.-C.); agvega@itagra.com (A.G.-V.)
[2] Area of Hydraulics and Hydrology, Department of Agroforestry Engineering, University of Valladolid, 34004 Palencia, Spain; juanfrancisco.fuentes@uva.es
[3] Sustainability Management, Auren Energia Companhia Energética de Sao Paulo—CESP, SP-613 Highway, km 78, Rosana 19273-000, Brazil; le_celestino@hotmail.com
[4] Grupo de Pesquisa em Tecnologia em Ecohidráulica e Conservaçao de Recursos Pesqueiros e Hídricos—GETECH, Programa de Pós-Graduaçao em Recursos Pesqueiros e Engenharia de Pesca, Universidade Estadual do Oeste do Paraná—UNIOESTE, Toledo 85903-000, Brazil; sergio.makrakis@unioeste.br
* Correspondence: jsanz@uva.es; Tel.: +34-979108358
† Presented at the IX Iberian Congress of Ichthyology, Porto, Portugal, 20–23 June 2022.
‡ Presenting author (Oral communication).

Citation: Bravo-Córdoba, F.J.; García-Vega, A.; Fuentes-Pérez, J.F.; Fernandes-Celestino, L.; Makrakis, S.; Sanz-Ronda, F.J. Two-Way Migration of a Potamodromous Cyprinid in a Small Hydropower Plant with a Pool Type Fishway. *Biol. Life Sci. Forum* **2022**, *13*, 38. https://doi.org/10.3390/blsf2022013038

Academic Editor: Alberto Teodorico Correia

Published: 6 June 2022

Publisher's Note: MDPI stays neutral with regard to jurisdictional claims in published maps and institutional affiliations.

Abstract: Most freshwater fish need to move freely through rivers to complete their life cycles. Thus, river barriers that hinder or block their longitudinal movement (e.g., dams, culverts, gauging stations), directly affect their reproductive, feeding, and habitat routes. A holistic solution to these barriers would need to allow directed, undistracted, and bidirectional fish migration between different habitats; that is to say, it would need to allow two-way migration. The most extended solution that would allow upstream fish migration is a fishway. However, for downstream migration fish have alternate routes such as spillways, turbines, or bypasses. Studies and discussions about two-way migration and bidirectional movement through a fishway have been focused on large dams and reservoirs; thus, there is a lack of available data on other environments, less popular species, or smaller dams and weirs. In this sense, it is possible to hypothesize that a fishway, especially in a smaller facility, could enhance two-way migration by allowing bidirectional movement. Therefore, as a first step to analyzing the possibility, we studied longitudinal connectivity (two-way migration and bidirectional movements) through a small run-of-river hydropower plant (HPP) with a step-pool type fishway, a common and representative configuration of several small HPPs around the world. A potamodromous cyprinid—the Iberian barbel (*Luciobarbus bocagei*)—was selected as the target species. In this study, radio and PIT tracking data were collected for four different years and combined to characterize movement in the full system: fishway, turbines/spillways, and the river reach downstream (up to 3 km) and upstream (up to 4 km) from the HPP. The results demonstrated the existence of several types of movement with inter-annual and intra-annual variability. Several fish even returned over the years. This suggests that, in this type of HPP facility, a fishway can provide bidirectional connectivity and two-way migration, thus ensuring that a great proportion of fish complete their life cycles.

Keywords: bidirectional; downstream migration; fish ladder; fish passes; PIT; potamodromous; radio tracking; run-of-river; upstream migration

Author Contributions: Conceptualization, F.J.B.-C., L.F.-C., S.M. and F.J.S.-R.; methodology, F.J.B.-C., F.J.S.-R., A.G.-V. and J.F.F.-P.; validation, A.G.-V. and J.F.F.-P.; formal analysis, F.J.B.-C.; data curation, F.J.B.-C.; writing—original draft preparation, F.J.B.-C.; writing—review and editing, All authors; visualization, F.J.B.-C. and J.F.F.-P.; supervision, L.F.-C., S.M. and F.J.S.-R.; project administration, F.J.S.-R.; funding acquisition, F.J.S.-R. All authors have read and agreed to the published version of the manuscript.

Funding: This research was partially funded by the European Union's H2020 research and innovation program under grant agreement No. 727830, FIThydro.

Institutional Review Board Statement: The study was conducted according to the guidelines of the Declaration of Helsinki, and approved by the Ethics Committee of University of Valladolid as well as the approval of the competent authorities, i.e. Regional Government on Natural Resources (Junta de Castilla y León) and Water Management Authority (Confederación Hidrográfica del Duero). (protocol code 7904309 and date of approval 3 august 2018).

Informed Consent Statement: Not applicable.

Data Availability Statement: Data are available upon reasonable request to the corresponding author.

Conflicts of Interest: The authors declare no conflict of interest.

biology and life sciences forum

MDPI

Abstract

Otolith Fingerprint and Shape of the Chub Mackerel (*Scomber colias*) in the Southwestern Atlantic Ocean [†]

Luiz Matsuda, Jr. [1,*], Felippe A. Daros [2,‡] and Paulo Schwingel [1]

[1] Laboratório de Ecossistemas Aquáticos e Pesqueiros, Escola do Mar, Ciência e Tecnologia, Universidade do Vale do Itajaí (UNIVALI), Rua Uruguai 458, Centro, Itajaí 88302-901, Brazil; schwingel@univali.br

[2] Coordenadoria de Curso de Engenharia de Pesca, Universidade Estadual Paulista "Júlio Mesquita Filho" (UNESP), Campus de Registro, Registro 11900-000, Brazil; felippe.daros@unesp.br

[*] Correspondence: matsuda_jr@hotmail.com

[†] Presented at the IX Iberian Congress of Ichthyology, Porto, Portugal, 20–23 June 2022.

[‡] Presenting author (Oral communication).

Citation: Matsuda, L., Jr.; Daros, F.A.; Schwingel, P. Otolith Fingerprint and Shape of the Chub Mackerel (*Scomber colias*) in the Southwestern Atlantic Ocean. *Biol. Life Sci. Forum* **2022**, *13*, 39. https://doi.org/10.3390/blsf2022013039

Academic Editor: Alberto Teodorico Correia

Published: 2 June 2022

Publisher's Note: MDPI stays neutral with regard to jurisdictional claims in published maps and institutional affiliations.

Abstract: The mackerel (*Scomber colias*, Gmelin 1789) is a shoal-forming scombrid that inhabits the coastal waters of the Atlantic Ocean. The species can be found at depths of up to 300 m, but often inhabits the warmer, shallower waters of coastal regions in its migration to feeding and spawning grounds. In the southeast and south of Brazil, the commercial capture of this species is achieved by purse seine fleets, with great fluctuations in the catch from one year to the next. In Brazil, a single stock of *S. colias* was considered for fishery management purposes. However, the species is not covered by any specific regulatory act in the Brazilian fisheries legislation. The aim of this study was to evaluate the homogeneity of fish stocks of the mackerel *S. colias* on the continental shelf of southeastern and southern Brazil, through an analysis of the shape and elemental chemical signatures of otoliths. The data used are from nine samples from fishing landings and scientific observers in the purse seine fleets that went out between 2008 and 2020. Multielemental signatures (^{44}Ca, ^{7}Li, ^{26}Mg, ^{55}Mn, ^{88}Sr, and ^{137}Ba) of whole otoliths was performed by Inductively Coupled Plasma Mass Spectrometry, and otolith shape patterns were obtained through wavelet coefficients and shape indices, for the Santa Catarina (SC), São Paulo (SP), and Rio de Janeiro (RJ) regions. The results of multivariate analysis (PERMANOVA, $p < 0.05$) for otolith chemistry showed differences between regions, which were confirmed in the pairwise test. In the Canonical Analysis of Principal Coordinates (CAP), with reclassification by the leave-one-out diagnosis, the individuals were assigned to their collection regions, with accuracies of 74% (SC), 90% (SP), and 65% (RJ), and global reclassification of 73%. The results for otolith shape alone showed no differences between the SC and SP samples, and individuals were assigned to their collection regions with lower precision (SC: 54%, SP: 70%, and RJ: 60%). When the otolith shape and chemical analyses were combined, the reliability of the results did not increase. This study indicates that mackerel stocks are not homogeneous in the continental shelf area of the southeast and south of Brazil.

Keywords: Scombridae; otolith microchemistry; otolith shape analysis

Author Contributions: All authors have read and agreed to the published version of the manuscript.

Funding: Coordenação de Aperfeiçoamento de Pessoal de Nível Superior (CAPES) and Fundo Brasileiro para a Biodiversidade (FUNBIO – Projeto Sardinha).

Institutional Review Board Statement: Not applicable.

Informed Consent Statement: Not applicable.

Data Availability Statement: Not applicable.

Conflicts of Interest: The authors declare no conflict of interest.

biology and life sciences forum

Abstract

Fishway Attraction Efficiency during Upstream and Down-Stream Migration: Field Tests in a Small Hydropower Plant with Run-of-the-River Configuration [†]

Francisco Javier Bravo-Córdoba [1,*], Julen Torrens [1,‡], Juan Francisco Fuentes-Pérez [2], Ana García-Vega [1] and Francisco Javier Sanz-Ronda [2]

[1] Centro Tecnológico Agrario y Agroalimentario—ITAGRA.CT, 34004 Palencia, Spain; julentorrens@gmail.com (J.T.); agvega@itagra.com (A.G.-V.)

[2] Area of Hydraulics and Hydrology, Department of Agroforestry Engineering, University of Valladolid, 34004 Palencia, Spain; juanfrancisco.fuentes@uva.es (J.F.F.-P.); jsanz@uva.es (F.J.S.-R.)

* Correspondence: fjbravo@itagra.com; Tel.:+34 979108358

† Presented at the IX Iberian Congress of Ichthyology, Porto, Portugal, 20–23 June 2022.

‡ Presenting author (Poster presentation).

Abstract: Understanding fishway attraction is one of the main open challenges in fishways research, and unraveling the mechanisms and relationships that trigger it is crucial to improve the performance of fishways. Furthermore, attraction is usually understood in terms of upstream migration; however, taking into account the possible bidirectional use of fishways, it is equally important to study this phenomenon during downstream migration, although this is usually considered negligible. Therefore, this study aims to advance our understanding of fishway attraction efficiency by considering both upstream and downstream movements in a key small hydropower plant scheme in the Iberian Peninsula. To achieve this, one of the most common Iberian fish species, the Iberian barbel (*Luciobarbus bocagei*, Steindachner), was monitored via telemetry in a stepped fishway. The studied fishway, considering the specialized literature, would be classified as poor in attraction, i.e., difficult to find due to its low competing discharge and the long distance between the main river flow and both fishway entrances. Fish were PIT tagged and released in different upstream and downstream locations and on different dates. The results showed that a significant proportion of the tagged barbels was able to successfully locate the fishway from both sides, in spite of the mentioned localization drawbacks, with inter-annual variability and with repeated events throughout the years. This suggests that even a fishway with a theoretical poor attraction can still be localized by fish, allowing their use as a two-way migration route, at least with species and HPP schemes such as those already studied.

Keywords: fish ladder; fish passes; PIT; potamodromous; cyprinids

Citation: Bravo-Córdoba, F.J.; Torrens, J.; Fuentes-Pérez, J.F.; García-Vega, A.; Sanz-Ronda, F.J. Fishway Attraction Efficiency during Upstream and Down-Stream Migration: Field Tests in a Small Hydropower Plant with Run-of-the-River Configuration. *Biol. Life Sci. Forum* **2022**, *13*, 40. https://doi.org/10.3390/blsf2022013040

Academic Editor: Alberto Teodorico Correia

Published: 2 June 2022

Publisher's Note: MDPI stays neutral with regard to jurisdictional claims in published maps and institutional affiliations.

Author Contributions: Conceptualization, F.J.B.-C. and F.J.S.-R.; methodology, F.J.B.-C., F.J.S.-R., A.G.-V. and J.F.F.-P.; validation, F.J.B.-C.; formal analysis, J.T.; data curation, J.T.; writing—original draft preparation, F.J.B.-C. and J.T.; writing—review and editing, All authors.; visualization, F.J.B.-C. and J.T.; supervision, F.J.B.-C. and F.J.S.-R.; project administration, F.J.S.-R.; funding acquisition, F.J.S.-R. All authors have read and agreed to the published version of the manuscript.

Funding: This research received no external funding.

Institutional Review Board Statement: The study was conducted according to the guidelines of the Declaration of Helsinki, and approved by the Ethics Committee of University of Valladolid as well as the approval of the competent authorities, i.e., Regional Government on Natural Resources (Junta de Castilla y León) and Water Management Authority (Confederación Hidrográfica del Duero). (protocol code 7904309 and date of approval 3 August 2018).

Informed Consent Statement: Not applicable.

Data Availability Statement: Data are available upon reasonable request to the corresponding author.

Conflicts of Interest: The authors declare no conflict of interest.

146

Abstract

Application of Machine Learning Methodologies for Unravelling the Philopatry and Dispersal Range of *Alosa* Species in the Eastern European Atlantic [†]

Alejandro Pico [1,*,‡], David José Nachón [1], Rufino Vieira-Lanero [1], Sandra Barca [1], María del Carmen Cobo [2], Rosa Crujeiras [3] and Fernando Cobo [1]

1   Department of Zoology, Genetics and Physical Anthropology, Campus Vida, University of Santiago de Compostela, 15782 Santiago de Compostela, Spain; davidjose.nachon@usc.es (D.J.N.); rufino.vieira@usc.es (R.V.-L.); sandra.barca@usc.es (S.B.); fernando.cobo@usc.es (F.C.)
2   Department of Biological Sciences, University of Alabama, Tuscaloosa, AL 35487, USA; mariadelcarmen.cobo@usc.es
3   Department of Statistics and Operations Research, Campus Vida, University of Santiago de Compostela, 15782 Santiago de Compostela, Spain; rosa.crujeiras@usc.es
*   Correspondence: alejandro.pico.calvo@rai.usc.es
†   Presented at the IX Iberian Congress of Ichthyology, Porto, Portugal, 20–23 June 2022.
‡   Presenting author (Oral communication).

Citation: Pico, A.; Nachón, D.J.; Vieira-Lanero, R.; Barca, S.; Cobo, M.d.C.; Crujeiras, R.; Cobo, F. Application of Machine Learning Methodologies for Unravelling the Philopatry and Dispersal Range of *Alosa* Species in the Eastern European Atlantic. *Biol. Life Sci. Forum* **2022**, *13*, 45. https://doi.org/10.3390/blsf2022013045

Academic Editor: Alberto Teodorico Correia

Published: 7 June 2022

Publisher's Note: MDPI stays neutral with regard to jurisdictional claims in published maps and institutional affiliations.

Abstract: Allis shad, *Alosa alosa* (Linnaeus, 1758), and twaite shad, *Alosa fallax* (Lacépède, 1803), are two anadromous fish species belonging to the family Clupeidae. Like all other diadromous species, their populations in European rivers are in decline and their scarcity has become an emerging problem. Anthropogenic activities, mainly large dams and other river obstacles have restricted their range in the eastern Atlantic and western Mediterranean. The studies carried out in Europe in the last 20 years provide a good overview of the conservation status of the shads, however, data on many biological characteristics of their life cycle, especially those corresponding to their development in the marine and coastal environment, are still very scarce. In the Iberian Peninsula, scientific data on their populations are scarcely available. In Galicia, there are studies on the populations of these species in the Ulla and Miño rivers; nevertheless, it is not known whether these species can reside in other Galician rivers suitable for their development. The microchemistry of the otoliths of diadromous fish is strongly correlated with the microchemistry of the water of their river of birth. In the present study, the microchemical composition of the otoliths of 95 adults, distributed between both species, caught at sea and in the Ulla and Miño rivers was analyzed and compared with that of the otoliths of 135 juveniles, distributed between both species, from seven Galician and French rivers. In this way, through this correlation it was possible to quantify the degree of philopatry of the specimens, specify their river of birth and establish their dispersal range during the marine phase. Thus, through the application of Machine Learning methodologies, we provide new information on the dispersal range of these species and the interactions between populations.

Keywords: machine learning; philopatry; shads; dispersal range; otoliths; anadromous species; European Atlantic

Author Contributions: Conceptualizacion, F.C., R.V.-L. and D.J.N.; methodology, A.P. and D.J.N.; software, A.P. and D.J.N.; validation, A.P., D.J.N., R.V.-L. and F.C.; formal analysis, A.P., D.J.N. and R.C.; investigation, A.P., D.J.N., R.V.-L., R.C. and F.C.; resources, A.P. and D.J.N.; data curation, A.P., D.J.N. and R.C.; writing—original draft preparation, A.P.; writing—review and editing, A.P., D.J.N., R.V.-L., S.B., M.d.C.C., R.C. and F.C.; visualization, A.P., D.J.N. and R.V.-L.; supervision, F.C., R.V.-L., R.C. and D.J.N.; project administration, F.C., R.V.-L. and D.J.N.; funding acquisition, F.C., R.V.-L. and D.J.N. All authors have read and agreed to the published version of the manuscript.

Funding: This project has been developed in the framework of the projects "Evaluación de las 'capturas incidentales' de *Alosa alosa* y *Alosa fallax* por la flota costera de Galicia: análisis del problema, sensibilización y proposición de medidas de gestión y protección (1MARDEALOSAS)", which has the collaboration of the Biodiversity Foundation of the Ministry for Ecological Transition and the Demographic Challenge, through the Pleamar Programme, co-financed by the FEMP and "Assessing and enhancing ecosystem services provided by diadromous fish in a climate change context (EAPA_18/2018-DiadES)", of the Interreg Atlantic Arc Programme.

Institutional Review Board Statement: Protocols used in this study conform to the ethical laws of the country and have been reviewed by the ethics committee of the University of Santiago de Compostela and the regional government (Xunta de Galicia).

Informed Consent Statement: Not applicable.

Data Availability Statement: Data from this research are available from the corresponding authors upon reasonable request.

Conflicts of Interest: The authors declare no conflict of interest.

biology and life sciences forum

Abstract

Habitat Use of *Gadiculus argenteus* (Pisces: Gadiformes) in the Galicia and Cantabrian Sea Waters [†]

Juan Carlos Arronte *,‡, José Manuel González-Irusta and Alberto Serrano

Instituto Español de Oceanografía (IEO-CSIC), C. O. de Santander, 39004 Santander, Spain; jmanuel.gonzalez@ieo.csic.es (J.M.G.-I.); alberto.serrano@ieo.csic.es (A.S.)

* Correspondence: jcarlos.arronte@ieo.csic.es; Tel.: +34-942-29-17-16

† Presented at the IX Iberian Congress of Ichthyology, Porto, Portugal, 20–23 June 2022.

‡ Presenting author (Oral communication).

Citation: Arronte, J.C.; González-Irusta, J.M.; Serrano, A. Habitat Use of *Gadiculus argenteus* (Pisces: Gadiformes) in the Galicia and Cantabrian Sea Waters. *Biol. Life Sci. Forum* **2022**, *13*, 49. https://doi.org/10.3390/blsf2022013049

Academic Editor: Alberto Teodorico Correia

Published: 7 June 2022

Publisher's Note: MDPI stays neutral with regard to jurisdictional claims in published maps and institutional affiliations.

Abstract: *Gadiculus argenteus*, is a quite common and relatively abundant fish present in the Galicia and Cantabrian Sea continental shelf and is one of the main trophic resources in the area. Despite its importance in the trophic ecosystem dynamics of the Spanish northern continental shelf, there is a general lack of knowledge about the ecological preferences of the species. The aims of this study are both to determine the importance of spatial, temporal, and oceanic environmental factors on the distribution of *G. argenteus* in this area and to generate, for the first time for the species, abundance maps that could help in the development of, for example, trophic models or marine management plans. In order to reach these goals, data on the abundance of this species from an annual bottom trawl survey (DEMERSALES) for the period 1998–2019, along with temporally invariant (depth, slope, sediment type, and percentage of organic matter) and annual (near-bottom temperature and salinity and chlorophyll-a concentration) environmental layers were modelled using delta generalised additive models (GAMs). The results helped us to identify the most suitable habitats for the species and which environmental factors have a significant effect on its distribution. According to our findings, the species showed higher abundances in the upper slope and a preference for muddy bottoms, with chlorophyll-a positively influencing its biomass. It aggregates mainly in the Galician waters and in the most eastern longitudes of the study area. The results of the models proved that most of the environmental variables chosen are relevant factors in the distribution of the species.

Keywords: *Gadiculus argenteus*; forage fish; distribution models; generalised additive models; delta models; essential fish habitat

Author Contributions: Conceptualization, J.C.A., J.M.G.-I. and A.S.; methodology, J.M.G.-I.; software, J.C.A. and J.M.G.-I.; formal analysis, J.C.A. and J.M.G.-I.; writing—original draft preparation, J.C.A. and J.M.G.-I.; writing—review and editing, J.C.A., J.M.G.-I. and A.S. All authors have read and agreed to the published version of the manuscript.

Funding: DEMERSALES survey is funded by the Instituto Español de Oceanografía (IEO) and European Maritime, Fisheries and Aquaculture Fund (EMFAF).

Institutional Review Board Statement: Not applicable.

Informed Consent Statement: Not applicable.

Data Availability Statement: Data not available.

Conflicts of Interest: The authors declare no conflict of interest.

Abstract

Population Structure of the Brazilian Carapeba *Eugerres brasilianus* in a Complex of Lagoon Systems from Southwest Atlantic Ocean Inferred from Otolith Elemental and Shape Signatures [†]

Paulo Almeida [1,2], Marcus Costa [2], Raiane Oliveira [2], Edgar Pinto [3,4], Agostinho Almeida [3], Rui Azevedo [3], Cassiano Monteiro-Neto [2] and Alberto Teodorico Correia [1,5,6,*,‡]

[1] Centro Interdisciplinar de Investigação Marinha e Ambiental (CIIMAR/CIMAR), 4450-208 Matosinhos, Portugal; prcalmeida@id.uff.br
[2] Departamento de Biologia Marinha, Universidade Federal Fluminense, Niterói 24001-970, Brazil; marcusrc@id.uff.br (M.C.); raiane1299@hotmail.com (R.O.); cmneto@id.uff.br (C.M.-N.)
[3] LAQV/REQUIMTE, Departamento de Ciências Químicas, Faculdade de Farmácia, Universidade do Porto, 4050-313 Porto, Portugal; ecp@ess.ipp.pt (E.P.); aalmeida@ff.up.pt (A.A.); ruiazevedo43@gmail.com (R.A.)
[4] Departamento de Saúde Ambiental, Escola Superior de Saúde do Politécnico do Porto (ESS/P.PORTO), Rua Dr. António Bernardino de Almeida 400, 4200-072 Porto, Portugal
[5] Faculdade de Ciências da Saúde, Universidade Fernando Pessoa (FCS/UFP), 4200-150 Porto, Portugal
[6] Instituto de Ciências Biomédicas Abel Salazar, Universidade do Porto (ICBAS-UP), 4050-313 Porto, Portugal
[*] Correspondence: atcorreia.ciimar@gmail.com
[†] Presented at the IX Iberian Congress of Ichthyology, Porto, Portugal, 20–23 June 2022.
[‡] Presenting author (Poster presentation).

Citation: Almeida, P.; Costa, M.; Oliveira, R.; Pinto, E.; Almeida, A.; Azevedo, R.; Monteiro-Neto, C.; Correia, A.T. Population Structure of the Brazilian Carapeba *Eugerres brasilianus* in a Complex of Lagoon Systems from Southwest Atlantic Ocean Inferred from Otolith Elemental and Shape Signatures. *Biol. Life Sci. Forum* **2022**, *13*, 60. https://doi.org/10.3390/blsf2022013060

Academic Editor: Felippe Daros

Published: 8 June 2022

Publisher's Note: MDPI stays neutral with regard to jurisdictional claims in published maps and institutional affiliations.

Abstract: The Brazilian mojarra, *Eugerres brasilianus*, is an economically important species for the artisanal fisheries that exist in the estuarine environments along the Southwest Atlantic Ocean. Despite this, knowledge about its population structure is scarce, and no management strategies have been applied to ensure the sustainability of *E. brasilianus* fisheries in Brazil. Thus, the present study intended to understand the population structure of *E. brasilianus* in a complex system of lagoons in the Southwest Atlantic Ocean. A total of 90 individuals were collected in the lagoons of Piratininga-Itaipu (IP), Saquarema (SQ) and Araruama (AR) between December 2019 and March 2020. For the analyses, 30 individuals per location from the same age group (2 years old), following age estimation by counting the annual growth increments, were used. The contour of the shape of each otolith was evaluated using elliptical Fourier descriptors (EFD). Multi-elemental signatures (MES) of the whole otoliths were obtained using solution-based inductively coupled plasma mass spectrometry. Data were analyzed using univariate and multivariate statistics to assess the degree of separation between individuals from different lagoons. EFD data showed differences between regions. MES exhibited distinct regional patterns, mainly driven by differences in Sr/Ca, Mg/Ca, Mn/Ca, Li/Ca and Cu/Ca ratios. Reclassification accuracy rates obtained from linear discriminant function analyses using both EFD and MES of otoliths were 100% (IP), 90% (SQ) and 97% (AR). Therefore, a clear distinction between the population groups was observed, probably related to the inherent characteristics of each lagoon system, their semi-restricted connectivity with the adjacent coastal zone, as well as the estuarine-opportunistic behavior of the species. Thus, the results suggest that these fisheries should be managed as different population-units.

Keywords: Gerreidae; Sagittae; natural tags; contour and chemical analyses

Author Contributions: Conceptualization, A.T.C.; Formal analysis, P.A. and A.T.C.; Investigation, P.A., M.C. and A.T.C.; Resources, M.C. and A.T.C.; Fish sampling, R.O. and C.M.-N.; Methodology, P.A., E.P., A.A., R.A. and A.T.C.; writing—original draft preparation, P.A.; Writing—review and editing, P.A., M.C. and A.T.C.; Supervision, M.C. and A.T.C.; Funding acquisition, M.C. and A.T.C. All authors have read and agreed to the published version of the manuscript.

Funding: This research was supported by the Strategic Funding UIDB/04423/2020 and UIDP/04423/2020 through Portuguese funds provided by FCT, and by Brazilian Funds provided by FUNBIO through the project Sistema Lagunares. Almeida P. has a PhD grant from CAPES.

Institutional Review Board Statement: Not applicable (animals obtained from commercial fisheries).

Informed Consent Statement: Not applicable.

Data Availability Statement: Datasets available upon request and after author's evaluation.

Conflicts of Interest: The authors declare no conflict of interest.

Abstract

Acoustic Telemetry Unravels Movements and Habitat Use Patterns of Juvenile Meagre (*A. regius*) in the Tagus Estuary [†]

João P. Marques [1,*,‡], Pedro R. Almeida [2,3], Pedro Moreira [1], Patrick Reis-Santos [4], Nuno Prista [5], José Lino Costa [1,6], Isabel Domingos [1,6], Carlos M. Alexandre [2], Catarina S. Mateus [2] and Bernardo R. Quintella [1,6,*]

[1] MARE—Marine and Environmental Sciences Centre, Faculty of Sciences, University of Lisbon, 1749-016 Lisbon, Portugal; pedroandremoreira.pt@gmail.com (P.M.); jlcosta@fc.ul.pt (J.L.C.); idomingos@fc.ul.pt (I.D.)

[2] MARE—Marine and Environmental Sciences Centre, University of Évora, 7004-516 Évora, Portugal; pmra@uevora.pt (P.R.A.); cmea@uevora.pt (C.M.A.); cspm@uevora.pt (C.S.M.)

[3] Department of Biology, School of Sciences and Technology, Universiy of Évora, 7004-516 Évora, Portugal

[4] Southern Seas Ecology Laboratories, School of Biological Sciences, The University of Adelaide, Adelaide 5005, Australia; patrick.santos@adelaide.edu.au

[5] Department of Aquatic Resources, Institute of Marine Research, Swedish University of Agricultural Sciences, 453 30 Lysekil, Sweden; nuno.prista@slu.se

[6] Department of Animal Biology, Faculty of Sciences, University of Lisbon, 1749-016 Lisbon, Portugal

[*] Correspondence: jpmarques@fc.ul.pt (J.P.M.); bsquintella@fc.ul.pt (B.R.Q.)

[†] Presented at the IX Iberian Congress of Ichthyology, Porto, Portugal, 20–23 June 2022.

[‡] Presenting author (Oral Communication).

Citation: Marques, J.P.; Almeida, P.R.; Moreira, P.; Reis-Santos, P.; Prista, N.; Costa, J.L.; Domingos, I.; Alexandre, C.M.; Mateus, C.S.; Quintella, B.R. Acoustic Telemetry Unravels Movements and Habitat Use Patterns of Juvenile Meagre (*A. regius*) in the Tagus Estuary. *Biol. Life Sci. Forum* **2022**, *13*, 64. https://doi.org/10.3390/blsf2022013064

Academic Editor: Alberto Teodorico Correia

Published: 8 June 2022

Publisher's Note: MDPI stays neutral with regard to jurisdictional claims in published maps and institutional affiliations.

Abstract: The meagre is among the largest *Sciaenidae* in the world (max: 230 cm, 103 kg), with a wide distribution range encompassing the NE and CE Atlantic Ocean and the Mediterranean Sea. The life cycle in Atlantic waters includes migratory movements from feeding and overwintering areas at sea to spawning and nursery areas in estuaries and coastal waters. However, significant spawning aggregations are only observed in five locations, among which is the Tagus estuary (Portugal). The meagre fishery that takes place within the Tagus estuary is significant, accounting for approximately two-thirds of Portuguese meagre catches. Despite its economic relevance, the meagre movements in that region remain largely unknown. The existence of a target fishery inside the estuary alongside a lack of routine biological data collection targeting the species and incipient fisheries control in the area, highlight an urgency to adopt innovative methodologies to unravel meagre migrations and its use of critical areas. We present the first insights of movement patterns and habitat use in the Tagus estuary using acoustic biotelemetry data collected between 2019 and 2021. The acoustic receiver array obtained a total of 142.183 registers from a total of 34 individuals tagged. From the tagged specimens, 33% revisited the Tagus estuary in subsequent years at least once, during the spring and summer, and 49% remained in the Tagus at least until autumn. Further analysis was conducted with juveniles tracked over 3 years to identify critical nursery areas using dynamic Brownian bridge movement models (utilization distribution estimations). The effects of abiotic conditions on the meagre behaviour were assessed using in situ sensor data (e.g., temperature and salinity) and other environmental predictors (e.g., photoperiod and tide cycle) and an explanatory model was developed that helps to understand the use of the Tagus estuary by juveniles. The information collected will be discussed in light of possible applications to promote sustainable management of meagre fisheries in the Tagus estuary and adjacent coastal areas.

Keywords: Tagus estuary; nursery habitat; site fidelity; acoustic biotelemetry; habitat use

Author Contributions: Conceptualization, B.R.Q., P.R.A. and J.P.M.; methodology, B.R.Q. and J.P.M.; formal analysis, J.P.M. and P.M.; writing—original draft preparation, J.P.M.; writing—review and editing, B.R.Q., P.R.A., P.R.-S., N.P., J.L.C., I.D., C.M.A. and C.S.M.; supervision, B.R.Q., P.R.A. and P.R.-S.; project administration, B.R.Q.; funding acquisition, B.R.Q., C.S.M., C.M.A., P.R.A., J.L.C. and I.D. All authors have read and agreed to the published version of the manuscript.

Funding: This research was funded by National Funds through FCT (Foundation for Science and Technology) via the project "MIGRACORV—Integrated approach to study the movement dynamics of the meagre *Argyrosomus regius*" (PTDC/BIA-BMA/30517/2017) and MARE's strategic programme (UID/MAR/04292/2020). Additional funding was provided by FCT through a PhD scholarship attributed to J.P.M. (2020.06801.BD), and individual contracts attributed to C.M.A. (CEECIND/02265/2018) and B.R.Q. (2020.02413.CEECIND). CoastNet (Portuguese Coastal Monitoring Network) research infrastructure, also funded by FCT and the European Regional Development Fund (FEDER), through LISBOA2020 and ALENTEJO2020 regional operational programs, in the framework of the National Roadmap of Research Infrastructures of strategic relevance (PINFRA/22128/2016), logistically supported this study.

Informed Consent Statement: Not applicable.

Conflicts of Interest: The authors declare no conflict of interest.

Abstract

Effectiveness Monitoring of Five Fish Ladders in Catalonia, NE of the Iberian Peninsula [†]

Quim Pou-Rovira [1,2,*,‡], Enric Aparicio [2], Rafel Rocaspana [3], Eloi Cruset [1], Guillem Llenas [1], Andreu Porcar [1], Blanca Font [1], Sergio Gaspar [4] and Clemente González [5]

[1] Sorelló, Estudis al medi aquatic SL, 17003 Girona, Spain; eloi.cruset@sorello.net (E.C.); guillem.llenas@sorello.net (G.L.); andreu.porcar@sorello.net (A.P.); blanca.font@sorellona.org (B.F.)
[2] Research Group on Ecology of Inland Waters (GRECO), University of Girona; 17003 Girona, Spain; enric.aparicio@gmail.com
[3] Gesna, Estudis ambientals SL, 25240 Linyola, Spain; rafel@gesna.net
[4] Summit Asesoria Ambiental, 31174 Etxauri, Spain; asesoriasummit@gmail.com
[5] ADRA ingenieria y gestion del medio, slp., 39608 Igollo de Camargo, Spain; adra@adraingenieria.com
* Correspondence: quim.pou@sorello.net
† Presented at the IX Iberian Congress of Ichthyology, Porto, Portugal, 20–23 June 2022.
‡ Presenting author (Oral communication).

Citation: Pou-Rovira, Q.; Aparicio, E.; Rocaspana, R.; Cruset, E.; Llenas, G.; Porcar, A.; Font, B.; Gaspar, S.; González, C. Effectiveness Monitoring of Five Fish Ladders in Catalonia, NE of the Iberian Peninsula. *Biol. Life Sci. Forum* **2022**, *13*, 70. https://doi.org/10.3390/blsf2022013070

Academic Editor: Alberto Teodorico Correia

Published: 9 June 2022

Publisher's Note: MDPI stays neutral with regard to jurisdictional claims in published maps and institutional affiliations.

Abstract: In recent years, a remarkable effort has been made to recover the connectivity for fish in the rivers of Catalonia, mainly through the construction of new fish ladders, or river connectors. However, evaluating the effectiveness of new fish ladders is not yet a common practice, although knowing how it works is a key aspect of improving the design of new river connectors in other infrastructures. For this reason, the Catalan Water Agency has launched a study to monitor the effectiveness of several recently built river connectors. A total of five fish ladders were selected in the rivers Anoia, Ripoll, Brugent, Llémena and Borró, placed in four different watersheds (Llobregat, Besós, Ter and Fluvià, respectively). These rivers have a similar hydrology (average river discharge inferior to 1 m³/s), and a similar potential fish assemblage (*Barbus haasi* or *B. meridionalis*, *Squalius laietanus*, *Anguilla anguilla*, and in some cases *Salaria fluviatilis*). Several exotic species are also present in some of these rivers: *Gambusia holbrooki*, *Lepomis gibbosus*, *Phoxinus* sp., *Barbatula* sp. During 2021, between two and three monitoring campaigns (spring, summer, fall) were carried out. The methodology used consisted first in the installation of permanent traps at the top exit of ladders to obtain direct estimates of pass rate by species, and on comparative fish sampling on each side of the barrier. Additionally, regular monitoring of several hydraulic variables (water velocity, draft, and elevation) was performed at a selection of internal points on each ladder. Finally, the expected mobility per species was also estimated using the package Fishmove (Radinger, 2013), to compare it with experimental estimations and determine the efficiency of ladders. The results demonstrate the effectiveness of the evaluated ladders, at least for some of the species present. Among native species, barbels (*B. haasi* and *B. meridionalis*) showed the highest rates of passage. Some exotic species were also able to occasionally use the ladders. The efficiency of the ladders is mostly high but decreases when there are internal points with excessive speeds, or other failures with respect to the optimal design.

Keywords: fish ladder effectiveness; river connectivity; ecological rehabilitation; Mediterranean rivers; *Barbus meridionalis*; *Barbus haasi*

Author Contributions: Conceptualization, Q.P.-R., E.A., R.R., S.G. and C.G.; methodology, all authors; formal analysis, Q.P.-R., E.A. and R.R.; data curation, Q.P.-R., E.A., E.C., G.L., A.P. and B.F.; writing—review and editing, Q.P.-R., E.A. and R.R. All authors have read and agreed to the published version of the manuscript.

Funding: This research was funded by Catalan Water Agency, through the contract CTN2000623.

Institutional Review Board Statement: Ethical review and approval were waived for this study due to the use of fish standard sampling methods, under the authorization on national authorities on nature conservation and animal welfare.

Informed Consent Statement: Not applicable.

Data Availability Statement: Not applicable.

Conflicts of Interest: The authors declare no conflict of interest.

Abstract

Use of Otolith Shape and Elemental Signatures to Infer the Population Structure of the Thicklip Grey Mullet *Chelon labrosus* in the Southern Bay of Biscay [†]

Anthony Nzioka [1], Ibon Cancio [1], Oihane Diaz De Cerio [1], Maren Ortiz-Zarragoitia [1], Edgar Pinto [2,3], Agostinho Almeida [2] and Alberto Teodorico Correia [4,5,6,*,‡]

[1] CBET Research Group, Department Zoology & Animal Cell Biology, Research Centre for Experimental Marine Biology and Biotechnology, University of the Basque Country, Areatza Hiribidea s/n, 48620 Plentzia, Basque Country, Spain; anthony.nzioka@ehu.eus (A.N.); ibon.cancio@ehu.eus (I.C.); oihane.diazdecerio@ehu.eus (O.D.D.C.); maren.ortiz@ehu.eus (M.O.-Z.)

[2] LAQV/REQUIMTE, Departamento de Ciências Químicas, Faculdade de Farmácia, Universidade do Porto, Rua de Jorge Viterbo Ferreira 228, 4050-313 Porto, Portugal; ecp@ess.ipp.pt (E.P.); aalmeida@ff.up.pt (A.A.)

[3] Departamento de Saúde Ambiental, Escola Superior de Saúde, P. Porto, Rua Dr. António Bernardino de Almeida 400, 4200-072 Porto, Portugal

[4] Centro Interdisciplinar de Investigação Marinha e Ambiental (CIIMAR), Terminal de Cruzeiros do Porto de Leixões, Avenida General Norton de Matos S/N, 4550-208 Matosinhos, Portugal

[5] Faculdade de Ciências da Saúde da Universidade Fernando Pessoa (FCS-UFP), Rua Carlos da Maia 296, 4200-150 Porto, Portugal

[6] Instituto de Ciências Biomédicas Abel Salazar da Universidade do Porto (ICBAS-UP), Rua de Jorge Viterbo Ferreira 228, 4050-313 Porto, Portugal

* Correspondence: atcorreia.ciimar@gmail.com

† Presented at the IX Iberian Congress of Ichthyology, Porto, Portugal, 20–23 June 2022.

‡ Presenting author (poster).

Citation: Nzioka, A.; Cancio, I.; De Cerio, O.D.; Ortiz-Zarragoitia, M.; Pinto, E.; Almeida, A.; Correia, A.T. Use of Otolith Shape and Elemental Signatures to Infer the Population Structure of the Thicklip Grey Mullet *Chelon labrosus* in the Southern Bay of Biscay. *Biol. Life Sci. Forum* **2022**, *13*, 71. https://doi.org/10.3390/blsf2022013071

Academic Editor: Alberto Correia

Published: 9 June 2022

Publisher's Note: MDPI stays neutral with regard to jurisdictional claims in published maps and institutional affiliations.

Abstract: Xenoestrogenic effects have been reported in thicklip grey mullet, *Chelon labrosus*, used as pollution sentinel organisms in estuaries in the Southeast Bay of Biscay with intersex gonads described in populations from some contaminated estuaries. Despite evidence of reproductive stress in this catadromous fish species, knowledge of mullet reproductive movements and connectivity between estuaries is lacking. This study investigates the population structure of *C. labrosus* using otolith shape and elemental signatures of 60 adult individuals collected from two estuaries found in the Southeast Bay of Biscay (Gernika and Plentzia). All samples were collected in June–July 2020. Otolith shape analysis was determined using elliptical Fourier descriptors, while elemental signatures (Sr:Ca, Li:Ca, Mg:Ca, Mn:Ca, Co:Ca, Ni:Ca, Cu:Ca and Ba:Ca) of whole sagittae were determined by inductively coupled plasma mass spectrophotometry. Both natural tags were analyzed with univariate and multivariate statistics to determine whether these signatures are geographically distinct and can be used to assess the degree of separation between individuals. The data showed significant differences in the otolith shape and elemental analyses, with canonical analysis of principal coordinates plots identifying two different groups, each one belonging to each estuary of origin. Differences in whole otolith elemental signatures between locations were driven by Sr:Ca, Li:Ca, and Ba:Ca. Sr:Ca and Li:Ca ratios were higher in Plentzia than in Gernika, while Ba:Ca was higher in Gernika. The high re-classification success rate using both tools obtained from stepwise linear discriminant function analysis supports these findings and suggests that Gernika and Plentzia individuals passed enough time in separated water compartments and should be regarded as two different population units. This could suggest that the intersex condition in mullets from Gernika is due to life-long exposure to xenoestrogens after homing during early larval development in that estuary, without migrations to other estuaries. Acknowledgements: Project funded by Basque Gov. (IT1302-19), Spanish MCIN and EU-FEDER/ERDF (PGC2018-101442-B-100) and EU H2020 (Assemble+ 730984).

Keywords: *Chelon labrosus*; otoliths; shape and elemental signatures; population structure

Author Contributions: Conceptualization, A.N., I.C. and A.T.C.; Formal analysis, A.N.; Investigation, A.N., I.C. and A.T.C.; Methodology, A.N., A.A., E.P., O.D.D.C., M.O.-Z. and A.T.C.; writing—original draft preparation, A.N.; Writing—review and editing, A.N., I.C. and A.T.C.; Supervision, I.C. and A.T.C.; Funding acquisition, I.C. and A.T.C. All authors have read and agreed to the published version of the manuscript.

Funding: This research was supported by the Strategic Funding UIDB/04423/2020 and UIDP/04423/2020 through Portuguese funds provided by FCT, and by the Basque Government (IT1302-19), Spanish Ministry of Science and Innovation (MCIN) and EU-FEDER/ERDF (PGC2018-101442-B-100) and EU H2020 (Assemble+ 730984).

Institutional Review Board Statement: Not applicable (animals obtained through fisheries).

Informed Consent Statement: Not applicable.

Data Availability Statement: Dataset available upon request and after author's evaluation.

Conflicts of Interest: The authors declare no conflict of interest.

biology and life sciences forum

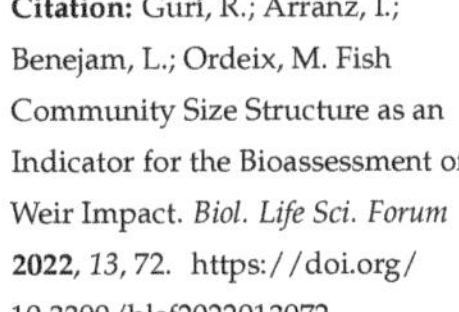

Abstract

Fish Community Size Structure as an Indicator for the Bioassessment of Weir Impact [†]

Rosa Gurí [1,*,‡], Ignasi Arranz [2], Lluís Benejam [3] and Marc Ordeix [1]

1 Centre d'Estudis dels Rius Mediterranis, University of Vic—Central University of Catalonia, Museu del Ter, 08560 Manlleu, Spain; marc.ordeix@uvic.cat
2 Laboratoire Evolution et Diversité Biologique (EDB), UMR5174, Université Toulouse III-Paul Sabatier, 31062 Toulouse, France; ignasiarranz@gmail.com
3 Aquatic Ecology Group, University of Vic—Central University of Catalonia, 08500 Vic, Spain; lluis.benejam@uvic.cat
* Correspondence: rosa.guri@uvic.cat; Tel.: +34-938515176
† Presented at the IX Iberian Congress of Ichthyology, Porto, Portugal, 20–23 June 2022.
‡ Presenting author (Oral communication).

Abstract: The ecological impacts of large dams on stream fish communities have been largely documented, but there is less research on the impacts of small hydropower plants (hereafter called weirs). Most of the studies that evaluate the impact of weirs have been focused on taxonomical approaches such as species richness or diversity. Size-based indicators can be used as alternative tools to evaluate the effects of several environmental changes and anthropogenic perturbations on riverine ecosystems because of the key role of body size in fish physiological rates (i.e., growth, reproduction, respiration). This work investigated the impact of 16 weirs of the upper Ter River basin (NE Iberian Peninsula) on fish community body size structure, comparing control reaches (distant from a weir) with reaches impacted by weirs (immediately downstream). We also controlled for the influence of environmental factors including altitudinal gradients, spatial connectivity, and stream depth. Additionally, we tested the usage of multiple size-based approaches under different sampling intensities from one pass to four passes with an electrofishing sampling design. The results revealed strong evidence that weirs have a negative effect on basic size metrics such as average length and median length: fish communities located in impacted sites showed smaller average and median body sizes than fish communities distant from weirs. In contrast, the size spectrum parameters and functional size diversity metrics showed weak responses to the impact of weirs. The results also showed that all size-based metrics exhibited consistent results under different sampling efforts, suggesting that one sampling pass provided a good representation of the community size structure. The results suggested that only basic size metrics such as average and median length could be useful indicators for the bioassessment of river flow alterations. Finally, size-based metrics can provide an alternative approach to characterize community fish structures by reducing the costs of fish surveys in management plans.

Keywords: biological assessment; body size distributions; environmental flows; fish assemblage; stream barrier impacts

Citation: Gurí, R.; Arranz, I.; Benejam, L.; Ordeix, M. Fish Community Size Structure as an Indicator for the Bioassessment of Weir Impact. *Biol. Life Sci. Forum* **2022**, *13*, 72. https://doi.org/10.3390/blsf2022013072

Academic Editor: Alberto Teodorico Correia

Published: 9 June 2022

Publisher's Note: MDPI stays neutral with regard to jurisdictional claims in published maps and institutional affiliations.

Author Contributions: Conceptualization, R.G., I.A., L.B. and M.O.; formal analysis, I.A. and R.G.; writing—original draft preparation, R.G.; writing—review and editing, R.G., I.A., L.B. and M.O.; supervision, I.A. and M.O.; funding acquisition, L.B. All authors have read and agreed to the published version of the manuscript.

Funding: This research was funded by the Spanish Ministry of Science, Innovation and Universities, grant number RTI2018-095363-B-I00 and the Catalan Water Agency (ACA), Government of Catalonia (CTN1001569).

Institutional Review Board Statement: Not applicable.

Informed Consent Statement: Not applicable.

Data Availability Statement: Data are available from the corresponding author R.G. upon reasonable request.

Conflicts of Interest: The authors declare no conflict of interest.

biology and life sciences forum

Abstract

Understanding the Impacts of River Network Fragmentation on Freshwater Fish Species †

Tamara Leite [1,2,*,‡], Florian Borgwardt [3] and Paulo Branco [1,2]

[1] Forest Research Centre (CEF)—School of Agriculture, University of Lisbon, 1349-017 Lisbon, Portugal; pjbranco@isa.ulisboa.pt
[2] Associated Laboratory TERRA, 1349-017 Lisbon, Portugal
[3] Institute of Hydrobiology and Aquatic Ecosystem Management (IHG), University of Natural Resources and Life Sciences, 1180 Vienna, Austria; florian.borgwardt@boku.ac.at
* Correspondence: tamaraleite@e-isa.ulisboa.pt
† Presented at the IX Iberian Congress of Ichthyology, Porto, Portugal, 20–23 June 2022.
‡ Presenting author (Poster presentation).

Citation: Leite, T.; Borgwardt, F.; Branco, P. Understanding the Impacts of River Network Fragmentation on Freshwater Fish Species. *Biol. Life Sci. Forum* **2022**, *13*, 73. https://doi.org/10.3390/blsf2022013073

Academic Editor: Alberto Teodorico Correia

Published: 9 June 2022

Publisher's Note: MDPI stays neutral with regard to jurisdictional claims in published maps and institutional affiliations.

Abstract: Longitudinal connectivity of freshwater systems allows upstream/downstream movements of aquatic migratory species and can be disrupted by natural or artificial barriers. Multiple obstacles, partially or totally blocking upstream movements, along the river promote cumulative non-additive impacts on freshwater fish. In addition to limiting fish migration, barriers can also affect habitat quality downstream by creating changes in flow regime, sediment and nutrient transport, and water temperature. Practically all European river corridors were already substantially modified by the building of dams and other barriers, resulting in at least 1.2 million instream barriers in 36 European countries. The impact of river network fragmentation is progressively intensifying, with the construction of additional barriers, and by the interaction of exiting barriers with other human-induced pressures such as water abstraction. Hence, it is important to assess the impacts of barriers on the structural and functional longitudinal connectivity of riverine systems and the consequences for fish species, assemblages, and diversity, to improve conservation measures and ecosystem management to halt biodiversity loss. The goal of this thesis project is to understand the impacts of riverine longitudinal connectivity loss, regarding potamodromous and diadromous fish species across Europe. For that, the research will be focused on: (i) determining connectivity loss caused by large dams in Europe, using graph theory analysis; (ii) understanding if we are over or underestimating the effects of artificial barriers by comparing the effects of natural to artificial fragmentation—regarding waterfalls, dams and smaller barriers, and the upstream water reservoir; (iii) distinguishing the effects of different types of barriers; and (iv) determining if functional resilience of fish communities can be improved by restoring connectivity. We expect that the fragmentation of freshwater systems will greatly reduce structural and functional longitudinal connectivity within river basins, which will ultimately have effects on fish species and communities, leading to a decrease of functional resilience.

Keywords: river network; river fragmentation; restoration; barriers; connectivity; fish community; functional resilience

Author Contributions: T.L.: Methodology, Investigation, Formal analysis, Data curation, Writing—original draft. P.B.: Conceptualization, Supervision, Methodology, Investigation, Writing—review & editing. F.B.: Supervision, Methodology, Investigation, Writing—review & editing. All authors have read and agreed to the published version of the manuscript.

Funding: This research was partially funded by the project Dammed Fish (PTDC/CTA-AMB/4086/2021). Tamara Leite was supported by a Ph.D. grant from the FLUVIO–River Restoration and Management programme funded by Fundação para a Ciência e a Tecnologia I. P. (FCT), Portugal

(UI/BD/15052/2021). Paulo Branco was financed by national funds via FCT, under "Norma Transitória—DL57/2016/CP1382/CT0020". Florian Borgwardt was financed by the MERLIN project (H2020-101036337). Forest Research Centre (CEF) is a research unit funded by Fundação para a Ciência e a Tecnologia I.P. (FCT), Portugal (UIDB/00239/2020).

Institutional Review Board Statement: Not applicable.

Informed Consent Statement: Not applicable.

Data Availability Statement: All data used is publicly available.

Conflicts of Interest: The authors declare no conflict of interest.

Abstract

Your Past Condemns You: Trace Elements of a Marine Catfish in Two Periods in an Altered Tropical Bay [†]

Taynara Pontes Franco [1,2,*,‡], Leonardo Caladrini Bruno [1], Rafael C. C. Rocha [3], Tatiana D. Saint Pierre [3] and Francisco Gerson Araújo [1]

[1] Laboratório de Ecologia de Peixes, Departamento de Biologia Animal, Universidade Federal Rural do Rio de Janeiro, Seropédica 23897-030, Brazil; leonardocalandrini@gmail.com (L.C.B.); gersonufrrj@gmail.com (F.G.A.)
[2] Centro Interdisciplinar de Investigação Marinha e Ambiental, 4450-208 Matosinhos, Portugal
[3] Laboratório de Espectrometria Atômica, Departamento de Química, Pontifícia Universidade Católica do Rio de Janeiro, Rio de Janeiro 22451-900, Brazil; rafaelccr@puc-rio.br (R.C.C.R.); tatispierre@puc-rio.br (T.D.S.P.)
* Correspondence: taynarafranco@hotmail.com; Tel.: +351-928-159-708
† Presented at the IX Iberian Congress of Ichthyology, Porto, Portugal, 20–23 June 2022.
‡ Presenting author (Poster presentation).

Abstract: Changes in coastal environments are usually caused by anthropogenic activities such as habitat degradation and nonpoint source pollution. Such changes reduce the life quality of the living organisms in altered environments, decreasing biotic descriptors and jeopardizing biodiversity. The aim of this study was to compare eventual changes in minor and trace elements in otoliths of the catfish *Genidens genidens* in two periods (1980s and 2017–2018) in an altered tropical bay (Sepetiba Bay, RJ, Brazil). The concentrations of 34 elements were determined by Inductively Coupled Plasma Mass Spectrometry (ICP-MS) in fish otoliths. Although some elements have shown higher concentrations in more contemporary periods (2017–2018) such as ^{7}Li, ^{11}Br, ^{85}Rb, and ^{205}Tl, and some did not differ between the two periods (^{208}Pb, ^{34}S, ^{44}Ca, ^{55}Mn, ^{82}Se, and ^{137}Ba), but most of the examined elements (such as ^{24}Mg, ^{59}Co, ^{60}Ni, ^{65}Cu, ^{105}Pd, ^{107}Ag, ^{66}Zn, ^{75}As, ^{69}Ga, ^{98}Mo, ^{53}Cr, ^{27}Al, ^{114}Cd, and ^{11}B) had higher concentrations in the 1980s compared to 2017-2018 (test w, $p < 0.05$). Although this result was the opposite of what was expected, this could be associated with the intensity of activities without any environmental control in the past and with the first dredging to deepen the access channel to the port created in 1982, allowing the operation of large ships, which promoted the resuspension of trace elements trapped in the sediment and the pollution carried into the bay. The stability of the incorporation of elements in otoliths, compared to other tissues, allows such records, which are consistent over time, to be used in order to better understand past and present variations in the quality of the environment. Such information can be useful in conservation programs as it provides a historical view of variations in the quality of the environment.

Keywords: otolith chemistry; coastal bay; pollution; environmental impact

Citation: Franco, T.P.; Caladrini Bruno, L.; Rocha, R.C.C.; Saint Pierre, T.D.; Araújo, F.G. Your Past Condemns You: Trace Elements of a Marine Catfish in Two Periods in an Altered Tropical Bay. *Biol. Life Sci. Forum* **2022**, *13*, 74. https://doi.org/10.3390/blsf2022013074

Academic Editor: Alberto Teodorico Correia

Published: 9 June 2022

Publisher's Note: MDPI stays neutral with regard to jurisdictional claims in published maps and institutional affiliations.

Author Contributions: Conceptualization, T.P.F., L.C.B. and F.G.A.; methodology, T.P.F. and R.C.C.R.; software, T.P.F. and R.C.C.R.; formal analysis, T.P.F.; investigation, T.P.F. and L.C.B.; resources, F.G.A. and T.D.S.P.; data curation, T.P.F.; writing—original draft preparation, T.P.F.; writing—review and editing, T.P.F., T.D.S.P. and F.G.A.; supervision, F.G.A.; project administration, F.G.A.; funding acquisition, F.G.A. All authors have read and agreed to the published version of the manuscript.

Funding: This work was partially supported by a compensatory measure established by the Term of Adjustment of Conduct of responsibility of PetroRio, conducted by the Ministério Público Federal-MPF/RJ, with the implementation of the Fundo Brasileiro para a Biodiversidade–FUNBIO (Grant #16/2017 to F.G.A.). Grants for TPF were provided by the Fundação Carlos Chagas Filho de Amparo à Pesquisa do Estado do Rio de Janeiro–FAPERJ (PDR10 E-26/202.301/2018 and E-26/202.302/2018).

Institutional Review Board Statement: All applicable international, national, and/or institutional guidelines for the care and use of animals were followed, including Universidade Federal Rural do Rio de Janeiro, Brazil, Animal Care Protocol (# 11874). This research was conducted under SISBIO Collection of Species Permit number 10707 issued by ICMBio, Brazilian Environmental Agency.

Conflicts of Interest: The authors declare no conflict of interest.

Abstract

Smart Fishways: A Sensor Network for the Assessment of Fishway Performance [†]

Juan Francisco Fuentes-Pérez [1,*], Ana García-Vega [2], Francisco Javier Bravo-Córdoba [2] and Francisco Javier Sanz-Ronda [1,‡]

[1] Area of Hydraulics and Hydrology, Department of Agroforestry Engineering, University of Valladolid, Avenida de Madrid 44, Campus La Yutera, 34004 Palencia, Spain; fjbravo@itagra.com

[2] Centro Tecnológico Agrario y Agroalimentario—ITAGRA.CT, Avenida de Madrid 44, Campus La Yutera, 34004 Palencia, Spain; jsanz@uva.es (A.G.-V.); agvega@itagra.com (F.J.B.-C.)

* Correspondence: juanfrancisco.fuentes@uva.es; Tel.: +34-979108358

† Presented at the IX Iberian Congress of Ichthyology, Porto, Portugal, 20–23 June 2022.

‡ Presenting author (Poster presentation).

Citation: Fuentes-Pérez, J.F.; García-Vega, A.; Bravo-Córdoba, F.J.; Sanz-Ronda, F.J. Smart Fishways: A Sensor Network for the Assessment of Fishway Performance. *Biol. Life Sci. Forum* **2022**, *13*, 76. https://doi.org/10.3390/blsf2022013076

Academic Editor: Alberto Teodorico Correia

Published: 9 June 2022

Abstract: River barriers cause the fragmentation of riverine habitats as well as changes in the ecology of freshwater systems, fish being one of the most affected organisms by these impacts. The most common solution to allow fish to move freely through river barriers and, thus, to complete their life cycles, are stepped fishways. However, they are currently far from an optimal solution as the natural variability of rivers (e.g., discharge, floating debris, etc.) modifies the hydraulic conditions within these structures, directly affecting the fish passage, i.e., their efficiency, and, thus, the continuous assessment and management of fishways becomes vital for guaranteeing fish migration. Smart Fishways is an EU-funded project which aims to assess the effect of hydrological variability on fishways and to develop a low-cost technological and methodological framework to monitor fishway performance in real-time. The main objective of this project is to combine fish biology, hydraulics, and sensor networks to create a new generation of smart fishways, capable of self-deciding their optimal management and configuration. The present work describes the first steps followed to develop the sensor network and the online platform for the Smart Fishways project, together with the results of an ongoing study in a field test in the Iberian Peninsula. The network follows a star architecture (one gateway controls all the nodes) with independent custom-made ultrasonic water level nodes and environmental sensors distributed through the fishway together with a fish detection system for a fish movement assessment, both managed remotely and autonomously by a central gateway. This work demonstrates how the network is able to optimize the timing of maintenance on a fishway in real time, as well as how it helps to detect those hydraulic configurations and environmental variables that maximize the fish passage.

Keywords: fishways; remote monitoring; smart management; environmental sensor network

Author Contributions: Conceptualization, J.F.F.-P. and F.J.S.-R.; methodology, J.F.F.-P.; software, J.F.F.-P.; validation, J.F.F.-P. and A.G.-V.; formal analysis, J.F.F.-P.; investigation, J.F.F.-P., A.G.-V., F.J.B.-C. and F.J.S.-R.; resources, J.F.F.-P. and F.J.S.-R.; data curation, J.F.F.-P.; writing—original draft preparation, J.F.F.-P.; writing—review and editing, J.F.F.-P., A.G.-V., F.J.B.-C. and F.J.S.-R.; visualization, J.F.F.-P.; supervision, F.J.S.-R.; project administration, J.F.F.-P. and F.J.S.-R.; funding acquisition, J.F.F.-P. and F.J.S.-R. All authors have read and agreed to the published version of the manuscript.

Funding: This research has received funding from the European Research Council (ERC) under the European Union's Horizon 2020 research and innovation program (Smart fishways-grant agreement n° 101032024).

Institutional Review Board Statement: The study was conducted according to the guidelines of the Declaration of Helsinki, and approved by the Ethics Committee of University of Valladolid as well as the approval of the competent authorities, i.e. Regional Government on Natural Resources (Junta de Castilla y León) and Water Management Authority (Confederación Hidrográfica del Duero).

Informed Consent Statement: Not applicable.

Data Availability Statement: Data are available upon reasonable request to the corresponding author.

Conflicts of Interest: The authors declare no conflict of interest.

Abstract

Evaluating Silver Eel Escapement at a Large Scale Using Eel Density Analysis (EDA) [†]

María Mateo [1,*,‡], Estibaliz Díaz [1], Laurent Beaulaton [2], Hilaire Drouineau [3], Carlos Antunes [4], Elsa Amilhat [5], Agnes Bardonnet [6], Maria João Correira [7], José Lino Costa [7], Anna Costarrosa [8], Ramón J. de Miguel Rubio [9], Isabel Domingos [7], Elisabeth Faliex [5], Carlos Fernandez Delgado [9], María Korta [1], Mathilde Labedan [10], Rui Monteiro [7], Ana Moura [4], Teresa Portela [7], Pierre Sagnes [10], Gaël Simon [5], Lluis Zamora [8] and Cédric Briand [11]

[1] AZTI, Sustainable Fisheries Management, Basque Research and Technology Alliance (BRTA), 48395 Sukarrieta, Spain; ediaz@azti.es (E.D.); mkorta@azti.es (M.K.)

[2] Service Conservation et Gestion Durable des Espèces Exploitées, Direction de la Recherche et de l'Appui Scientifique, OFB & Management of Diadromous Fish in their Environment, OFB, INRAE, Institute Agro, E2S UPPA, 35000 Rennes, France; laurent.beaulaton@ofb.gouv.fr

[3] INRAE, UR EABX, 33612 Cestas, France; hilaire.drouineau@inrae.fr

[4] CIIMAR—Interdisciplinary Centre of Marine and Environmental Research, Universidade do Porto, 4450-208 Matosinhos, Portugal; cantunes@ciimar.up.pt (C.A.); anacatarinamoura@hotmail.com (A.M.)

[5] UMR 5110 CNRS-UPVD, Laboratoire CEFREM, Université of Perpignan Via Domitia, 66100 Perpignan, France; elsa.amilhat@univ-perp.fr (E.A.); faliex@univ-perp.fr (E.F.); gsimon@univ-perp.fr (G.S.)

[6] UMR 1224, Ecobiop, Aquapôle, Pôle Gest'Aqua, Inrae/UPPA, 64310 St Pée sur Nivelle, France; agnes.bardonnet@inrae.fr

[7] Departamento de Biologia Animal, Centro de Ciências do Mar e do Ambiente (MARE), Faculdade de Ciências, Universidade de Lisboa, 1600-548 Lisboa, Portugal; mjcorreia@fc.ul.pt (M.J.C.); jlcosta@fc.ul.pt (J.L.C.); idomingos@fc.ul.pt (I.D.); rmmonteiro@fc.ul.pt (R.M.); tmportela@fc.ul.pt (T.P.)

[8] Grupo de Investigación en Ecología Acuática Continental (GRECO), Instituto de Ecología Acuática, Facultad de Ciencias, Universidad de Girona, 17003 Girona, Spain; acostarrosa@gmail.com (A.C.); lluis.zamora@udg.edu (L.Z.)

[9] Departamento de Zoología, Universidad de Córdoba, 14071 Córdoba, Spain; a92mirur@uco.es (R.J.d.M.R.); ba1fedec@uco.es (C.F.D.)

[10] OFB Pôle de Recherche et Développement en Ecohydraulique OFB-IMFT, 31400 Toulouse, France; mathilde.labedan@toulouse-inp.fr (M.L.); pierre.sagnes@ofb.gouv.fr (P.S.)

[11] EPTB Vilaine, 56130 La Roche Bernard, France; cedric.briand@eptb-vilaine.fr

* Correspondence: mmateo@azti.es

† Presented at the IX Iberian Congress of Ichthyology, Porto, Portugal, 20–23 June 2022.

‡ Presenting author (oral communication).

Citation: Mateo, M.; Díaz, E.; Beaulaton, L.; Drouineau, H.; Antunes, C.; Amilhat, E.; Bardonnet, A.; Correira, M.J.; Costa, J.L.; Costarrosa, A.; et al. Evaluating Silver Eel Escapement at a Large Scale Using Eel Density Analysis (EDA). *Biol. Life Sci. Forum* **2022**, 13, 95. https://doi.org/10.3390/blsf2022013095

Academic Editor: Alberto Teodorico Correia

Published: 15 June 2022

Publisher's Note: MDPI stays neutral with regard to jurisdictional claims in published maps and institutional affiliations.

Abstract: The European eel population has declined over the last 50 years and is outside safe biological limits. In 2007, the European Commission enforced a regulation to ensure that all Member States implement Management Plans to achieve an escapement rate of 40% for silver eels. However, escapement is assessed using diverse methods, and therefore, estimates of different countries are not comparable. Thus, European scientists and managers face the challenge of finding a method that could be applied all along the European eel distribution range using widely available data. During the SUDOANG project, the Eel Density Analysis model (EDA) was used to estimate the silver eel escapement in Spain, France, and Portugal. EDA can estimate silver eel escapement at different scales (basin, EMU, country, international) using data from routine electrofishing surveys, such as those collected for the Water Framework Directive (WFD). EDA is based on open-source software and is widely applicable to European rivers. A chained river network in the three countries was built, including the hydrological attributes, the location, and the characteristics of 10,574 obstacles. A total of 46,118 electrofishing operations from 1985 to 2019 were also compiled and projected to the hydrographic network. Finally, EDA was implemented through the sub-models on eel presence–absence, density, size structure, and silvering. EDA estimated a total fluvial population of 12.3, 11.7, and 5.8 million silver eels for 2015 in France, Spain, and Portugal, respectively. Eel presence and abundance decreased as the altitude and the distance to the sea increased. Finally, by setting the cumulated height of dams to zero, EDA provided an estimate of the effect of habitat loss due to dams on the eel population. These results correspond to the first assessment of silver eels carried out simultaneously and with the same methodologies in three European countries.

Keywords: European eel; silver escapement; EDA model; habitat loss

Author Contributions: Conceptualization, C.B., M.M., E.D., H.D. and L.B.; methodology, C.B., M.M., H.D. and L.B.; software, C.B., M.M. and L.B.; validation, M.K., E.D., L.B., H.D., C.A., E.A., A.B., I.D., C.F.D., P.S. and L.Z.; formal analysis, C.B. and M.M.; investigation, C.B., M.M., E.D., H.D. and L.B.; resources, C.A., E.A., A.B., I.D., C.F.D., P.S. and L.Z.; data curation, M.M., C.B., C.A., E.A., A.B., M.J.C., J.L.C., A.C., R.J.d.M.R., I.D., E.F., C.F.D., M.L., R.M., A.M., T.P., P.S., G.S. and L.Z.; writing—original draft preparation, M.M., C.B. and E.D.; writing—review and editing, M.M., C.B., E.D., L.B., H.D., C.A., E.A., A.B., M.J.C., J.L.C., A.C., R.J.d.M.R., I.D., E.F., C.F.D., M.L., R.M., A.M., T.P., P.S., G.S. and L.Z.; visualization, M.M., C.B., M.K. and H.D.; supervision, C.B., L.B. and E.D.; project administration, E.D.; funding acquisition, E.D. All authors have read and agreed to the published version of the manuscript.

Funding: This research was co-financed by the INTERREG SUDOE Programme through the European Regional Development Fund (ERDF).

Institutional Review Board Statement: Not applicable.

Informed Consent Statement: Not applicable.

Data Availability Statement: The data presented in this study are available in Zenodo: Mateo, M.; Drouineau, H.; Pella, H.; Beaulaton, L.; Amilhat, E.; Bardonnet, A.; Domingos, I.; Fernández-Delgado, C.; De Miguel Rubio, R.; Herrera, M.; Korta, M.; Zamora, L.; Díaz, E.; Briand, C. (2021). Atlas of European Eel Distribution (*Anguilla anguilla*) in Portugal, Spain and France (1.0.0) [Data set]. Zenodo. https://doi.org/10.5281/zenodo.6021838. Briand, C.; Mateo, M.; Drouineau, H.; Korta, M.; Díaz, E.; Beaulaton, L. (2022). Electrofishing data for eel (*Anguilla anguilla*) in the SUDOE area (1.0.0) [Data set]. Zenodo. https://doi.org/10.5281/zenodo.6023561. The database is also available for download on the project website: https://sudoang.eu/en/.

Conflicts of Interest: The authors declare no conflict of interest.

Abstract

River Network Connectivity—An Holistic Approach to Improve the Sustainability of Fish Populations [†]

Paulo Branco [1,2,*,‡], Pedro Segurado [1,2], José Maria Santos [1,2], Susana D. Amaral [1], Gonçalo Duarte [1,2], Filipe Romão [3], Tamara Leite [1,2], António Pinheiro [3] and Maria T. Ferreira [1,2]

[1] Forest Research Centre, School of Agriculture, University of Lisbon, 1349-017 Lisbon, Portugal; psegurado@isa.ulisboa.pt (P.S.); jmsantos@isa.ulisboa.pt (J.M.S.); samaral@isa.ulisboa.pt (S.D.A.); goncalofduarte@isa.ulisboa.pt (G.D.); tamaraleite@e-isa.ulisboa.pt (T.L.); terferreira@isa.ulisboa.pt (M.T.F.)
[2] Associated Laboratory TERRA, 1349-017 Lisbon, Portugal
[3] CERIS—Civil Engineering for Research and Innovation for Sustainability, Técnico, University of Lisbon, 1049-001 Lisboa, Portugal; filipe.romao@tecnico.ulisboa.pt (F.R.); antonio.pinheiro@tecnico.ulisboa.pt (A.P.)
[*] Correspondence: pjbranco@isa.ulisboa.pt
[†] Presented at the IX Iberian Congress of Ichthyology, Porto, Portugal, 20–23 June 2022.
[‡] Presenting author (oral communication).

Citation: Branco, P.; Segurado, P.; Santos, J.M.; Amaral, S.D.; Duarte, G.; Romão, F.; Leite, T.; Pinheiro, A.; Ferreira, M.T. River Network Connectivity—An Holistic Approach to Improve the Sustainability of Fish Populations. *Biol. Life Sci. Forum* **2022**, *13*, 105. https://doi.org/10.3390/blsf2022013105

Academic Editor: Alberto Teodorico Correia

Published: 16 June 2022

Publisher's Note: MDPI stays neutral with regard to jurisdictional claims in published maps and institutional affiliations.

Abstract: Rivers are intrinsically linked with human settlement and civilization development. This has forced upon rivers a multitude of pressures. Arguably, one of the most pervasive pressures is river network fragmentation. The consequent loss of longitudinal connectivity has long-lasting disruptive effects on ecosystem functioning. Fish are among the most affected organisms, as they are unable to freely disperse along river networks, affecting the tenuous population and meta-community balance. To tackle fragmentation problems, it is important to evaluate the degree of fragmentation, understand the impacts, develop and apply cost-effective prioritization procedures of connectivity restoration, and design connectivity enhancement solutions that may be applied to different barrier types. In this work, we demonstrate how one should take a holistic approach to river network connectivity studies by presenting the key findings and take-home messages of a 15-year research path focused on river network connectivity. This encompasses theoretical and laboratory-controlled experiments and fieldwork, ranging from historical fish occurrences to predictions of future distributions and from fish passage research to management and planning solutions to enhance connectivity. This global analysis intends to demonstrate that the optimal way to address river network connectivity issues is to establish a holistic perspective, taking overarching approaches at multiple spatial and temporal scales.

Keywords: fishways; river barriers; graph theory; functional connectivity; ecohydraulics

Author Contributions: P.B.: Writing, Development, concept, funding, coordination; P.S.: Concept, corrdination, development, writing; J.M.S.: Writing, concept, funding, coordination; S.D.A.: Development, writing; G.D.: Development, concept, writing; F.R.: Development, writing; T.L.: Development; A.P.: Concept, writing, coordination, funding; M.T.F.: Concept, coordination, funding. All authors have read and agreed to the published version of the manuscript.

Funding: Tamara Leite was supported by a Ph.D. grant from the FLUVIO–River Restoration and Management programme funded by Fundação para a Ciência e a Tecnologia I.P. (FCT), Portugal (UI/BD/15052/2021). Gonçalo Duarte has been financed by national funds via FCT–Fundação para a Ciência e a Tecnologia, I.P., within the project UIDP/00239/2020. Paulo Branco was financed by national funds via FCT, under "Norma Transitória—DL57/2016/CP1382/CT0020". The study was partially funded by the project Dammed Fish (PTDC/CTA-AMB/4086/2021). Forest Research Centre (CEF) is a research unit funded by Fundação para a Ciência e a Tecnologia I.P. (FCT), Portugal (UIDB/00239/2020), and the Associate Laboratory TERRA (LA/P/0092/2020) is also funded by FCT.

Institutional Review Board Statement: Not Applicable.

Informed Consent Statement: Not Applicable.

Data Availability Statement: Data is available upon request.

Conflicts of Interest: The authors declare no conflits of interest.

biology and life sciences forum

Abstract

Contribution of the Nursery Areas to the Major Fishing Grounds of the Brazilian Sardine (*Sardinella Brasiliensis*) in Southeastern Brazilian Bight through Otolith Fingerprinting [†]

Rafael Schroeder [1,2,3,‡], Paulo Ricardo Schwingel [2], Richard Schwarz [1], Felippe Alexandre Daros [4], Taynara Pontes Franco [3,5], Natasha Travenisk Hoff [6], Ana Méndez-Vicente [7], Jorge Pisonero Castro [7], André Martins Vaz-dos-Santos [8] and Alberto Teodorico Correia [3,9,10,*]

[1] Laboratório de Estudos Marinhos Aplicados, Escola do Mar, Ciência e Tecnologia, Universidade do Vale do Itajaí (UNIVALI), Rua Uruguai 458, Itajaí 88302-901, Brazil; schroederichthys@gmail.com (R.S.); ricschwarz@gmail.com (R.S.)
[2] Laboratório de Ecossistemas Aquáticos e Pesqueiros, Escola do Mar, Ciência e Tecnologia, Universidade do Vale do Itajaí (UNIVALI), Rua Uruguai 458, Itajaí 88302-901, Brazil; schwingel@univali.br
[3] Centro Interdisciplinar de Investigação Marinha e Ambiental (CIIMAR), Terminal de Cruzeiros do Porto de Leixões, Avenida General Norton de Matos S/N, 4550-208 Matosinhos, Portugal; taynarafranco@hotmail.com
[4] Coordenadoria de Curso de Engenharia de Pesca, Universidade Estadual Paulista "Júlio Mesquita Filho" (UNESP), Campus de Registro, Registro 11900-000, Brazil; felippe.daros@unesp.br
[5] Laboratório de Ecologia de Peixes, Departamento de Biologia Animal, Universidade Federal Rural do Rio de Janeiro, Seropédica 23897-030, Brazil
[6] Instituto Oceanográfico, Universidade de São Paulo, São Paulo 05508-120, Brazil; natasha.hoff@usp.br
[7] Department of Physics, Faculty of Science, University of Oviedo, 33007 Oviedo, Spain; mendezana@uniovi.es (A.M.-V.); pisonerojorge@uniovi.es (J.P.C.)
[8] Laboratório de Esclerocronologia, Setor Palotina, Universidade Federal do Paraná, R. Pioneiro 2153, Palotina 85950-000, Brazil; andrevaz@gmail.com
[9] Faculdade de Ciências, Universidade Fernando Pessoa, Rua Carlos da Maia 296, 4200-150 Porto, Portugal
[10] Instituto de Ciências Biomédicas Abel Salazar, Universidade do Porto (ICBAS-UP), Rua de Jorge Viterbo Ferreira 228, 4050-313 Porto, Portugal
* Correspondence: atcorreia.ciimar@gmail.com
† Presented at the IX Iberian Congress of Ichthyology, Porto, Portugal, 20–23 June 2022.
‡ Presenting author (oral communication).

Citation: Schroeder, R.; Schwingel, P.R.; Schwarz, R.; Daros, F.A.; Franco, T.P.; Hoff, N.T.; Méndez-Vicente, A.; Castro, J.P.; Vaz-dos-Santos, A.M.; Correia, A.T. Contribution of the Nursery Areas to the Major Fishing Grounds of the Brazilian Sardine (*Sardinella Brasiliensis*) in Southeastern Brazilian Bight through Otolith Fingerprinting. *Biol. Life Sci. Forum* 2022, 13, 113. https://doi.org/10.3390/blsf2022013113

Academic Editor: Pedro Miguel Guerreiro

Published: 17 June 2022

Publisher's Note: MDPI stays neutral with regard to jurisdictional claims in published maps and institutional affiliations.

Abstract: In the late 1970s, studies on the population structure of *S. brasiliensis* suggested the existence of two stocks, considering distinct regional somatic growth rates and spawning areas (23–25° S and 26–28° S). This scenario was further confirmed by geochemical signatures of whole otoliths combined with basic biological data regarding 2-year-old sardines collected in SW-S Brazil. However, information about sardine movements and connectivity between their main juvenile recruitment areas and the adult fishing grounds is currently limited. In this study, natal otolith elemental fingerprints (core section) of young-of-year (age 0+) and adult (age 2+) individuals were collected, respectively, in the main spawning areas (2019) and fishing grounds (2021) and evaluated. Elemental signatures of recruits were compared with those of adult fish from the same cohort to estimate connectivity between juvenile recruitment areas (RJ-22° S, SP-23° S and SC-26° S) and regional adult populations captured in the major fishing grounds (22–23° S, 24–25° S and 26–27° S). Uni- and multi-elemental chemical signatures showed significant differences for age 0+ and for age 2+. Pairwise comparisons associated age 0+ and age 2+ with the northern distribution area (RJ + SP) and differentiated them from those of SC. The leave-one-out reclassification matrix combining chemical fingerprints and reassigned the individuals to their original areas with moderate-to-high accuracy: RJ 0+ (85%), SP 0+ (80%), SC 0+ (85%), and from RJ 2+ (80%), SP 2+ (70%), SC 2+ (75%). This variability was driven by Ba/Ca, Fe/Ca, Mg/Ca, Mn/Ca, and Sr/Ca ratios. Maximum likelihood analysis suggested for the 2019 cohort that replenishment of adult populations of *S. brasiliensis* along the Brazilian coast was mostly derived from the northern recruitment area (RJ + SP = 64%). Nonetheless, an important contribution from the southern counterparts to the northern stock was detected (36%), supporting the hypothesis of meta-population structure.

Keywords: pelagic fish; fisheries; otolith microchemistry; rational management

172

Author Contributions: Conceptualization, R.S. (Rafael Schroeder) and A.T.C.; Statistical analysis, R.S. (Rafael Schroeder); Investigation, R.S. (Rafael Schroeder), A.T.C., P.R.S., A.M.V.-d.-S.; Fish Sampling, R.S. (Richard Schwarz), F.A.D., T.P.F. and N.T.H.; Analytical analysis, R.S. (Rafael Schroeder), A.T.C., A.M.-V. and J.P.C.; Writing—original draft preparation, R.S. (Rafael Schroeder); writing—review and editing, R.S. (Rafael Schroeder) and A.T.C.; Supervision, A.T.C.; Funding acquisition, A.T.C. and P.R.S. All authors have read and agreed to the published version of the manuscript.

Funding: Rafael Schroeder received a PhD scholarship from the Coordination for the Improvement of Higher Education Personnel (CAPES). Alberto T. Correia was supported by national funds through FCT—Foundation for Science and Technology within the scope of UIDB/04423/2020 and UIDP/04423/2020. This study was also funded by the Brazilian Biodiversity Fund (FUNBIO).

Institutional Review Board Statement: Not applicable (seafood product).

Informed Consent Statement: Not applicable.

Data Availability Statement: Not available. Only upon request.

Conflicts of Interest: The authors declare no conflict of interest.

Abstract

Restocking Trials with Hatchery-Reared Dusky Groupers in a Marine Protected Area of the Southwestern Portuguese Coast [†]

Ana Filipa Silva [1,*,‡], Bárbara Horta e Costa [2], José Lino Costa [1,3], Esmeralda Pereira [4], João Pedro Marques [1], João J. Castro [4,5,6], Pedro G. Lino [7], Ana Candeias-Mendes [7], Pedro Pousão-Ferreira [7], Inês Sousa [2], Luís Bentes [2], Jorge M. S. Gonçalves [2], Pedro Raposo de Almeida [4,5] and Bernardo Ruivo Quintella [1,3,*]

1   MARE—Centro de Ciências do Mar e do Ambiente, Faculdade de Ciências, Universidade de Lisboa, 1749-016 Lisboa, Portugal; jlcosta@fc.ul.pt (J.L.C.); jpmarques@fc.ul.pt (J.P.M.)
2   CCMAR—Centro de Ciências do Mar, Universidade do Algarve, 8005-139 Faro, Portugal; bbcosta@ualg.pt (B.H.e.C.); isousa@ualg.pt (I.S.); lbentes@ualg.pt (L.B.); jgoncal@ualg.pt (J.M.S.G.)
3   Departamento de Biologia Animal, Faculdade de Ciências, Universidade de Lisboa, 1749-016 Lisboa, Portugal
4   MARE—Centro de Ciências do Mar e do Ambiente, Universidade de Évora, 7002-554 Évora, Portugal; ecdp@uevora.pt (E.P.); jjc@uevora.pt (J.J.C.); pmra@uevora.pt (P.R.d.A.)
5   Departamento de Biologia, Escola de Ciências e Tecnologia, Universidade de Évora, 7002-554 Évora, Portugal
6   Laboratório de Ciências do Mar, Escola de Ciências e Tecnologia, Universidade de Évora, 7521-903 Évora, Portugal
7   IPMA/EPPO—Instituto Português do Mar e da Atmosfera, 8700-194 Olhão, Portugal; plino@ipma.pt (P.G.L.); ana.mendes@ipma.pt (A.C.-M.); pedro.pousao@ipma.pt (P.P.-F.)
*   Correspondence: afmsilva@fc.ul.pt (A.F.S.); bsquintella@fc.ul.pt (B.R.Q.)
†   Presented at the IX Iberian Congress of Ichthyology, Porto, Portugal, 20–23 June 2022.
‡   Presenting author (Oral communication).

Citation: Silva, A.F.; Horta e Costa, B.; Costa, J.L.; Pereira, E.; Marques, J.P.; Castro, J.J.; Lino, P.G.; Candeias-Mendes, A.; Pousão-Ferreira, P.; Sousa, I.; et al. Restocking Trials with Hatchery-Reared Dusky Groupers in a Marine Protected Area of the Southwestern Portuguese Coast. *Biol. Life Sci. Forum* **2022**, *13*, 117. https://doi.org/10.3390/blsf2022013117

Academic Editor: Alberto Teodorico Correia

Published: 17 June 2022

Publisher's Note: MDPI stays neutral with regard to jurisdictional claims in published maps and institutional affiliations.

Abstract: Marine Protected Areas (MPAs) have played an important role in the protection of endangered species such as the dusky grouper (*Epinephelus marginatus*), and the no-take areas have been particularly crucial for this purpose. Yet, despite the establishment of no-take areas and the legislation banning dusky groupers' catches since 2011 in a southwestern Portuguese MPA (SACVMP—'Sudoeste Alentejano' and 'Costa Vicentina' Marine Park), there is still no evidence of this population's recovery. In the face of this, the present work aimed to monitor the experimental hatchery-reared adult dusky groupers' restocking of two no-take areas of SACVMP with acoustic biotelemetry. In 2019 and 2021, thirty groupers tagged with acoustic transmitters were released in two no-take areas, and the site attachment and their movements were assessed. None of the tagged fish set residency in either of the releasing areas, mostly leaving there at dusk and night. A rarely reported event for this species was also observed, as some individuals moved for more than a hundred kilometers along the Portuguese coast. At least some of those ranging movements were performed close to the rocky shore, which may point out the importance of coastal MPAs in promoting the connectivity of fish species associated with rocky reef habitats. Future studies must focus on the conditions that promote the site attachment and the site fidelity of released hatchery-reared dusky groupers so that large-scale restocking programs can be successfully implemented in MPAs with appropriate habitats.

Keywords: acoustic biotelemetry; *Epinephelus marginatus*; no-take areas; ranging behavior; serranidae

Author Contributions: Conceptualization, A.F.S., B.R.Q., P.R.d.A., B.H.e.C.; methodology, A.F.S., B.R.Q., J.L.C., P.R.d.A., B.H.e.C.; software, A.F.S. and J.P.M.; formal analysis, A.F.S. and J.P.M.; investigation, A.F.S., J.P.M., E.P., B.R.Q., B.H.e.C., I.S., L.B., A.C.-M., P.G.L.; resources, P.P.-F., B.R.Q., J.M.S.G.; data curation, A.F.S.; writing—original draft preparation, A.F.S.; writing—review and editing, B.H.e.C., J.L.C., E.P., J.P.M., J.J.C., P.G.L., A.C.-M., P.P.-F., I.S., L.B., J.M.S.G., P.R.d.A. and B.R.Q.; visualization, A.F.S.; supervision, B.R.Q.; funding acquisition, B.R.Q., J.J.C. and J.M.S.G. All authors have read and agreed to the published version of the manuscript.

Funding: This research was co-funded by PO SEUR—Operational Programme for Sustainability and Efficient Use of Resources, Fundo Ambiental (Portuguese Environmental Fund) and by Aljezur, Odemira and Vila do Bispo Municipalities, through project "Sistemas de Informação e Monitorização da Biodi-versidade Marinha das Áreas Classificadas do Sudoeste Alentejano e Costa Vicentina—MARSW", with the grant number POSEUR-03-2215-FC-000046.

Institutional Review Board Statement: This study was carried out in strict accordance with the recommendations present in the Guide for the Care and Use of Laboratory Animals of the European Union—in Portugal represented by the Decree-Law Nr. 129/92, Ordinance Nr. 1005/92.

Informed Consent Statement: Not applicable.

Data Availability Statement: The data will be available after proper publication on the ETN (European Tracking Network) data portal.

Conflicts of Interest: The authors declare no conflict of interest.

Abstract

LIFE Agueda—Gaining Habitat for Migratory Fish in the Vouga River Basin [†]

Sílvia Pedro [1,*,‡], Carlos M. Alexandre [1], Catarina S. Mateus [1], Bernardo R. Quintella [2,3], Maria João Lança [4], Ana F. Belo [1], Esmeralda Pereira [1], Ana S. Rato [1], Inês Oliveira [1], Sara Silva [1] and Pedro R. Almeida [1,5]

[1] MARE—Marine and Environmental Sciences Centre/ARNET—Aquatic Research Network, Universidade de Évora, 7002-554 Évora, Portugal; cmea@uevora.pt (C.M.A.); cspm@uevora.pt (C.S.M.); affadb@uevora.pt (A.F.B.); ecdp@uevora.pt (E.P.); assrato@uevora.pt (A.S.R.); icso@uevora.pt (I.O.); ssrs@uevora.pt (S.S.); pmra@uevora.pt (P.R.A.)

[2] MARE—Marine and Environmental Sciences Centre/ARNET—Aquatic Research Network, Faculdade de Ciências, Universidade de Lisboa, 1749-016 Lisboa, Portugal; bsquintella@fc.ul.pt

[3] Departamento de Biologia Animal, Faculdade de Ciências, Universidade de Lisboa, 1749-016 Lisboa, Portugal

[4] MED-Instituto Mediterrânico para a Agricultura, Ambiente e Desenvolvimento & Departamento de Zootecnia, Escola de Ciências e Tecnologia, Universidade de Évora, 7002-554 Évora, Portugal; mjlanca@uevora.pt

[5] Departamento de Biologia, Escola de Ciências e Tecnologia, Universidade de Évora, 7002-554 Évora, Portugal

* Correspondence: ssfp@uevora.pt; Tel.: +351-217-500-148

† Presented at the IX Iberian Congress of Ichthyology, Porto, Portugal, 20–23 June 2022.

‡ Presenting author (oral communication).

Citation: Pedro, S.; Alexandre, C.M.; Mateus, C.S.; Quintella, B.R.; Lança, M.J.; Belo, A.F.; Pereira, E.; Rato, A.S.; Oliveira, I.; Silva, S.; et al. LIFE Agueda—Gaining Habitat for Migratory Fish in the Vouga River Basin. *Biol. Life Sci. Forum* **2022**, *13*, 118. https://doi.org/10.3390/blsf2022013118

Academic Editor: Alberto Teodorico Correia

Published: 17 June 2022

Publisher's Note: MDPI stays neutral with regard to jurisdictional claims in published maps and institutional affiliations.

Abstract: Habitat loss and overfishing are the most significant threats to diadromous fish, most of them of high socioeconomic and conservationist importance, such as *Alosa alosa*, *Alosa fallax*, *Petromyzon marinus* and *Anguilla anguilla*. The main objective of the LIFE Águeda project (LIFE16 ENV/PT/000411) is the removal of hydro-morphological pressures towards the reestablishment of conditions for a good ecological status, as required by the Water Framework Directive (WFD) and associated River Basin Management Plans. Actions to achieve the project's objectives include the restoration of river morphology through the construction of nature-like fish passes, removal of river obstacles and re-naturalization of the riverbed. Aside from these interventions, the project also contemplates riparian habitat restoration, design and operation of a pilot translocation program directed to European eel juveniles, management of recreational and commercial fisheries, and stakeholders' engagement, safeguarding compatibility of ecosystem uses. To reestablish longitudinal connectivity in rivers Agueda and Alfusqueiro, a total of five fish passes (two modular and temporary vertical-slot and three nature-like fish passes) will be installed in obstacles where current uses need to be secured and removal is not an option. Obsolete or illegally built obstacles are to be completely or partially removed, in a total of eight interventions in both rivers. By placing PIT antennas in the most upstream fishways planned to be built in both rivers, and by monitoring the movements of tagged fish from target species, the efficiency of these habitat restoration actions, and the reestablishment of longitudinal connectivity, will be assessed.

Keywords: connectivity; fish passes; obstacle removal; fish migration

Author Contributions: Conceptualization, P.R.A., B.R.Q., C.M.A. and C.S.M.; methodology, P.R.A., B.R.Q., C.M.A., C.S.M. and S.P.; writing—original draft preparation, S.P.; writing—review and editing—S.P., C.M.A., C.S.M., B.R.Q., M.J.L., A.F.B., E.P., A.S.R., I.O., S.S. and P.R.A.; supervision, P.R.A.; project administration, P.R.A. and S.P.; funding acquisition, P.R.A., B.R.Q., C.M.A. and C.S.M. All authors have read and agreed to the published version of the manuscript.

Funding: This research was funded by EU's LIFE Programme for the Environment and Climate action (LIFE16 ENV/PT/000411), and FCT—Fundação para a Ciência e a Tecnologia (strategic project UIDB/04292/2020 awarded to MARE and project LA/P/0069/2020 granted to Associate Laboratory ARNET). FCT also supported this study through the individual contracts attributed to Carlos M. Alexandre (CEECIND/02265/2018) and to Bernardo R. Quintella (2020.02413.CEECIND), and the PhD scholarships attributed to Sara Silva (2021.05558.BD) and Ana S. Rato (2021.05339.BD). Co-funded by EDP-Gestão da Produção de Energia, S.A. and The Navigator Company.

Institutional Review Board Statement: Not Applicable.

Informed Consent Statement: Not Applicable.

Data Availability Statement: Not Applicable.

Conflicts of Interest: The authors declare no conflict of interest.

Abstract

Dammed Fish: Impact of Structural and Functional River Network Connectivity Losses on Fish Biodiversity—Optimizing Management Solutions [†]

Paulo Branco [1,2,*,‡], Florian Borgwardt [3], Jesse O'Hanley [4], Rui Figueira [5,6,7], Gonçalo Duarte [1,2], José Maria Santos [1,2], Pedro Segurado [1,2], Tamara Leite [1,2] and Maria T. Ferreira [1,2]

[1] Forest Research Centre, School of Agriculture, University of Lisbon, 1649-004 Lisbon, Portugal; goncalofduarte@isa.ulisboa.pt (G.D.); jmsantos@isa.ulisboa.pt (J.M.S.); psegurado@isa.ulisboa.pt (P.S.); tamaraleite@e-isa.ulisboa.pt (T.L.); terferreira@isa.ulisboa.pt (M.T.F.)

[2] Associated Laboratory TERRA, 1349-017 Lisbon, Portugal

[3] Institute of Hydrobiology and Aquatic Ecosystem Management (IHG), University of Natural Resources and Life Sciences, 1180 Vienna, Austria; florian.borgwardt@boku.ac.at

[4] Kent Business School, University of Kent, Canterbury CT2 7NT, UK; j.ohanley@kent.ac.uk

[5] Linking Landscape, Environment, Agriculture and Food (LEAF), School of Agriculture, University of Lisbon, 1649-004 Lisbon, Portugal; ruifigueira@isa.ulisboa.pt

[6] Research Centre in Biodiversity and Genetic Resources (CIBIO/InBIO), School of Agriculture, University of Lisbon, 1649-004 Lisbon, Portugal

[7] Research Centre in Biodiversity and Genetic Resources (CIBIO/InBIO), Campus A, University of Porto, 4200-450 Porto, Portugal

[*] Correspondence: pjbranco@isa.ulisboa.pt

[†] Presented at the IX Iberian Congress of Ichthyology, Porto, Portugal, 20–23 June 2022.

[‡] Presenting author (Poster presentation).

Citation: Branco, P.; Borgwardt, F.; O'Hanley, J.; Figueira, R.; Duarte, G.; Santos, J.M.; Segurado, P.; Leite, T.; Ferreira, M.T. Dammed Fish: Impact of Structural and Functional River Network Connectivity Losses on Fish Biodiversity—Optimizing Management Solutions. *Biol. Life Sci. Forum* **2022**, *13*, 119. https://doi.org/10.3390/blsf2022013119

Academic Editor: Alberto Teodorico Correia

Published: 17 June 2022

Publisher's Note: MDPI stays neutral with regard to jurisdictional claims in published maps and institutional affiliations.

Abstract: Rivers have always been linked to society development, as such they are amongst the most impacted ecosystems in the world. One of the most impairing impacts is the one promoted by instream barriers that fragment river network connectivity and impede fish movements. Barriers are especially deleterious to freshwater dependent fish species that see their dispersion ability severely affected with possible consequences for population and species maintenance, altering genetic flow and community balance. With this in mind, the European Biodiversity Strategy has the specific goal of rehabilitating 25,000 km of free-flowing rivers. Additionally, the European Water Framework Directive (WFD) also determines that longitudinal connectivity re-establishment is vital to achieve the goal of good ecological status. To implement adequate measures for river network connectivity enhancement, connectivity has to be quantified at the basin scale accounting for the cumulative impacts of all the dams present in a given system. Dammed Fish is a Europe-wide research project aiming to evaluate and propose solutions and tools to inform river network connectivity management, to improve fish biodiversity and enhance the biotic quality of European rivers. The project is structured into five interconnected tasks to evaluate how dams, by themselves and combined with other pressures, affect river network connectivity, biodiversity loss, species range contraction and species turnover in (riverine) fish. Results will contribute to further research and improved management of river network connectivity by developing three free tools: RivFish—to link fish data and river networks; RivConnect—to calculate basin-wide network connectivity; and RivOpt—to optimize basin-wide connectivity management solutions considering conflicting management goals. Thus, Dammed Fish proposes an overall and integrative approach to connectivity management over large spatial extents.

Keywords: river barriers; graph theory; functional connectivity; software development; Europe

Author Contributions: P.B.: Writing, Development, concept, funding, coordination; P.S.: Concept, writing; J.M.S.: Writing, concept; G.D.: Concept, writing; F.B.: Concept, writing; T.L.: Development; J.O.: Concept, writing; R.F.: Concept, writting; M.T.F.: Concept. All authors have read and agreed to the published version of the manuscript.

Funding: Dammed Fish is a project funded by Fundação para a Ciência e a Tecnologia I.P. (FCT), Portugal.

Institutional Review Board Statement: Not applicable.

Informed Consent Statement: Not Applicable.

Data Availability Statement: Not applicable.

Conflicts of Interest: The authors declare no conflict of interest.

Abstract

Migration Patterns and Behaviour of Trout (*Salmo trutta* L.) in the Southern Limit of the Species Distribution [†]

Sara Silva [1,*,‡], Carlos M. Alexandre [1], Ana R. Ribeiro [1], Andreia Domingues [1], Ana S. Rato [1], Joana Pereira [1], Catarina S. Mateus [1], Bernardo R. Quintella [2,3] and Pedro R. Almeida [1,4]

[1] MARE—Centro de Ciências do Mar e do Ambiente/ARNET—Rede de Investigação Aquática, Universidade de Évora, 7004-516 Évora, Portugal; cmea@uevora.pt (C.M.A.); anacarita_ribeiro@hotmail.com (A.R.R.); afdomingues@uevora.pt (A.D.); assrato@uevora.pt (A.S.R.); joana_gomespereira@hotmail.com (J.P.); cspm@uevora.pt (C.S.M.); pmra@uevora.pt (P.R.A.)

[2] MARE—Centro de Ciências do Mar e do Ambiente/ARNET—Rede de Investigação Aquática, Faculdade de Ciências, Universidade de Lisboa, 1749-016 Lisboa, Portugal; bsquintella@fc.ul.pt

[3] Departamento de Biologia Animal, Faculdade de Ciências, Universidade de Lisboa, 1749-016 Lisboa, Portugal

[4] Departamento de Biologia, Escola de Ciências e Tecnologias, Universidade de Évora, 7002-516 Évora, Portugal

[*] Correspondence: ssrs@uevora.pt; Tel.: +351-963537867

[†] Presented at the IX Iberian Congress of Ichthyology, Porto, Portugal, 20–23 June 2022.

[‡] Presenting author (oral communication).

Abstract: The trout *Salmo trutta* L. is an iconic fish species very well studied across most of its range. However, there is a lack of information about the biology and ecology of *S. trutta* populations from southern European rivers, which coincide with the southern limit of its global distribution. This study aims to analyse the movement patterns and habitat use of *S. trutta* in the Mondego River basin, Central Portugal, and relate them with the environmental factors to which the species is exposed. Biotelemetry represents an important tool to obtain temporal and spatial specific details about the behaviour of target species and, in this work, we used a set of complementary techniques, namely acoustics, radio and PIT telemetry. A total of 114 trout specimens were tagged with PIT-tags, to be identified in future recaptures or detected at an antenna installed at Coimbra dam fish pass. From these, 18 were also tagged with Dual Mode transmitters, that include radio and acoustic telemetry signals, allowing to track the species' movements from the estuary to the upstream freshwater sections. Results show the existence of a migratory peak between November and January that coincides with the reproduction season, while reinforcing the importance of Alva River to spawning *S. trutta*, one of the main tributaries in the study area. River fragmentation in the study area, particularly in the tributaries, is still limiting the vital area of the tracked individuals, although recent restoration actions provided easier access to some important areas in this river for distinct trout life-stages. This study aims to improve the knowledge of southern European trout populations, contributing to enhancing efforts for restoring and managing these populations inhabiting areas under severe climate change effects.

Keywords: *Salmo trutta*; Mediterranean rivers; biotelemetry; fisheries management

Citation: Silva, S.; Alexandre, C.M.; Ribeiro, A.R.; Domingues, A.; Rato, A.S.; Pereira, J.; Mateus, C.S.; Quintella, B.R.; Almeida, P.R. Migration Patterns and Behaviour of Trout (*Salmo trutta* L.) in the Southern Limit of the Species Distribution. *Biol. Life Sci. Forum* **2022**, *13*, 121. https://doi.org/10.3390/blsf2022013121

Academic Editor: Alberto Teodorico Correia

Published: 17 June 2022

Publisher's Note: MDPI stays neutral with regard to jurisdictional claims in published maps and institutional affiliations.

Author Contributions: Conceptualization, C.M.A., B.R.Q. and P.R.A.; methodology, S.S., C.M.A., B.R.Q. and P.R.A.; formal analysis, C.M.A., B.R.Q., S.S., A.R.R. and A.D.; investigation, S.S., C.M.A., C.S.M., B.R.Q., A.R.R., A.S.R., J.P. and A.D.; resources, C.M.A. and P.R.A.; data curation, S.S., A.D., A.R.R. and C.M.A.; writing—original draft preparation, S.S.; writing—review and editing, C.M.A., A.R.R., A.D., A.S.R., J.P., C.S.M., B.R.Q. and P.R.A.; supervision, C.M.A., C.S.M., B.R.Q. and P.R.A.; project administration, C.M.A. and P.R.A.; funding acquisition, C.M.A., C.S.M., B.R.Q. and P.R.A. All authors have read and agreed to the published version of the manuscript.

Funding: The present study was supported by DiadES project (EAPA_18/2018, Assessing and enhancing ecosystem services provided by diadromous fish in a climate change context), which is co-financed by the Interreg Atlantic Area Programme through the European Regional Development Fund. Funding was also provided by the Portuguese Science Foundation through the strategy plan for MARE (Marine and Environmental Sciences Centre), via project UIDB/04292/2020, and under the project LA/P/0069/2020 granted to the Associate Laboratory ARNET. FCT also supported this study through the individual contracts attributed to Carlos M. Alexandre (CEECIND/02265/2018) and to Bernardo R. Quintella (2020.02413.CEECIND), and the PhD scholarships attributed to Sara Silva (2021.05558.BD), Ana S. Rato (2021.05339.BD) and Andreia Domingues (2021.05644.BD).

Institutional Review Board Statement: Not applicable.

Informed Consent Statement: Informed consent was obtained from all subjects involved in the study.

Data Availability Statement: Data is available from correspondence author, upon reasonable request.

Conflicts of Interest: The authors declare no conflict of interest.

Abstract

Assessment of Fish Passage and Behaviour through a Tidal Weir Using an Underwater Sonar [†]

Roberto Oliveira [1,*,‡], Carlos M. Alexandre [1], Bernardo R. Quintella [2], Ana S. Rato [1], Sílvia Pedro [1] and Pedro R. Almeida [1,3]

1   MARE—Marine and Environmental Sciences Centre/ARNET—Aquatic Research NETwork, Universidade de Évora, 7002-554 Évora, Portugal; cmea@uevora.pt (C.M.A.); assrato@uevora.pt (A.S.R.); ssfp@uevora.pt (S.P.); pmra@uevora.pt (P.R.A.)
2   MARE—Marine and Environmental Sciences Centre/ARNET—Aquatic Research NETwork, Departamento de Biologia Animal, Faculdade de Ciências, Universidade de Lisboa, 1749-016 Lisboa, Portugal; bsquintella@fc.ul.pt
3   Departamento de Biologia, Universidade de Évora, 7004-516 Évora, Portugal
*   Correspondence: rlpo@uevora.pt
†   Presented at the IX Iberian Congress of Ichthyology, Porto, Portugal, 20–23 June 2022.
‡   Presenting author (Oral communication).

Abstract: The "Rio Novo do Principe" temporary tidal weir is built annually in the brackish section of the Vouga river to prevent saline intrusion during the low-flow summer period, thus securing freshwater abstraction for agricultural and industrial uses. Compatibilization between these objectives and successful fish migration, coupled with adequate biological monitoring of this infrastructure, is essential, since it is located on a Nature 2000 site (Ria de Aveiro: PTZPE0004), an area considered important as a migratory corridor for diadromous fish species. In 2019, an experimental fishway was added to the weir, and a monitoring program has been ongoing since then, using an underwater acoustic camera (ARIS 1800 Sonar) to study fish behaviour upon facing and passing this obstacle. This monitoring was carried out between July and November of 2019 and 2020, on a weekly or fortnightly basis, for a complete 24-h cycle, spanning 12-h intervals, downstream and upstream of the weir. The number of fish (e.g., grey mullets and sand smelts) that successfully used the fishway in each monitoring session, varied between 1.02 fishes/min in 2019, and 0.86 fishes/min in 2020, depending on environmental conditions. An extrapolation of the number of fish recorded in the function of the lunar phase for the complete operation period of the structure (142 days in 2019 and 126 days in 2020) resulted in 158,207 individuals in 2019 and 154,961 individuals in 2020. GLM analysis with the fish counts as response variable showed that the environmental predictors that significantly influence the experimental fishway use were salinity, tidal phase, and the moon phase, for both years. Compatibilization between the prevention of saltwater intrusion and successful fish migration may be hard to achieve, but results from this study provide insights into fish behaviour when facing such obstacles and can help promote the optimization of fishways solutions in tidal areas.

Keywords: fishway; tidal weir; ARIS sonar monitoring

Citation: Oliveira, R.; Alexandre, C.M.; Quintella, B.R.; Rato, A.S.; Pedro, S.; Almeida, P.R. Assessment of Fish Passage and Behaviour through a Tidal Weir Using an Underwater Sonar. *Biol. Life Sci. Forum* **2022**, *13*, 123. https://doi.org/10.3390/blsf2022013123

Academic Editor: Alberto Teodorico Correia

Published: 17 June 2022

Publisher's Note: MDPI stays neutral with regard to jurisdictional claims in published maps and institutional affiliations.

Author Contributions: Conceptualization, P.R.A., B.R.Q., C.M.A., S.P.; methodology, P.R.A., B.R.Q., C.M.A., S.P., R.O.; writing—original draft preparation, R.O.; writing—review and editing—R.O., S.P., C.M.A., B.R.Q., A.S.R., P.R.A.; supervision, P.R.A.; project administration, P.R.A. and C.M.A.; funding acquisition, P.R.A., B.R.Q., C.M.A., S.P. All authors have read and agreed to the published version of the manuscript.

Funding: This research was co-funded by EU's LIFE Programme for the Environment and Climate action (LIFE16 ENV/PT/000411) and The Navigator Company. This study was also supported by the Portuguese Science Foundation (FCT) though the strategy plan for MARE (Marine and Environmental Sciences Centre), via project UIDB/04292/2020, and under the project LA/P0069/2020 granted to the Associate Laboratory ARNET FCT also supports this study through individual contracts attributed to Carlos M. Alexandre (CEECIND/02265/2018) and to Bernardo R. Quintella (2020.02413.CEECIND), and the PhD scholarship attributed to Ana S. Rato (2021.05339.BD).

Institutional Review Board Statement: Not applicable.

Informed Consent Statement: Not applicable.

Data Availability Statement: Data is available from correspondence author, upon reasonable request.

Conflicts of Interest: The authors declare no conflict of interest.

Abstract

Migratory Patterns of Two Potamodromous Fish Species Assessed through Fish-Pass Monitoring in Mondego River, Portugal [†]

Ana S. Rato [1,*,‡], Carlos M. Alexandre [1], Sílvia Pedro [1], Catarina S. Mateus [1], Esmeralda Pereira [1], Ana F. Belo [1], Bernardo R. Quintella [2,3], Maria F. Quadrado [4], Ana Telhado [4], Carlos Batista [4] and Pedro R. Almeida [1,5]

[1] MARE—Centro de Ciências do Mar e do Ambiente/ARNET—Rede de Investigação Aquática, Universidade de Évora, 7004-516 Évora, Portugal; cmea@uevora.pt (C.M.A.); ssfp@uevora.pt (S.P.); cspm@uevora.pt (C.S.M.); ecdp@uevora.pt (E.P.); affadb@uevora.pt (A.F.B.); pmra@uevora.pt (P.R.A.)

[2] MARE—Centro de Ciências do Mar e do Ambiente, Faculdade de Ciências, Universidade de Lisboa, 1749-016 Lisboa, Portugal; bsquintella@fc.ul.pt

[3] Departamento de Biologia Animal, Faculdade de Ciências, Universidade de Lisboa, 1749-016 Lisboa, Portugal

[4] Departamento de Recursos Hídricos e Departamento do Litoral e Proteção Costeira, Agência Portuguesa do Ambiente, I.P., 2610-124 Amadora, Portugal; maria.quadrado@apambiente.pt (M.F.Q.); ana.telhado@apambiente.pt (A.T.); carlos.batista@apambiente.pt (C.B.)

[5] Departamento de Biologia, Escola de Ciências e Tecnologias, Universidade de Évora, 7002-554 Évora, Portugal

[*] Correspondence: assrato@uevora.pt; Tel.: +351-912-749-847

[†] Presented at the IX Iberian Congress of Ichthyology, Porto, Portugal, 20–23 June 2022.

[‡] Presenting author (oral communication).

Citation: Rato, A.S.; Alexandre, C.M.; Pedro, S.; Mateus, C.S.; Pereira, E.; Belo, A.F.; Quintella, B.R.; Quadrado, M.F.; Telhado, A.; Batista, C.; et al. Migratory Patterns of Two Potamodromous Fish Species Assessed through Fish-Pass Monitoring in Mondego River, Portugal. *Biol. Life Sci. Forum* **2022**, *13*, 127. https://doi.org/10.3390/blsf2022013127

Academic Editor: Alberto Teodorico Correia

Published: 17 June 2022

Publisher's Note: MDPI stays neutral with regard to jurisdictional claims in published maps and institutional affiliations.

Abstract: The Iberian barbel (*Luciobarbus bocagei* Steindachner, 1864) and the Iberian nase (*Pseudochondrostoma polylepis* Steindachner, 1864) are two potamodromous species that migrate upstream in freshwater environments to reproduce. Thus, river fragmentation is a major threat to these species, and fish passes are one of the most-used mitigation measures to restore the longitudinal connectivity of impounded rivers, enabling these species to reach upstream spawning sites. Since 2013, the fish pass installed in the Coimbra dam (Mondego River) has been equipped with a video-recording system to continuously monitor fish passage. Based on visual count data between 2013 to 2015, a total of 61,965 movements of Iberian barbel (up- and downstream) and a total of 138,207 movements of Iberian nase (up- and downstream) were registered, with the migratory upstream movements of nase occurring over a wider period (i.e., January to December) relative to what is described in the literature. The analysis conducted to evaluate the temporal variability in the size of fish using the fish pass showed significant differences between the studied months for both species in both migratory directions; upstream-moving barbel showed a bigger body length in May, and nase showed bigger body lengths in the months of May, June and November. Boosted Regression Trees were used to identify the environmental variables that triggered these movements, with water temperature and flow being, overall, two of the most important variables for both species in both migratory directions. This study updates the relatively scarce available information concerning these species migrations, including movement activity and the associated peaks, size-structure characterization during the migratory periods, and the identification of environmental variables that seem to trigger Iberian barbel and nase movements.

Keywords: potamodromy; migratory peak; Iberian barbel; Iberian nase

Author Contributions: Conceptualization, C.M.A., C.S.M., P.R.A.; methodology, C.M.A., S.P., C.S.M., E.P., A.F.B., B.R.Q., P.R.A.; formal analysis, A.S.R., C.M.A., S.P.; investigation, A.S.R., C.M.A., S.P., C.S.M., E.P., A.F.B., B.R.Q.; resources, M.F.Q., A.T., C.B., P.R.A.; data curation, A.S.R., C.S.M., M.F.Q., A.T., C.B.; writing—original drafts presentation, A.S.R.; writing—review and editing, C.M.A.,

S.P., C.S.M., E.P., A.F.B., B.R.Q., M.F.Q., A.T., C.B., P.R.A.; supervision, C.M.A., S.P., C.S.M., P.R.A.; project administration, M.F.Q., A.T., C.B., P.R.A.; funding acquisition, B.R.Q., P.R.A. All authors have read and agreed to the published version of the manuscript.

Funding: The present study was supported by the project ANADROMOS–Plano Operacional de Monitorização e Gestão de Peixes Anádromos em Portugal (MAR-01.03.02-FEAMP-0002) funded by the European Fisheries Fund, and by the Portuguese Environment Agency (APA) through the "Coimbra fish pass monitoring program". Funding was also provided by the Portuguese Science Foundation (FCT) through the strategy plan for MARE (Marine and Environmental Sciences Centre), via project UIDB/04292/2020, and under the project LA/P/0069/2020 granted to Associate Laboratory ARNET. FCT also supported this study through the individual contracts attributed to Carlos M. Alexandre (CEECIND/02265/2018) and to Bernardo R. Quintella (2020.02413.CEECIND), and the PhD scholarship attributed to Ana S. Rato (2021.05339.BD).

Institutional Review Board Statement: Not applicable.

Informed Consent Statement: Not applicable.

Data Availability Statement: Data is available from correspondence author, upon reasonable request.

Conflicts of Interest: The authors declare no conflict of interest.

Abstract

Is Cantharidin Able to Reduce the Inflammation Produced by λ-Carrageenin in Head-Kidney Leucocytes from Gilthead Seabream (*Sparus aurata*)? †

Jose Carlos Campos-Sánchez, Francisco Antonio Guardiola and María Ángeles Esteban *,‡

Immunobiology for Aquaculture Group, Department of Cell Biology and Histology, Faculty of Biology, Campus Regional de Excelencia Internacional "Campus Mare Nostrum", University of Murcia, 30100 Murcia, Spain; josecarlos.campos@um.es (J.C.C.-S.); faguardiola@um.es (F.A.G.)
* Correspondence: author: aesteban@um.es
† Presented at the IX Iberian Congress of Ichthyology, Porto, Portugal, 20–23 June 2022.
‡ Presenting author (Oral communication).

Citation: Campos-Sánchez, J.C.; Guardiola, F.A.; Esteban, M.Á. Is Cantharidin Able to Reduce the Inflammation Produced by λ-Carrageenin in Head-Kidney Leucocytes from Gilthead Seabream (*Sparus aurata*)? *Biol. Life Sci. Forum* **2022**, *13*, 8. https://doi.org/10.3390/blsf2022013008

Academic Editor: Alberto Teodorico Correia

Published: 2 June 2022

Publisher's Note: MDPI stays neutral with regard to jurisdictional claims in published maps and institutional affiliations.

Abstract: Inflammation is a well-characterized process in mammals, but it has been poorly studied in fish. Among the diverse methods to study inflammation, carrageenin, a mucopolysaccharide obtained from the cell walls of the red algae (Rhodophyceae family), has been used for decades as a model of acute inflammation in rodents. Otherwise, cantharidin, a toxic vesicant terpene secreted by male blister beetles of the Meloidae and Oedemeridae families, has been used in low doses in both folk and traditional Chinese medicine due to its therapeutical properties. The present study aims to evaluate the effects of cantharidin on gilthead seabream (*Sparus aurata*) head-kidney leucocytes (HKLs) after their stimulation with λ-carrageenin. In this study, specimens of gilthead seabream were anesthetized with clove oil and bled from the caudal vein. Head-kidney leucocytes were isolated and incubated with combinations of final solutions of 1000, 1, and 0 μg mL^{-1} (PBS diluted in sRPMI; control) of λ-carrageenin and 5 and 0 μg mL^{-1} (DMSO diluted in sRPMI; control) of cantharidin for 24 h. After incubation, the following parameters were analyzed: viability, morphologic alterations by transmission electron microscopy, and the gene expression of inflammatory-related genes (*il1b*, *tnfa*, *il6*, *il10*, and *tgfb*). Results evidenced a decrease in viability in HKLs incubated with 1000 μg mL^{-1} of λ-carrageenin and 5 μg mL^{-1} of cantharidin compared with control HKLs. The morphologic study revealed evident signs of cell death in HKLs after being exposed to 5 μg mL^{-1} of cantharidin and 1000 μg mL^{-1} of λ-carrageenin. In addition, the expression of the *il1b* and *il10* genes were up-regulated and down-regulated, respectively, after the HKLs incubation with 1000 μg mL^{-1} of λ-carrageenin and 0 μg mL^{-1} of cantharidin in comparison to the control. In contrast, the gene expression of *il1b* was down-regulated in HKLs incubated with 1000 μg mL^{-1} of λ-carrageenin and 5 μg mL^{-1} of cantharidin in comparison to those incubated with the same concentration of λ-carrageenin and 0 μg mL^{-1} of cantharidin. Our results corroborate the effects of carrageenin as an inductor of inflammation in fish cells and point at cantharidin as a possible natural anti-inflammatory product able to reduce exacerbated inflammatory responses.

Keywords: λ-carrageenin; cantharidin; gilthead seabream (*Sparus aurata*); aquaculture

Author Contributions: Conceptualization, methodology, J.C.C.-S. and F.A.G.; software, validation, formal analysis, investigation, resources, data curation, writing—original draft preparation, J.C.C.-S.; writing—review and editing; visualization, J.C.C.-S., F.A.G. and M.Á.E.; supervision, F.A.G. and M.Á.E.; project administration, funding acquisition, M.Á.E. All authors have read and agreed to the published version of the manuscript.

Funding: This research was funded by the Spanish Ministry of Economy and Competitiveness (MINECO), co-funded by the European Regional Development Funds (ERDF/FEDER) (grant no. AGL-2017-83370-C3-1-R).

Institutional Review Board Statement: The study was conducted in accordance with the Ethical Committee of the University of Murcia (Permit Number: A13150104), following the guidelines of the European Union for animal handling (2010/63/EU).

Data Availability Statement: All the data supporting the findings of this study are available on request from the corresponding author.

Conflicts of Interest: The authors declare no conflict of interest.

Abstract

Transcriptomic Analysis of Differentially Expressed Genes in Kidney and Intestine of *Dicentrarchus labrax* Fed Different Nutritional Amounts of Inorganic Phosphate [†]

Carolina Vargas-Lagos [1,‡], Sandra Silva [1], Laura Guerrero [2], Marco Montes de Oca [3], Bruno Louro [1], Alexandra Alves [1], Josep Rotllant [2] and Pedro M. Guerreiro [1,*]

[1] Centro de Ciências do Mar (CCMAR), Universidade do Algarve, 8005-139 Faro, Portugal; carolinavargaslagos@gmail.com (C.V.-L.); scsilva@ualg.pt (S.S.); blouro@ualg.pt (B.L.); alcalves@ualg.pt (A.A.)

[2] Department of Biotechnology and Aquaculture, Institute of Marine research (IMM-CSIC), 36208 Vigo, Spain; lguerrero@iim.csic.es (L.G.); rotllant@iim.csic.es (J.R.)

[3] Escuela de Medicina, Universidad de Magallanes, Punta Arenas 01855, Chile; emontesdeocas@gmail.com

[*] Correspondence: pmgg@ualg.pt

[†] Presented at the IX Iberian Congress of Ichthyology, Porto, Portugal, 20–23 June 2022.

[‡] Presenting author (Poster presentation).

Citation: Vargas-Lagos, C.; Silva, S.; Guerrero, L.; Montes de Oca, M.; Louro, B.; Alves, A.; Rotllant, J.; Guerreiro, P.M. Transcriptomic Analysis of Differentially Expressed Genes in Kidney and Intestine of *Dicentrarchus labrax* Fed Different Nutritional Amounts of Inorganic Phosphate. *Biol. Life Sci. Forum* **2022**, *13*, 17. https://doi.org/10.3390/blsf2022013017

Academic Editor: Alberto Teodorico Correia

Published: 6 June 2022

Publisher's Note: MDPI stays neutral with regard to jurisdictional claims in published maps and institutional affiliations.

Abstract: Phosphorus (P), in the form of inorganic phosphate (Pi), is one of the most important macronutrients for all organisms, including fish. It is indispensable for the formation of hard tissues such as bones, but also for cell signalling and cell membrane formation, and energy transduction, among many other functions and is kept under well-controlled conditions, since its deficiency or overload may lead to skeletal malformation or ectopic calcification, disturbances of intermediary metabolism, growth and function impairment, endocrine dysfunction, and eventually death. Fish feeds used in aquaculture are therefore P-rich but excess/unused/excreted P in the effluents can lead to eutrophication and a consequent deleterious change in the aquatic ecosystem. The objective of this study was to evaluate the expression profiles and transcripts modified by dietary P, to identify pathways and mechanisms involved in P transport and regulation in the kidney and intestine. Juvenile *Dicentrarchus labrax* were fed using a commercial feed (1.1% P) or tailored-made feeds containing 0.05%, 1.1%, or 3% Pi. Fish (duplicate tanks, $n = 10$) were fed for 70 days and weighed periodically to evaluate growth changes. Kidney and intestine were used for RNA extraction. Next-Generation Sequencing and RNAseq library preparation were performed in an Illumina system following the manufacturer's recommendations. Annotation was performed using the available sea bass genome assembly. Bioinformatic analysis showed significant differences in expression patterns among the three conditions tested in both tissues. In the kidney, increased P led to a total of 135 differentially expressed genes (DEGs; 82 up and 53 down), while only 54 (11 up and 43 down) genes responded to P restriction. In the intestine, high P affected the expression of 50 genes (16 up and 34 down) whereas only 26 (6 up and 20 down) were modified by low P. However, DEGs between high and low P were 156 in kidney and 154 in intestine. Preliminary analysis suggests the most affected pathways were those involved in cellular metabolism and phosphorylation but also on the structure of cell membranes, either for maintaining membrane integrity or in genes related to transmembrane ion transport. We expect this research to reveal the molecular implications of dietary P imbalance looking at specific targets such as membrane transporters and regulatory factors, but also to the larger metabolic pathways affected in these two key organs for P uptake and excretion.

Keywords: inorganic phosphate; fish feeds; nutritional physiology; transcriptomics; aquaculture; environmental protection

Author Contributions: Conceptualization, P.M.G.; methodology, S.S., B.L., L.G., P.M.G.; formal analysis, C.V-L., S.S., B.L., L.G., P.M.G., M.M.d.O.; investigation, S.S., A.A., L.G., C.V.-L.; resources,

P.M.G., J.R.; data curation, C.V.-L., S.S., B.L., L.G.; writing—original draft preparation, P.M.G., C.V.-L.; writing—review and editing, P.M.G., C.V.-L.; supervision, P.M.G., J.R.; project administration, P.M.G.; funding acquisition, P.M.G. All authors have read and agreed to the published version of the manuscript.

Funding: Funded by Fundação para a Ciência e Tecnologia (FCT), grant number PTDC/BIA-ANM/4225/2012 and FCT-UIDB/04326/2021.

Institutional Review Board Statement: Not applicable.

Informed Consent Statement: Not applicable.

Data Availability Statement: Not applicable.

Conflicts of Interest: The authors declare no conflict of interest.

Abstract

Metabolic Responses and Resilience to Environmental Challenges in the Sedentary Batrachoid *Halobatrachus didactylus* [†]

Juan M. Molina [1,2], Andreas Kunzmann [2], João Reis [3] and Pedro M. Guerreiro [3,*,‡]

1. Conicet Bahia Blanca, Universidad Nacoinal del Sur, Bahia Blanca B8000, Argentina; jmmolina@criba.edu.ar
2. Leibniz-Zentrum für Marine Tropenforschung (ZMT), 28359 Bremen, Germany; akunzmann@leibniz-zmt.de
3. Centro de Ciências do Mar (CCMAR), Universidade do Algarve, 8005-139 Faro, Portugal; jreis@ualg.pt
* Correspondence: pmgg@ualg.pt
† Presented at the IX Iberian Congress of Ichthyology, Porto, Portugal, 20–23 June 2022.
‡ Presenting author (Poster presentation).

Citation: Molina, J.M.; Kunzmann, A.; Reis, J.; Guerreiro, P.M. Metabolic Responses and Resilience to Environmental Challenges in the Sedentary Batrachoid *Halobatrachus didactylus* . Biol. Life Sci. Forum **2022**, 13, 50. https://doi.org/10.3390/blsf2022013050

Academic Editor: Alberto Teodorico Correia

Published: 2 June 2022

Publisher's Note: MDPI stays neutral with regard to jurisdictional claims in published maps and institutional affiliations.

Abstract: The Lusitanian toadfish, *Halobatrachus didactylus* is a marine teleost found in coastal lagoons and river estuaries, often exposed to important changes in salinity, temperature and reduced oxygen. Sedentary species, with strong site fidelity and low migratory ability along the temperature gradient such as this may be especially impacted by climate change. We aimed at establishing the tolerance limits to acute temperature and oxygen changes, and evaluate respiratory and metabolic responses in chronic control, warm and hypoxic (35% O_2) conditions. Critical temperature maximum (CTmax) was determined in 12 individuals exposed to a temperature ramp of 3 °C per hour starting at 18 °C, and was found to be 34.8 ± 0.66 °C. Critical oxygen level (PO_2crit) was determined in 8 fish at 18 °C while performing intermittent respirometry and oxygen depletion was created by nitrogen injection in the tank. PO_2crit was calculated as the inflexion point between oxyregulation and oxyconformation, which was found to be around 1.2 mgO_2/L, but fish survived down to 3% O_2, recovering from 0.2 mgO_2/L but showing increased hematocrit (Hct), red blood cell (RBC) counts and blood pH. We also quantified routine aerobic scope and daily activity patterns, finding this fish to be extremely sedentary. *H. didactylus* showed one of the lowest daytime basal metabolic rates (MR) found in the literature but activity increased significantly at night (over two-fold when closed inside the metabolic chambers). The effect of temperature on metabolic rate (MR) was evaluated using a temperature ramp ranging from 8 to 32 °C (1 °C/h). Acute temperature changes resulted in a steady increase in MR up to circa 29 °C, beyond which MR become increasingly variable, especially among smaller individuals. Indeed, small fish appear to show high- and low-MR groups, and were more susceptible to heat and hypoxia than larger individuals. In chronic acclimation, the MR was increased by 3- and 4-fold (hypoxia vs. normoxia) in fish at 28 °C in relation to those at 12 °C. Standard MR were not statistically different between normoxia and hypoxia at 12 °C, but maximum MR in hypoxia was only about 2/3 of that in normoxia. Fish in high temperature lost weight (mean −3.1%) and had higher metabolism, while in low temperature, weight increased (mean +9.3%) and metabolism was low, and HIS was significantly lower in high temperature groups. Fish in hypoxic conditions showed consistently high Hct but not RBC or hemoglobin (Hb). Overall this study indicates that *H. didactylus* is highly tolerant to hypoxia and temperature variations. It remains to be seen if other populations along the Atlantic coast show similar metrics. The measured CTmax is close to the actual maximum temperature possible to experience in Ria Formosa ponds during summer, and it would not be unexpected to find this species establishing stable populations in other regions if climate change forces it out of its actual distribution.

Keywords: respirometry; metabolic rates; hypoxia; temperature; sedentary species; climate change; physiology

Funding: Funded by FCT-UIDB/04326/2020 & AssemblePlus 8307, EMBRC.PT ALG-01-0145-FEDER-022121.

Institutional Review Board Statement: Not applicable.

Informed Consent Statement: Not applicable.

Data Availability Statement: Not applicable.

Conflicts of Interest: The authors declare no conflict of interest.

Abstract

Ecotoxicological Effects in Gilthead Seabream (*Sparus aurata*) Exposed to Environmentally Realistic Concentrations of Nickel Nanoparticles [†]

Eduardo Motta [1,2,‡], Alberto Teodorico Correia [1,3,4,*], José Fernando Gonçalves [3], David Daniel [5], Bruno Nunes [5] and José Neves [1,3]

[1] Centro Interdisciplinar de Investigação Marinha e Ambiental (CIIMAR), 4550-208 Matosinhos, Portugal; edumottabiomedico@gmail.com (E.M.); jneves@ufp.edu.pt (J.N.)

[2] Faculdade de Ciência e Tecnologia, Universidade Fernando Pessoa (FCT-UFP), 4249-004 Porto, Portugal

[3] Faculdade de Ciências da Saúde, Universidade Fernando Pessoa (FCS-UFP), 4200-150 Porto, Portugal; jfmg@icbas.up.pt

[4] Instituto Ciências Biomédicas Abel Salazar (ICBAS), 4050-313 Porto, Portugal

[5] Centro de Estudos do Ambiente e do Mar (CESAM), 3810-193 Aveiro, Portugal; d.daniel@ua.pt (D.D.); nunes.b@ua.pt (B.N.)

[*] Correspondence: atcorreia.ciimar@gmail.com

[†] Presented at the IX Iberian Congress of Ichthyology, Porto, Portugal, 20–23 June 2022.

[‡] Presenting author (Oral communication).

Citation: Motta, E.; Correia, A.T.; Gonçalves, J.F.; Daniel, D.; Nunes, B.; Neves, J. Ecotoxicological Effects in Gilthead Seabream (*Sparus aurata*) Exposed to Environmentally Realistic Concentrations of Nickel Nanoparticles. *Biol. Life Sci. Forum* **2022**, *13*, 59. https://doi.org/10.3390/blsf2022013059

Academic Editor: Felippe Daros

Published: 8 June 2022

Publisher's Note: MDPI stays neutral with regard to jurisdictional claims in published maps and institutional affiliations.

Abstract: Metalic nanoparticles (NPs) are emerging microcontaminants that have had, in recent years, increasing use in various sectors of the economy and society. Consequently, there is an urgent need to understand the environmental health consequences of the entry of these contaminants into the aquatic compartment. The objective of this study was to evaluate the ecotoxicological effects resulting from chronic exposure (28 days) to nickel NPs (Ni-NPs) at environmentally realistic concentrations (0.05 mg/L; 0.5 mg/L 5 mg/L), including a negative control (0.00 mg/L), in gills and liver of *Sparus aurata*. Antioxidant defense (Catalase, CAT), phase II metabolic detoxification (Glutathione S-Transferases, GSTs) enzymes, and lipid peroxidation (thiobarbituric acid reactive species, TBARS) were evaluated. Although the data showed that gills did not show significant differences in GST and CAT activities among the experimental goups, the group exposed to the highest dose (5 mg/L) showed a higher concentration of TBARS compared to the control. Regarding the liver, significant inhibition of catalase was observed for the different groups exposed to different concentrations of Ni-NPs. The assays performed suggest that the nanoparticles could promote biochemical alterations in the livers and gills of the exposed individuals, but more biomarkers of oxidative stress are needed to reveal the mechanistic pathways of Ni-NPs.

Keywords: nanoparticles; ecotoxicity; fish

Author Contributions: Conceptualization, A.T.C. and J.N.; Methodology, E.M., A.T.C., D.D., B.N. and J.N.; Aquarium facilities, J.F.G.; Statistical analysis, A.T.C. and E.M.; Investigation, E.M., A.T.C. and J.N.; Writing—original draft preparation, E.M.; writing—review and editing, E.M., A.T.C. and J.N.; Supervision, A.T.C. and J.N.; Funding acquisition, A.T.C. All authors have read and agreed to the published version of the manuscript.

Funding: This research was supported by the Strategic Funding UIDB/04423/2020 and UIDP/04423/2020 through Portuguese funds provided by FCT and by the project Ocean3R (NORTE-01-0145-FEDER-000064) supported by North Portugal Regional Operational Programme (NORTE 2020), under the PORTUGAL 2020 Partnership Agreement, through the European Regional Development Fund (ERDF).

Institutional Review Board Statement: Experiments were prior authorized by the Ethical Committee of the host institution (ORBEA/CIIMAR), including the euthanasia procedure. Furthermore, this work took into consideration the Portuguese animal welfare testing regulations (Decree-Law 113/2013).

Informed Consent Statement: Not apllicable.

Data Availability Statement: The data presented in this study are available on request from the corresponding author.

Conflicts of Interest: The authors declare no conflict of interest.

biology and life sciences forum

Abstract

Changes in Morphology of Hepatocytes and Tissue-Biodistribution in Iron-Overload Gilthead Seabream (*Sparus aurata*) [†]

Jhon A. Serna-Duque [1], José Carlos Campos-Sánchez [1,‡], Cristóbal Espinosa Ruiz [1], Salvadora Martínez-López [2] and Maria Ángeles Esteban [1,*]

[1] Immunobiology for Aquaculture Group, Department of Cell Biology and Histology, Faculty of Biology, University of Murcia, 30100 Murcia, Spain; jhonalberto.sernad@um.es (J.A.S.-D.); josecarlos.campos@um.es (J.C.C.-S.); cer48658@um.es (C.E.R.)

[2] Department of Agricultural Chemistry, Geology and Pedology, Faculty of Chemistry, University of Murcia, 30100 Murcia, Spain; salvadora.martinez@um.es

[*] Correspondence: aesteban@um.es

[†] Presented at the IX Iberian Congress of Ichthyology, Porto, Portugal, 20–23 June 2022.

[‡] Presenting author (Poster presentation).

Abstract: Iron is an essential metallic micronutrient for life, but its bioavailability is especially scarce in marine areas. We have studied iron accumulation in gilthead seabream (*Sparus aurata*) to improve the knowledge of iron metabolism in fish. For this aim, acute iron overload was induced by intraperitoneally injection of iron-dextran. Nine different tissues were collected at 72 h after injection. Iron determination was performed in all samples by atomic absorption spectroscopy and liver was stained with hematoxylin–eosin for histologic study. Iron overload provoked big changes in the organization of the liver and, in particular, in the hepatocytes, which presented iron-deposits as hemosiderin and an increase of vacuolation by metabolic stress. Likewise, iron concentration in the studied samples indicated great changes in iron-distribution in tissues. In the control group, intestine, gill and blood are the predominant tissues in total proportion of iron in fish body, while iron overload produced a large increase in liver and spleen. In this way, the liver and spleen are storage organs of iron, and this is accumulated in the cytoplasm of hepatocytes. Hence, the study of iron accumulation has contributed to understand the iron-physiology and open doors to applications in aquaculture.

Keywords: iron metabolism; metal accumulation; gilthead seabream; teleost

Citation: Serna-Duque, J.A.; Campos-Sánchez, J.C.; Espinosa Ruiz, C.; Martínez-López, S.; Esteban, M.Á. Changes in Morphology of Hepatocytes and Tissue-Biodistribution in Iron-Overload Gilthead Seabream (*Sparus aurata*). *Biol. Life Sci. Forum* **2022**, *13*, 82. https://doi.org/10.3390/blsf2022013082

Academic Editor: Alberto Teodorico Correia

Published: 13 June 2022

Publisher's Note: MDPI stays neutral with regard to jurisdictional claims in published maps and institutional affiliations.

Author Contributions: Term, J.A.S.-D.; Conceptualization, M.Á.E.; methodology, J.A.S.-D., C.E.R. and S.M.-L.; software, J.A.S.-D. and S.M.-L.; validation, C.E.R. and M.Á.E.; formal analysis, C.E.R.; investigation, J.A.S.-D., C.E.R. and S.M.-L.; resources, M.Á.E.; data curation, C.E.R.; writing—original draft preparation, all authors; writing—review and editing, all authors; visualization, J.A.S.-D.; supervision, C.E.R. and M.Á.E.; project administration, M.Á.E.; funding acquisition, M.Á.E. All authors have read and agreed to the published version of the manuscript.

Funding: J.A.S.-D. has a grant (FPU19/02192). Project PID2020-113637RB-C21 of 346 research funded by MCIN/AEI/10.13039/501100011033.

Institutional Review Board Statement: All animal studies were carried out in accordance with the European Union regulations for animal experimentation and the Bioethical Committee of the University of Murcia.

Informed Consent Statement: Not applicable.

Conflicts of Interest: The authors declare no conflict of interest.

Abstract

A Comparison of Olfactory Sensitivity in Seawater- and Freshwater-Adapted Bass, *Dicentrarchus labrax* [†]

Zélia Velez [*,‡], Peter C. Hubbard and Pedro M. Guerreiro

Centro de Ciências do Mar (CCMAR), 8005-139 Faro, Portugal; phubbard@ualg.pt (P.C.H.); pmgg@ualg.pt (P.M.G.)
* Correspondence: zvelez@ualg.pt
† Presented at the IX Iberian Congress of Ichthyology, Porto, 20–23 June 2022.
‡ Presenting author (oral communication).

Abstract: Fish rely heavily on olfaction for many aspects of their lives including foraging, defense, migration, and reproduction. Olfactory receptor neurons in the olfactory epithelium are in direct contact with the water, and are, therefore, exposed to changes in water chemistry. The European seabass, *Dicentrarchus labrax*, uses estuaries as feeding grounds and migrates between seawater and brackish water; but some can be found in 100% freshwater. However, little is known about how the olfactory system adjusts to waters of such different ionic composition and whether this affects its function and ability to discriminate between odorants. The aim of this study was, therefore, to compare olfactory sensitivity in seabass adapted to either seawater (SW) or freshwater (5 ppt; FW), to odorants conveyed at different salinities, using multi-unit recording from the olfactory nerve. In SW-adapted fish, olfactory sensitivity to amino acids (AA) was consistently higher when AA were presented in seawater (SW-AA) than when presented in freshwater (FW-AA), whereas in FW-adapted fish, olfactory sensitivity to FW-AA was either equal or slightly lower to SW-AA. SW-adapted fish responded to decreases in external $[Ca^{2+}]$ and to increases in external $[Na^+]$. FW-adapted fish responded to increases of both ions. In SW-adapted fish, Ca^{2+}-free artificial seawater (ASW) completely inhibited olfactory responses to amino acids, whereas Na^+-free ASW had no effect. However, in FW-adapted fish, lack of either ion in the water had no effect. Taken together, these results suggest that, as a primarily marine species, the olfactory system of the seabass is more sensitive in seawater; however, it can still function in freshwater, albeit with reduced sensitivity. Furthermore, in seawater, the olfactory transduction process is likely mediated by influx of external Ca^{2+}, but not Na^+. In FW-adapted fish, the transduction process relies on neither external Ca^{2+} nor Na^+, suggesting that the process of hyperosmoregulatory ability to adjust to life in ion-poor water. Further work is needed to clarify how changes in salinity affect olfactory sensitivity, and the mechanisms by which euryhaline species are able to adapt to such changes when moving between media of different ionic composition and variable pH.

Keywords: olfaction; salinity; odorants; calcium; sodium; amino acids

Citation: Velez, Z.; Hubbard, P.C.; Guerreiro, P.M. A Comparison of Olfactory Sensitivity in Seawater- and Freshwater-Adapted Bass, *Dicentrarchus labrax*. *Biol. Life Sci. Forum* **2022**, *13*, 125. https://doi.org/10.3390/blsf2022013125

Academic Editor: Alberto Teodorico Correia

Published: 17 June 2022

Publisher's Note: MDPI stays neutral with regard to jurisdictional claims in published maps and institutional affiliations.

Funding: This study received Portuguese national funds from FCT—Foundation for Science and Technology through projects UIDB/04326/2020, UIDP/04326/2020, LA/P/0101/2020 and EXPL/BIA-ECO/1161/2021.

Institutional Review Board Statement: Not applicable.

Informed Consent Statement: Not applicable.

Data Availability Statement: Not applicable.

Conflicts of Interest: The authors declare no conflict of interest.

Abstract

Historical Food-Web Changes in Invaded Fish Communities in the Lower Guadiana Basin [†]

Christos Gkenas [1,*,‡], Joana Martelo [1,2], Julien Cucherousset [3], Filipe Ribeiro [1], João Gago [1,4], Maria Judite Alves [2,5], Diogo Ribeiro [1,4], Gisela Cheoo [6] and Maria Filomena Magalhães [2]

1 Centro de Ciências do Mar e do Ambiente (MARE), Faculdade de Ciências, Campo Grande, Universidade de Lisboa, 1749-016 Lisboa, Portugal; jmmartelo@fc.ul.pt (J.M.); fmvribeiro@gmail.com (F.R.); joao.gago@esa.ipsantarem.pt (J.G.); diogorrribeiro@hotmail.com (D.R.)
2 Centro de Ecologia, Alterações Ambientais e Evolução (cE3c), Faculdade de Ciências, Campo Grande, Universidade de Lisboa, 1749-016 Lisboa, Portugal; mjalves@museus.ulisboa.pt (M.J.A.); mfmagalhaes@fc.ul.pt (M.F.M.)
3 UMR5174 Laboratoire Évolution & Diversité Biologique (EDB), Centre National de la Recherche Scientifique (CNRS), Université Paul Sabatier, ENFA, 118 Route de Narbonne, 31062 Toulouse, France; julien.cucherousset@univ-tlse3.fr
4 Escola Superior Agrária, Instituto Politécnico de Santarém, 2001-904 Santarém, Portugal
5 Museu Nacional de História Natural e da Ciência, Universidade de Lisboa, 1250-102 Lisboa, Portugal
6 Faculdade de Ciências, Campo Grande, Universidade de Lisboa, 1749-016 Lisboa, Portugal; giselacheoo@gmail.com
* Correspondence: chrisgenas@gmail.com
† Presented at the IX Iberian Congress of Ichthyology, Porto, Portugal, 20–23 June 2022.
‡ Presenting author (Oral presentation).

Citation: Gkenas, C.; Martelo, J.; Cucherousset, J.; Ribeiro, F.; Gago, J.; Alves, M.J.; Ribeiro, D.; Cheoo, G.; Magalhães, M.F. Historical Food-Web Changes in Invaded Fish Communities in the Lower Guadiana Basin. *Biol. Life Sci. Forum* **2022**, *13*, 2. https://doi.org/10.3390/blsf2022013002

Academic Editor: Alberto Teodorico Correia

Published: 1 June 2022

Publisher's Note: MDPI stays neutral with regard to jurisdictional claims in published maps and institutional affiliations.

Abstract: Freshwater ecosystems are increasingly being reshaped by biological invasions, leading to biotic homogenization and biodiversity loss. However, the extent to which novel species may drive changes in food-web structure over time remains poorly understood. Clarifying changes in historical ecological processes is critical to inform conservation and restoration efforts in recipient ecosystems. Here, we address food-web changes associated with fish invasions in the Lower Guadiana Basin (LGB) over the past 40 years, by contrasting feeding relationships between museum-archived and contemporary specimens, using stable carbon (δ13C) and nitrogen (δ15N) ratios. Specifically, trophic niches of museum-archived fishes sampled throughout 1978–1987 and 1999–2004 corresponding to the initial establishment and spread of non-native fishes, respectively, were compared with those of fishes sampled in 2019, characterizing the integration of non-native species in the recipient ecosystem. We focused on five native species (*Anaecypris hispanica*, *Cobitis paludica*, *Iberochondrostoma lemmingii*, *Squalius pyrenaicus* and *Squalius alburnoides*) and four non-native species (*Lepomis gibbosus*, *Australoheros facetus*, *Micropterus salmoides* and *Gambusia holbrooki*) with potential to cover multiple trophic positions in the food-webs. We approached historical baseline resources using prey items in gut contents of the museum-archived fishes and characterized primary producers and macroinvertebrates in 2019. Prior to analysis, samples were normalized for high lipid content and corrected for preservation. We found considerable asymmetries in niche partitioning among species as invasion progressed. Over time, native species tended to be displaced to lower trophic levels, while non-native species showed significantly higher trophic niches, driven mainly by increases in trophic (δ15N) range. Our study highlights that stable isotopes may provide important insights on historical food-web structure and particularly on processes underpinning ecological changes associated with anthropogenetic pressures on freshwater ecosystems.

Keywords: food-webs; stable isotopes; museum specimens; trophic niche; non-native species

Author Contributions: Conceptualization, C.G., J.C. and M.F.M.; methodology, C.G., J.C. and M.F.M.; investigation, C.G., J.M., J.C., F.R., J.G., M.J.A., D.R., G.C. and M.F.M.; writing—original draft preparation, C.G.; writing—review and editing, C.G. and M.F.M.; visualization, C.G.; project adminis-tration, C.G.; funding acquisition, C.G. All authors have read and agreed to the published version of the manuscript.

Funding: This study was conducted in the frame of the project ISO-INVA co-funded by international funds through Lisboa 2020—Programa Operacional Regional de Lisboa, in its FEDER component (Project ref. LISBOA- 01-0145-FEDER-029105) and national funds through FCT—Fundação para a Ciência e a Tecnologia, I.P (Project ref. PTDC/CTA-AMB/29105/2017).

Institutional Review Board Statement: Not applicable.

Informed Consent Statement: Not applicable.

Data Availability Statement: The data presented in this study are available on request from the corresponding author. The data are not publicly available due to privacy restrictions.

Conflicts of Interest: The authors declare no conflict of interest.

Abstract

Feeding Habits of the Invasive Weakfish (*Cynoscion regalis*) in the Gulf of Cadiz [†]

Gala González [1,2,*,‡], Jose A. Cuesta [1,2], Cesar Vilas [2,3], Francisco Baldó [4], Carlos Fernández Delgado [5] and Enrique González-Ortegón [1,2]

[1] Institute of Marine Science of Andalusia (CSIC-ICMAN), 11519 Cádiz, Spain; jose.cuesta@csic.es (J.A.C.); e.gonzalez.ortegon@csic.es (E.G.-O.)
[2] Associated Unit IFAPA-CSIC "Crecimiento Azul", 11519 Cádiz, Spain; cesar.vilas@juntadeandalucia.es
[3] Andalusia Research and Training Institute for Fisheries and Agriculture (IFAPA), 11500 Cádiz, Spain
[4] Spanish Institute of Oceanography (IEO-Cádiz), 11006 Cadiz, Spain; francisco.baldo@ieo.es
[5] Department of Zoology, University of Córdoba, 14071 Córdoba, Spain; ba1fedec@uco.es
* Correspondence: gala.gonzalez@csic.es; Tel.: +34-637183648
† Presented at the IX Iberian Congress of Ichthyology, Porto, Portugal, 20–23 June 2022.
‡ Presenting author (Oral communication).

Citation: González, G.; Cuesta, J.A.; Vilas, C.; Baldó, F.; Delgado, C.F.; González-Ortegón, E. Feeding Habits of the Invasive Weakfish (*Cynoscion regalis*) in the Gulf of Cadiz. *Biol. Life Sci. Forum* **2022**, *13*, 3. https://doi.org/10.3390/blsf2022013003

Academic Editor: Alberto Teodorico Correia

Published: 1 June 2022

Publisher's Note: MDPI stays neutral with regard to jurisdictional claims in published maps and institutional affiliations.

Abstract: Weakfish (*Cynoscion regalis*) has been present in Iberian waters since at least 2011, when it was first recorded in the Guadalquivir estuary. Little is known about the preferences and feeding strategies of weakfish outside of its native range; therefore, in this work, we carried out a comprehensive study between March 2021 and September 2021 to elucidate these matters. In total, the stomach contents of 300 fish were examined. The fish were collected in spring and summer in the Gulf of Cadiz (Spain), with individuals caught ranging from 185 to 590 mm in total length. Due to the sampling period and size range of individuals, ontogenic and seasonal (spring–summer) variations in the diet were also explored. Overall, fish and crustaceans were the dominant groups consumed by weakfish. The European anchovy (*Engraulis encrasicolus*) and caramote prawn (*Penaeus kerathurus*) were the most abundant prey in each group. While no differences were found in the percentage of occurrence of fish in the non-empty stomachs analyzed in spring and summer (83%), a small increase was found in the percentage of occurrence of crustaceans from spring (20%) to summer (29%). In addition, the analysis of the results also suggested that weakfish of smaller sizes feed more on crustaceans, while bigger individuals feed mainly on fish. Our findings indicate that weakfish is a fully carnivorous species with a preference for fish. All this is in line with the diet and strategy that weakfish exhibits in its native area and with the first assessment made with the existing population in the Sado Estuary (Portugal). In recent years, due to the appearance of non-native species, there is increasing concern among local fishermen and the local authorities about the reduction in captures of certain species such as *Penaeus kerathurus*; therefore, this study could help researchers to understand the role of weakfish concerning this matter. The present study contributes to the ecological knowledge about this species in the Iberian Peninsula and the Gulf of Cadiz, helping the local authorities to establish, if needed, an appropriate management program to cope with this recently introduced species.

Keywords: weakfish; stomach content analysis; fish feeding; Iberian Peninsula; non-native species

Author Contributions: Conceptualization, E.G.-O. and J.A.C.; methodology, G.G., E.G.-O. and J.A.C.; formal analysis, G.G. and E.G.-O.; investigation, All authors; resources, E.G.-O. and J.A.C.; writing—original draft preparation, G.G.; writing—review and editing, All authors; funding acquisition, E.G.-O. and J.A.C. All authors have read and agreed to the published version of the manuscript.

Funding: This research was funded by the Regional Government of Andalusia with grant number P18-RT-5010.

Institutional Review Board Statement: Not applicable.

Informed Consent Statement: Not applicable.

Data Availability Statement: Not applicable.

Conflicts of Interest: The authors declare no conflict of interest.

Abstract

Lepomis gibbosus European Invasion Process: Niche Differentiation and Future Climate Scenarios [†]

Angela Lambea-Camblor [1], Felipe Morcillo [2], Jesús Muñoz [3] and Anabel Perdices [1,*,‡]

[1] Museo Nacional de Ciencias Naturales-CSIC, 28006 Madrid, Spain; angela.lambea@gmail.com
[2] Departamento de Biodiversidad, Ecología y Evolución, Facultad de Ciencias Biológicas, Universidad Complutense de Madrid, 28040 Madrid, Spain; fmorcill@ucm.es
[3] Real Jardín Botánico-CSIC, 28014 Madrid, Spain; jmunoz@rjb.csic.es
* Correspondence: aperdices@mncn.csic.es
[†] Presented at the IX Iberian Congress of Ichthyology, Porto, Portugal, 20–23 June 2022.
[‡] Presenting author (poster).

Citation: Lambea-Camblor, A.; Morcillo, F.; Muñoz, J.; Perdices, A. *Lepomis gibbosus* European Invasion Process: Niche Differentiation and Future Climate Scenarios. *Biol. Life Sci. Forum* **2022**, *13*, 6. https://doi.org/10.3390/blsf2022013006

Academic Editor: Alberto Teodorico Correia

Published: 2 June 2022

Abstract: In recent years, ecological studies have highlighted the importance of niche in the invasion of new habitats by invasive species, stating that introduction into a similar niche is critical for rapid expansion into the new habitat. In this study, we focus on *Lepomis gibbosus* or Pumpkinseed sunfish, a centrarchid native to eastern North America that is among the top 10 introduced freshwater fishes with the greatest ecological impact globally. We compared their native and introduced niches in Europe based on occurrence data from three time periods (1900–1959, 1960–1989 and 1990–2021) using niche identity and background similarity tests. In addition, through MaxEnt software, we modelled present and future potential distribution of Pumpkinseed sunfish under climate change conditions. Our results show that *L. gibbosus* has significantly modified its niche in the process of invasion through Europe, highlighting the great adaptability of this species to higher temperatures and irregular water regimes. This niche differentiation clearly distinguished the European niche occupied by the invasive populations from the North-American-native niche, supporting previous studies that suggested that European *L. gibbosus* populations would not be able to survive in its native area. Our modelling results based on different future climatic conditions pointed out that temperature and precipitation are the most influential variables that could facilitate the establishment of sunfish on almost all continents. Globally, it is expected that Pumpkinseed sunfish will soon colonize the African continent, where there are already species that cohabit with this centrarchid. This reinforces the need to develop new prevention measures before its presence becomes irreversible, taking into account the role that climate change will play in its establishment.

Keywords: aquatic systems; biological invasions; climate change; pumpkinseed sunfish; species distribution models

Author Contributions: Conceptualization, A.L.-C., F.M., J.M. and A.P.; methodology, A.L.-C., J.M. and A.P.; writing-review and editing, A.L.-C., F.M., J.M. and A.P.; supervision, F.M., J.M. and A.P. All authors have read and agreed to the published version of the manuscript.

Funding: This research received no external funding.

Institutional Review Board Statement: Not applicable.

Informed Consent Statement: Not applicable.

Conflicts of Interest: The authors declare no conflict of interest.

biology and life sciences forum

Abstract

The Course of Natural Colonization of the Toadfish *Halobatrachus didactylus* (Batrachoididae) in Galician Waters (NW Spain) †

Rafael Bañón [1,*,‡], Bruno Almón [1,2], Mariña Fabeiro [3], Sonia Rábade [3], Alexandre Alonso-Fernández [3], Sandra Mosquera [4], José Carlos Mariño [5], María Berta Ríos [6] and Alejandro de Carlos [7]

[1] Grupo de Estudio do Medio Mariño (GEMM), 15960 Ribeira, Spain; brunoalmon2@yahoo.es
[2] Instituto Español de Oceanografía, Centro Oceanográfico de Vigo (IEO, CSIC), 36390 Vigo, Spain
[3] Instituto de Investigaciones Marinas (IIM-CSIC), 36208 Vigo, Spain; fabeiro@iim.csic.es (M.F.); soniaru@iim.csic.es (S.R.); alex@iim.csic.es (A.A.-F.)
[4] Cofradía de Pescadores San Juan de Redondela, 36693 Redpndela, Spain; asistenciatecnicaredondela@gmail.com
[5] Cofradía de Pescadores San Antonio de Cambados, 36630 Cambados, Spain; maicarril@gmail.com
[6] Cofradía de Pescadores San José de Cangas do Morrazo, 36940 Cangas, Spain; berta@cofradiadecangas.org
[7] Departamento de Bioquímica, Xenética e Inmunoloxía, Universidade de Vigo, 36310 Vigo, Spain; adcarlos@uvigo.es
* Correspondence: anoplogaster@yahoo.es; Tel.: +34-655-220-949
† Presented at the IX Iberian Congress of Ichthyology, Porto, 20–23 June 2022.
‡ Presenting author (Poster presentation).

Citation: Bañón, R.; Almón, B.; Fabeiro, M.; Rábade, S.; Alonso-Fernández, A.; Mosquera, S.; Mariño, J.C.; Ríos, M.B.; de Carlos, A. The Course of Natural Colonization of the Toadfish *Halobatrachus didactylus* (Batrachoididae) in Galician Waters (NW Spain). *Biol. Life Sci. Forum* **2022**, *13*, 35. https://doi.org/10.3390/blsf2022013035

Academic Editor: Alberto Teodorico Correia

Published: 6 June 2022

Publisher's Note: MDPI stays neutral with regard to jurisdictional claims in published maps and institutional affiliations.

Abstract: Climate change is causing a northward shift of fish species and the tropicalisation of temperate zones. In this context, the toadfish *Halobatrachus didactylus* was found for the first time in 2018 in Galician waters (NW Spain), where a total of 39 specimens have been recorded so far in the southernmost part of those known as Rías Baixas. Preliminary analyses of the specimens showed a diet based mainly on crustaceans and molluscs. The length composition varied between 17 and 40 cm TL, and the estimated ages of seven individuals, with sizes ranging from 18 to 35 cm TL, ranged from 3 to 7 years. Histological reproductive analysis of 17 specimens showed an unbalanced sex ratio 2.4:1, favourable to males. Out of 12 males found, 2 (17%) were immature, ranging in size between 18 and 26.6 cm TL, whereas the other 10 (83%) were mature and greater than 25.8 cm TL. Furthermore, six of them were in the active spawning phase. Only five females were found, one (20%) being in the developing phase, and the remaining four (80%) in the spawning capable phase, with oocytes in the advanced vitellogenesis stage but without evidence of imminent spawning. These results suggest that this species is reproductively active in Galician waters. From a molecular point of view, this population has been compared with others along the Portuguese coast using the nucleotide sequences of a putative fragment of the mitochondrial control region and of the first intron of the ribosomal protein S7 gene. In the first case, a single haplotype is detected, which is the same as that occurring in all individuals captured further south in the Tagus and Sado estuaries. In the case of the S7 intron, no trend towards lower genetic diversity that could indicate lineage selection was detected. These data seem to support a very recent colonisation event.

Keywords: non-native species; toadfish; range extension; invasion process; fish biology

Author Contributions: Conceptualization, R.B. and A.d.C.; methodology, R.B., B.A., A.A.-F. and A.d.C.; formal analysis, R.B., B.A., S.R., M.F. A.A.-F. and A.d.C.; resources, R.B., S.M., J.C.M. and M.B.R.; data curation, R.B., B.A., A.d.C.; writing—original draft preparation, R.B., B.A. and A.d.C.; writing—review and editing, R.B., B.A., S.R., M.F., A.A.-F., S.M., J.C.M., M.B.R. and A.d.C.; funding acquisition, A.d.C.; All authors have read and agreed to the published version of the manuscript.

Funding: A.d.C. was funded by the GRC Program from the Xunta de Galicia, Spain, grant code ED431C 2019/28.

Institutional Review Board Statement: Not applicable.

Informed Consent Statement: Not applicable.

Data Availability Statement: DNA sequences and their corresponding metadata, such as specimen photographs, capture coordinates and dates, are part of the project "Marine Fishes from Galicia" (code FIGAL), deposited in the Barcode of Life database (www.boldsystems.org).

Conflicts of Interest: The authors declare no conflict of interest.

*biology and life
sciences forum*

Abstract

Upstream Movement Capacity of Invasive Signal Crayfish (*Pacifastacus leniusculus*) under Different Environmental and Biometric Factors [†]

Francisco Javier Bravo-Córdoba [1,‡], Juan Francisco Fuentes-Pérez [2], Carlos Escudero-Ortega [1], Ana García-Vega [1] and Francisco Javier Sanz-Ronda [2,*]

[1] Centro Tecnológico Agrario y Agroalimentario—ITAGRA.CT, 34004 Palencia, Spain; fjbravo@itagra.com (F.J.B.-C.); carlose1209@gmail.com (C.E.-O.); agvega@itagra.com (A.G.-V.)
[2] Area of Hydraulics and Hydrology, Department of Agroforestry Engineering, University of Valladolid, 34004 Palencia, Spain; juanfrancisco.fuentes@uva.es
[*] Correspondence: jsanz@uva.es; Tel.: +34-979108358
[†] Presented at the IX Iberian Congress of Ichthyology, Porto, Portugal, 20–23 June 2022.
[‡] Presenting author (Poster presentation).

Abstract: The spread of invasive crayfish species is a major threat to endemic species worldwide. This threat affects native crayfish as well as flora and fauna species in general. In order to limit their dispersal, different methods have been used, the most promising of which are those related to physical barriers. For their design, it is essential to know the limits in the capacity of crayfish to move under different hydraulic scenarios, although to date, there are few studies on this topic. The present work analyzes the volitional upstream movement capacity of the signal crayfish (*Pacifastacus leniusculus*) in a laboratory open flume, with different configurations of environmental and hydraulic variables (bed roughness, flow velocity, water temperature, times of day) and accounting for the possible effect of biometric factors (carapace length, sex). Twenty-four different trials with five individuals per trial were carried out, tracking all crayfish movements individually by visual tags and with a video monitoring system. Data were analyzed using survival analysis techniques and parametric models were developed, considering as response variables the maximum distance traveled and the movement speed. The results showed that the combination of bed roughness and flow velocity were the best predictors to explain crayfish movement performance, with a flow velocity greater than 0.8 m/s together on a non-rough bed being the limiting factor; the water temperature and the sex also have a significant effect. This information can serve as a basis for the design of future barriers to the dispersal of invasive crayfish species in the Iberian Peninsula.

Keywords: invasive species; alien species; dispersal barriers; survival analysis

Citation: Bravo-Córdoba, F.J.; Fuentes-Pérez, J.F.; Escudero-Ortega, C.; García-Vega, A.; Sanz-Ronda, F.J. Upstream Movement Capacity of Invasive Signal Crayfish (*Pacifastacus leniusculus*) under Different Environmental and Biometric Factors. *Biol. Life Sci. Forum* **2022**, *13*, 36. https://doi.org/10.3390/blsf2022013036

Academic Editor: Alberto Teodorico Correia

Published: 6 June 2022

Publisher's Note: MDPI stays neutral with regard to jurisdictional claims in published maps and institutional affiliations.

Author Contributions: Conceptualization, F.J.B.-C., J.F.F.-P. and F.J.S.-R.; methodology, F.J.B.-C., F.J.S.-R., C.E.-O., A.G.-V. and J.F.F.-P.; validation, F.J.B.-C.; formal analysis, C.E.-O.; data curation, C.E.-O.; writing—original draft preparation, C.E.-O.; writing—review and editing, All authors; visualization, F.J.B.-C. and C.E.-O.; supervision, F.J.B.-C., J.F.F.-P. and F.J.S.-R.; project administration, F.J.S.-R.; funding acquisition, F.J.S.-R. All authors have read and agreed to the published version of the manuscript.

Funding: This research received no external funding.

Institutional Review Board Statement: The study was conducted according to the guidelines of the Declaration of Helsinki, and approved by the Ethics Committee of University of Valladolid as well as the approval of the competent authorities, i.e. Regional Government on Natural Resources (Junta de Castilla y León) and Water Management Authority (Confederación Hidrográfica del Duero).

Informed Consent Statement: Not applicable.

Data Availability Statement: Data are available upon reasonable request to the corresponding author.

Conflicts of Interest: The authors declare no conflict of interest.

biology and life sciences forum

MDPI

Abstract

Activity Patterns and Tridimensional Space Use of the European Catfish on a Reservoir in River Tagus (Portugal) [†]

Gil S. Santos [1,*,‡], Bernardo R. Quintella [1,2], Esmeralda Pereira [3], Ana Filipa Silva [1], Pedro R. Almeida [3,4], Diogo Ribeiro [1] and Filipe Ribeiro [1]

[1] MARE—Centro de Ciências do Mar e do Ambiente, Faculdade de Ciências, Universidade de Lisboa, 1749-016 Lisboa, Portugal; bsquintella@fc.ul.pt (B.R.Q.); afmsilva@fc.ul.pt (A.F.S.); diogorrribeiro@hotmail.com (D.R.); fmvribeiro@gmail.com (F.R.)
[2] Departamento de Biologia Animal, Faculdade de Ciências, Universidade de Lisboa, 1749-016 Lisboa, Portugal
[3] MARE—Centro de Ciências do Mar e do Ambiente, Universidade de Évora, 7521-903 Évora, Portugal; ecdp@uevora.pt (E.P.); pmra@uevora.pt (P.R.A.)
[4] Departamento de Biologia, Escola de Ciências e Tecnologia, Universidade de Évora, 7002-554 Évora, Portugal
[*] Correspondence: gilsaraivasantos@gmail.com
[†] Presented at the IX Iberian Congress of Ichthyology, Porto, Portugal, 20–23 June 2022.
[‡] Presenting author (Oral communication).

Citation: Santos, G.S.; Quintella, B.R.; Pereira, E.; Silva, A.F.; Almeida, P.R.; Ribeiro, D.; Ribeiro, F. Activity Patterns and Tridimensional Space Use of the European Catfish on a Reservoir in River Tagus (Portugal). *Biol. Life Sci. Forum* **2022**, *13*, 61. https://doi.org/10.3390/blsf2022013061

Academic Editor: Alberto Teodorico Correia

Published: 8 June 2022

Publisher's Note: MDPI stays neutral with regard to jurisdictional claims in published maps and institutional affiliations.

Abstract: The European catfish (*Silurus glanis*) is a non-native species with invasive character to Iberian freshwaters. Being the largest fish species in those invaded water bodies, with high fecundity rates, a large life expectancy and an extraordinary predatory potential, *S. glanis* has all the indicators that it could be exerting a dangerous pressure on native fish communities. Albeit there are some studies regarding the activity and depth use of this catfish, many of them are restricted to its native range and do not describe the circadian and annual behaviours in detail. Moreover, no studies have compared the differences in habitat use and movement ranges between adults and juveniles in a recently invaded territory. To fill these knowledge gaps, this study resorts to acoustic biotelemetry to track 25 fish (10 adults and 15 juveniles) in a Tagus river reservoir, the Belver dam, through an array of fixed acoustic receivers. The fish were internally tagged with acoustic transmitters, which, in the case of the adults, including a 3D-accelerometer and pressure sensors that allow obtaining information on activity and depth use for over a year. The results show that *S. glanis* is active throughout the year but with higher activity levels in summer and minimal in autumn, and with a crepuscular and nocturnal increase in activity. This species occupies shallower depths in spring/summer, while in autumn/winter roams at relatively deeper waters. Circadian vertical movement patterns were identified; however, they vary seasonally and have some individual variability. The areas used by the adults are larger than the juveniles' and increase in warmer months. Adult preferences in the use of specific areas across the year and a possible migration to a spawning site were identified. Such findings will support the development of more effective control measures, for instance, by providing information on how to allocate the fishing efforts in space and time to maximize the efficiency of mass removal actions of this invasive fish.

Keywords: *Silurus glanis*; invasive species; habitat use; movement patterns; vertical migration; acoustic biotelemetry

Author Contributions: Conceptualization, B.R.Q. and F.R.; methodology, B.R.Q., F.R. and G.S.S.; software, G.S.S.; formal analysis, G.S.S.; investigation, G.S.S., F.R., B.R.Q., E.P., A.F.S., P.R.A. and D.R.; resources, F.R. and B.R.Q.; writing—original draft preparation, G.S.S.; writing—review and editing, F.R., B.R.Q., E.P., A.F.S., P.R.A. and D.R.; visualization, G.S.S.; supervision, F.R. and B.R.Q.; project administration, F.R.; funding acquisition, F.R. and B.R.Q. All authors have read and agreed to the published version of the manuscript.

Funding: This study was conducted in the frame of the projects: *"FRISK—Freshwater fish invasions risk assessment: identifying invasion routes"* financed by the Foundation for Science and Technology (FCT ref. PTDC/AAGMAA/0350/2014); *"SONICINVADERS—Sounds of Invasion—Detecting Invasive Fish in Freshwaters Ecosystems with Passive Acoustics"* financed by the Foundation for Science and Technology (FCT ref. PTDC/CTAAMB/28782/2017) and by the European Regional Development Fund (FEDER) (ALT20-03-0145-FEDER-028782); *"MEGAPREDATOR—A giant on the water: from predation pressure to population control of the European catfish (Silurus glanis)"* (FCT ref. PTDC/ASP-PES/4181/2020) and *"CoastNet—Portuguese Coastal Monitoring Network"* financed by the Foundation for Science and Technology (FCT) and the European Regional Development Fund (FEDER), through LISBOA2020 and ALENTEJO2020 regional operational programs, in the framework of the National Roadmap of Research Infrastructures of strategic relevance (PINFRA/22128/2016).

Institutional Review Board Statement: Not applicable.

Informed Consent Statement: Not applicable.

Data Availability Statement: The data will be available after proper publication on the ETN (European Tracking Network) data portal.

Conflicts of Interest: The authors declare no conflict of interest.

Abstract

Presence of Invasive Alien Species (IAS) in Impounded Waters of Navarre (Spain) Using Multi-Mesh Gillnets [†]

Gonzalo Moncada [1], Julen Torrens [1,2], Javier Oscoz [2], Enrique Baquero [2,3], Nora Escribano [2], Enrique Miranda [2], Imanol Miqueleiz [2], Ignacio Ruiz de la Cuesta [2], Ibon Tobes [4] and Rafael Miranda [2,3,*]

[1] Sociedad Ibérica de Ictiología SIBIC, Irunlarrea 1, 31008 Pamplona, Spain; gmoncada@alumni.unav.es (G.M.); julentorrens@gmail.com (J.T.)

[2] Biodiversity Data Analytics and Environmental Quality Group, Department of Environmental Biology, Universidad de Navarra, Irunlarrea 1, 31008 Pamplona, Spain; javioscoz@gmail.com (J.O.); ebaquero@unav.es (E.B.); nescribano@unav.es (N.E.); emirandagar@alumni.unav.es (E.M.); imiqueleiz@alumni.unav.es (I.M.); iruiz.19@alumni.unav.es (I.R.d.l.C.)

[3] BIOMA Instituto de Biodiversidad y Medioambiente, Universidad de Navarra, Irunlarrea 1, 31008 Pamplona, Spain

[4] Centro de Investigación en Biodiversidad y Cambio Climático (BioCamb), Facultad de Ciencias del Medio Ambiente, Universidad Tecnológica Indoamérica, Quito EC170103, Ecuador; ibontobes@uti.edu.ec

* Correspondence: rmiranda@unav.es

† Presented at the IX Iberian Congress of Ichthyology, Porto, Portugal, 20–23 June 2022.

Citation: Moncada, G.; Torrens, J.; Oscoz, J.; Baquero, E.; Escribano, N.; Miranda, E.; Miqueleiz, I.; de la Cuesta, I.R.; Tobes, I.; Miranda, R. Presence of Invasive Alien Species (IAS) in Impounded Waters of Navarre (Spain) Using Multi-Mesh Gillnets. *Biol. Life Sci. Forum* **2022**, *13*, 62. https://doi.org/10.3390/blsf2022013062

Academic Editor: Alberto Teodorico Correia

Published: 8 June 2022

Publisher's Note: MDPI stays neutral with regard to jurisdictional claims in published maps and institutional affiliations.

Abstract: The introduction of invasive alien species (IAS) is a severe problem in ecosystems worldwide, heavily impacting biodiversity and especially endemic species. This situation is especially worrying in the Iberian Peninsula, since Spain and Portugal's rivers and lakes host an outstanding richness of endemic freshwater species. Ignorance about IAS presence and distribution is a serious problem that hampers its management. Regarding invasive fish species, difficulties in sampling and studying the ichthyofauna of lentic and deep waters, where many IAS inhabit, comprise some of the reasons for this lack of knowledge. In this study, we sampled the fish community of ten impounded waters in Navarre (Ebro River Basin, Spain) using multi-mesh gillnets. Four sampling points were dams located in rivers, and the remaining points were ponds. One of these ponds had a direct connection with a water channel, and another was found in a flood plain, so it connects with a nearby river in floods. The remaining ponds did not have a direct connection to any major river. A total of 14 fish species were detected (9 of which were IAS (64.3%)), with 3383 specimens collected (56% IAS). Only one of the analyzed dams did not contain IAS. Numerically, the most abundant fish was the exotic bleak (*Alburnus alburnus*) (almost 44% of the captures), followed by the native Ebro nase (*Parachondrostoma miegii*) (23.4%). The most widely distributed IAS were the common carp (*Cyprinus carpio*) and the bleak, which appeared in 70% and 60% of the sampling stations, respectively. IAS in dams accounted for 50% of the total species found and represented 21.3% of fish abundance. On the other hand, 64.3% of species in ponds were IAS, reaching 68.3% of fish abundance. This percentage increased when analyzing only the ponds without a natural connection with rivers or canals, where IAS introduction would be anthropic. IAS species in these points represented 71.4% of total fish species and 92.5% of the abundance of fish.

Keywords: invasive alien species (IAS); impounded waters; multi-mesh gillnets; Navarre; Ebro Basin; Spain

Author Contributions: Conceptualization, G.M. and R.M.; formal analysis, G.M.; investigation, G.M., J.T., J.O., E.B., N.E., E.M., I.M., I.R.d.l.C. and I.T.; writing—original draft preparation, R.M.; writing—review and editing, G.M. and R.M.; project administration, G.M. and J.O.; funding acquisition, R.M. All authors have read and agreed to the published version of the manuscript.

Funding: This research was funded by European Commission LIFE program, project LIFE17 GIE/ES/000515, in collaboration with Government of Navarra.

Institutional Review Board Statement: Not applicable.

Informed Consent Statement: Not applicable.

Data Availability Statement: Not applicable.

Conflicts of Interest: The authors declare no conflict of interest.

biology and life sciences forum

Abstract

Effect of the Introduction of Catfish (*Silurus glanis*) on the Native Fish Fauna in the Torrejón Reservoir (Cáceres, Spain) [†]

Carlos Orduna [1,*,‡], Amadora Rodríguez-Ruiz [2], África Oliver-Blanco [1], Juan Ramón Cid-Quintero [1] and Lourdes Encina [2]

[1] EcoFishUS Research, 41009 Seville, Spain; africa@ecofishus.com (Á.O.-B.); juancid@ecofishus.com (J.R.C.-Q.)
[2] Departamento de Biología Vegetal y Ecología, Universidad de Sevilla, 41080 Seville, Spain; dora@us.es (A.R.-R.); lencina@us.es (L.E.)
* Correspondence: carlos@ecofishus.com
† Presented at the IX Iberian Congress of Ichthyology, Porto, Portugal, 20–23 June 2022.
‡ Presenting author (oral communication).

Citation: Orduna, C.; Rodríguez-Ruiz, A.; Oliver-Blanco, Á.; Cid-Quintero, J.R.; Encina, L. Effect of the Introduction of Catfish (*Silurus glanis*) on the Native Fish Fauna in the Torrejón Reservoir (Cáceres, Spain). *Biol. Life Sci. Forum* **2022**, *13*, 122. https://doi.org/10.3390/blsf2022013122

Academic Editor: Alberto Teodorico Correia

Published: 17 June 2022

Publisher's Note: MDPI stays neutral with regard to jurisdictional claims in published maps and institutional affiliations.

Abstract: The introduction of exotic fish species is a global ecological and conservation problem that has caused the reduction and extinction of numerous native species. In Spain, the introduction of exotic freshwater fish is one of the main factors threatening the survival of their native fish species. One such species that recently arrived in Spanish freshwaters is the catfish (*Silurus glanis*). This study was conducted from 2010 until 2021 in the Torrejón reservoir in the Tagus basin. The results obtained show that the introduction of catfish has led to a significant decrease in the abundance of fish in the reservoir, as well as a clear alteration in the fish assemblage, causing changes in the proportion of species and their biomass. The common barbel (*Luciobarbus bocagei*), the only native species that still maintained an abundant population in the reservoir, has clearly been the most disadvantaged species. The size structure of the species present in the reservoir allows us to differentiate adult catfish specimens from the rest of the species using hydroacoustic methods and thus be able to estimate their density, biomass, and spatio-temporal distribution. Based on the results obtained, we can affirm that the species is active in the reservoir throughout the year and that, despite its fame as a "bottom dweller", catfish use surface habitats more than previously thought. It has shown a great capacity of adaptation to exploit the new environments and resources, together with a great tolerance against adverse conditions, such as low oxygen concentration (even anoxia) or high concentrations of ammonium, that are characteristic of the hypolimnion of Torrejón during the stratification period. The insights are an important contribution both to proving the effect of the introduction of catfish in native fish populations and to the implementation of alternative tools, such as hydroacoustic methodologies, in future management programs for this exotic species in our freshwater ecosystems.

Keywords: *Silurus glanis*; catfish; exotic; invasive

Author Contributions: Conceptualization, methodology, A.R.-R., L.E. and C.O.; software, J.R.C.-Q.; validation, A.R.-R. and L.E.; formal analysis, C.O.; investigation, J.R.C.-Q., Á.O.-B. and C.O.; resources, C.O.; data curation, Á.O.-B.; writing—original draft preparation, C.O.; writing—review and editing, visualization, supervision, A.R.-R. and L.E.; project administration, funding acquisition, C.O. All authors have read and agreed to the published version of the manuscript.

Funding: This research received no external funding.

Institutional Review Board Statement: Not applicable.

Informed Consent Statement: Not applicable.

Data Availability Statement: Not applicable.

Conflicts of Interest: The authors declare no conflict of interest.

Abstract

Using DNA Metabarcoding to Uncover the Diet of the Invasive Mozambique Tilapia (*Oreochromis mossambicus*) in Mangroves from the Island of São Tomé [†]

Sofia Nogueira [1,*,‡], Manuel Curto [1], Pedro M. Félix [1], Joshua Heumüller [1], Filipa Afonso [1], Diogo Dias [1], Paula Chainho [1,2], Ana Brito [1,3], Ricardo Lima [4] and Filipe Ribeiro [1]

[1] Centro de Ciências do Mar e do Ambiente (MARE), Faculdade de Ciências da Universidade de Lisboa, Campo Grande, 1749-016 Lisboa, Portugal; macurto@fc.ul.pt (M.C.); pmfelix@fc.ul.pt (P.M.F.); jaheumuller@fc.ul.pt (J.H.); fmafonso@fc.ul.pt (F.A.); dmlvmd@gmail.com (D.D.); pmchainho@fc.ul.pt (P.C.); acbrito@fc.ul.pt (A.B.); fmvribeiro@gmail.com (F.R.)

[2] Centre for Energy and Environment Research (CINEA), Campus do IPS—Estefanilha, Polytechnic Institute of Setúbal, 2910-761 Setúbal, Portugal

[3] Departamento de Biologia Vegetal, Faculdade de Ciências da Universidade de Lisboa, Campo Grande 016, 1749-016 Lisboa, Portugal

[4] Departamento de Biologia Animal (DBA) & Centro de Ecologia, Evolução e Alterações Ambientais (cE3c), Faculdade de Ciências da Universidade de Lisboa, Campo Grande, 1749-016 Lisboa, Portugal; rfaustinol@gmail.com

* Correspondence: asnogueirac@gmail.com

† Presented at the IX Iberian Congress of Ichthyology, Porto, Portugal, 20–23 June 2022.

‡ Presenting author (Oral communication).

Citation: Nogueira, S.; Curto, M.; Félix, P.M.; Heumüller, J.; Afonso, F.; Dias, D.; Chainho, P.; Brito, A.; Lima, R.; Ribeiro, F. Using DNA Metabarcoding to Uncover the Diet of the Invasive Mozambique Tilapia (*Oreochromis mossambicus*) in Mangroves from the Island of São Tomé. *Biol. Life Sci. Forum* **2022**, *13*, 132. https://doi.org/10.3390/blsf2022013132

Academic Editor: Alberto Teodorico Correia

Published: 20 June 2022

Publisher's Note: MDPI stays neutral with regard to jurisdictional claims in published maps and institutional affiliations.

Abstract: Biological invasions are considered one of the main extinction drivers of native species worldwide. Invasive species have detrimental effects on local ecosystems by means of competition, predation, habitat modification and nutrient cycling, as well as disease spreading. Along with ecological impacts, there are socio-economic consequences to human populations dependent on the services provided by these ecosystems. One of the most fundamental steps towards understanding the influence of invasive species is to determine their role on the local food web. The Mozambique tilapia (*Oreochromis mossambicus*) is an extremely aggressive opportunistic feeder that has a high biological and ecological plasticity, including flexible reproductive strategies (including mouthbrooding) and tolerance to a wide range of temperature, salinity and dissolved oxygen conditions. Its biological traits coupled with being used in aquaculture have made it a successful invader, widely distributed outside its native range. Yet, its populations have rarely been studied in the wild; hence, its potential impacts remain largely unknown. In this study, we investigate the diet of the invasive Mozambique tilapia in two mangroves of the oceanic island of São Tomé. We applied metabarcoding to simultaneously identify multiple taxa in tilapia gut content samples, using high-throughput sequencing. To achieve a greater taxonomic coverage, we combined the use of two barcodes, the 18S ribosomal RNA and the cytochrome C oxidase subunit I genes, to target phytoplankton and animal species, respectively. We found a total of 251 amplicon sequence variants belonging to 96 taxa. The results revealed diet differences between specimens from the two mangrove locations, not only regarding the level of biodiversity, but also in the frequencies of the occurrence of certain functional groups. Some taxa, such as diatoms, green algae and rotifers were found in the gut contents at both mangroves, whereas others, such as arthropods and mollusks, were almost exclusive to one of them. These findings provide useful insights into the ecological implications of the biological invasion of vulnerable island ecosystems, offering some specific guidance on how to minimize the impact of tilapia on the mangroves of the São Tomé Island.

Keywords: biological invasions; trophic niche; island; diet; COI; 18S

Funding: This study was funded by the Critical Ecosystem Partnership Fund (CEPF) "Participatory Management of Malanza and Praia das Conchas Mangroves in São Tomé" (GFWA-2018-LG-02); This study had the support of national funds through Fundação para a Ciência e Tecnologia (FCT), under the project LA/P/0069/2020 granted to the Associate Laboratory ARNET and through the Centre for Ecology, Evolution and Environmental Changes's (cE3c) Unit funding (UIDB/00329/2020). A. B. was supported by the Scientific Employment Stimulus Programme (CEECIND/00095/2017), funded by Fundação para a Ciência e a Tecnologia (FCT).

Institutional Review Board Statement: Not applicable.

Informed Consent Statement: Not applicable.

Data Availability Statement: The data presented in this study are available on request from the corresponding author. The data are not publicly available due to privacy restrictions.

Conflicts of Interest: The authors declare no conflict of interest.

Abstract

Anguillicola crassus Infection in Different Ecosystems of the Southwestern European Area [†]

Elisabeth Faliex [1,*,‡], Carlos Antunes [2], Agnès Bardonnet [3], Anna Costarrosa [4], Estibaliz Diaz [5,‡], Ramón De Miguel [6], Isabel Domingos [7], Carlos Fernandez-Delgado [6], Mercedes Herrera [6], Maria Korta [5], Raphaël Lagarde [1], Manon Mercader [1], Rui Monteiro [7], Ana Moura [2], Teresa Portela [7], François Prellwitz [1], Noemie Regli [1], Jacques Rives [3], Gaël Simon [1], Lluís Zamora Hernández [4] and Elsa Amilhat [1]

[1] UMR 5110 CNRS-UPVD, Laboratoire CEFREM, Université of Perpignan Via Domitia, 66100 Perpignan, France; raphael.lagarde@univ-perp.fr (R.L.); mercader.manon@gmail.com (M.M.); francois.prellwitz@migado.fr (F.P.); noemie.regli@gmail.com (N.R.); gsimon@univ-perp.fr (G.S.); elsa.amilhat@univ-perp.fr (E.A.)

[2] CIIMAR—Interdisciplinary Centre of Marine and Environmental Research, Universidade do Porto, 4450-208 Matosinhos, Portugal; cantunes@ciimar.up.pt (C.A.); anacatarinamoura@hotmail.com (A.M.)

[3] UMR 1224, Ecobiop, Aquapôle, Pôle Gest'Aqua, Inrae/ UPPA, 64310 St Pée sur Nivelle, France; agnes.bardonnet@inrae.fr (A.B.); jacques.rives@inrae.fr (J.R.)

[4] GRECO, Instituto de Ecología Acuática, Facultad de Ciencias, Universidad de Girona, 17003 Girona, Spain; anna.costarrosa@udg.edu (A.C.); lluis.zamora@udg.edu (L.Z.H.)

[5] AZTI, Sustainable Fisheries Management, Basque Research and Technology Alliance (BRTA), 48395 Sukarrieta, Spain; ediaz@azti.es (E.D.); mkorta@azti.es (M.K.)

[6] Departamento de Zoología, Universidad de Córdoba, 14071 Córdoba, Spain; ramon@guadalictio.es (R.D.M.); ba1fedec@uco.es (C.F.-D.); zo2hearm@uco.es (M.H.)

[7] Departamento de Biologia Animal, Centro de Ciências do Mar e do Ambiente (MARE), Faculdade de Ciências, Universidade de Lisboa, 1600-548 Lisboa, Portugal; idomingos@fc.ul.pt (I.D.); rmmonteiro@fc.ul.pt (R.M.); tportelas@fc.ul.pt (T.P.)

* Correspondence: faliex@univ-perp.fr

† Presented at the IX Iberian Congress of Ichthyology, Porto, Portugal, 20–23 June 2022.

‡ Presenting author (Poster).

Citation: Faliex, E.; Antunes, C.; Bardonnet, A.; Costarrosa, A.; Diaz, E.; De Miguel, R.; Domingos, I.; Fernandez-Delgado, C.; Herrera, M.; Korta, M.; et al. *Anguillicola crassus* Infection in Different Ecosystems of the Southwestern European Area. *Biol. Life Sci. Forum* **2022**, *13*, 141. https://doi.org/10.3390/blsf2022013141

Academic Editor: Alberto Teodorico Correia

Published: 21 June 2022

Publisher's Note: MDPI stays neutral with regard to jurisdictional claims in published maps and institutional affiliations.

Abstract: The European eel is critically endangered (IUCN, 2020) due to several factors (climate change, barriers to migration, pollution, unsustainable exploitation, illegal trading and poaching, and pathogens). However, the relative effect of these factors can hardly be quantified. *Anguillicola crassus* is a nematode known as one of the most harmful parasites for the European eel. Being hematophagous and inducing serious alterations to the swimbladder, it can reduce swimming performances and thus may impair silver eel migration and reproduction. Since its introduction (in the early 1980s) from Asia to Europe, this invasive parasite has spread rapidly through its new host's geographic range. In that context, an extensive study was conducted in 2018 in seven pilot basins (one lagoon and two rivers on the Mediterranean side and four rivers on the Atlantic side) in the SUDOANG project to investigate the epidemiological status of *A. crassus* and the parameters influencing its distribution. The eel population was monitored in a harmonized way through a sampling network using standardized data collection methods. We sampled 36 to 96 eels (yellow and silver stages) in each pilot basin by electrofishing or fyke nets. We identified significant differences of *A. crassus* occurrence between basins, which were partly explained by eel life traits (density, feeding activity) and habitat characteristics (temperature, salinity). This study contributes to quantifying and better understanding the mechanisms of infection of *A. crassus* in different ecosystems and to increasing the awareness of its possible negative effects on the eels of the Sudoe Area.

Keywords: *Anguillicola crassus*; European eel; swimbladder invasive parasite; prevalence; SDI; Southwestern European area

Author Contributions: Conceptualization, E.F., E.A., G.S. and E.D.; methodology, E.F., E.A. and G.S.; field work, R.L., M.M., F.P., N.R., G.S., E.A., E.F., C.A., A.M., E.D., M.K., A.B., J.R., R.D.M., C.F.-D., M.H., A.C., L.Z.H., I.D., R.M. and T.P.; validation, E.F., E.A., G.S. and E.D.; formal analysis, E.F., E.A. and R.L.; investigation, R.L., M.M., F.P., N.R., G.S., E.A., E.F., C.A., A.M., E.D., M.K., A.B., J.R., R.D.M., C.F.-D., M.H., A.C., L.Z.H., I.D., R.M. and T.P.; resources, E.F., E.A., C.A., E.D., A.B., C.F.-D., A.C. and I.D.; data curation, E.F., E.A., C.A., E.D., A.B., C.F.-D., A.C. and I.D.; writing—original draft preparation, E.F., E.A.; writing—review and editing, R.L., M.M., F.P., N.R., G.S., E.A., E.F., C.A., A.M., E.D., M.K., A.B., J.R., R.D.M., C.F.-D., M.H., A.C., L.Z.H., I.D., R.M. and T.P.; visualization, E.F., E.A., and R.L.; supervision, E.F., E.A. and E.D.; project administration, E.D.; funding acquisition, E.D. All authors have read and agreed to the published version of the manuscript.

Funding: This research was co-financed by the INTERREG SUDOE Programme through the European Regional Development Fund (ERDF).

Institutional Review Board Statement: Procedures were conducted in accordance with the guidelines of the French Ethical Committee for animal experiments by the licensed scientist working under the authority and approval of the certificat d'experimenter sur les animaux vertébrés vivants (experimental animal certificate) number 66.0801 (Elisabeth Faliex) of the CEFREM, University of Perpignan.

Informed Consent Statement: Not applicable.

Data Availability Statement: The datasets generated during and/or analysed during the current study are available from the corresponding author on reasonable request.

Conflicts of Interest: The authors declare no conflict of interest.

Abstract

On the Efficacy of a Pre-Filtering Density Separation Method for Microplastic Analysis [†]

Sara Feiteira [1,2,‡], Felipe Abrunhosa [2], Luana S. Corona [2,3], Joana I. Robalo [2] and Rita Castilho [2,3,*]

[1]   MARE—Marine and Environmental Sciences Centre, ISPA Instituto Universitário, Rua Jardim do Tabaco 34, 1149-041 Lisboa, Portugal; feiteirasara@gmail.com
[2]   Centre of Marine Sciences (CCMAR), Campus de Gambelas, 8005-139 Faro, Portugal; fe.abrunhosa@gmail.com (F.A.); luanacorona02@gmail.com (L.S.C.); jrobalo@ispa.pt (J.I.R.)
[3]   Campus de Gambelas, University of the Algarve, 8005-139 Faro, Portugal
*   Correspondence: rcastil@ualg.pt
†   Presented at the IX Iberian Congress of Ichthyology, Porto, Portugal, 20–23 June 2022.
‡   Presenting author (poster).

Abstract: Microplastics (particles with a diameter between 1 and 5 mm) in the marine environment are a growing concern due to their involuntary ingestion by fish and other marine species. The small microplastic size makes them easily consumed in the water and hence readily introduced into the marine food chain, with yet unknown bioaccumulative and toxic consequences. The proximity to urban areas, industrial activities, and sewage disposal potentially increases the presence of microplastics in the marine environment. The gastrointestinal tract (GIT) of some species contains high quantities of debris, sediment, and non-digestible materials such as calcium carbonate resulting from their dietary or behavioral habits. This study aims to assess the efficacy of a pre-filtering density separation method using a hypersaline solution to facilitate the subsequent filtration procedure. This additional step is expected to accelerate the procedure as a whole, improving the filtering process and ensuring a more accurate detection of microplastics.

Keywords: filtration; hypersaline solution; non-digestible materials; gastrointestinal tract

Citation: Feiteira, S.; Abrunhosa, F.; Corona, L.S.; Robalo, J.I.; Castilho, R. On the Efficacy of a Pre-Filtering Density Separation Method for Microplastic Analysis. *Biol. Life Sci. Forum* **2022**, *13*, 4. https://doi.org/10.3390/blsf2022013004

Academic Editor: Alberto Teodorico

Published: 1 June 2022

Publisher's Note: MDPI stays neutral with regard to jurisdictional claims in published maps and institutional affiliations.

Author Contributions: Conceptualization, R.C. and J.I.R.; methodology, F.A., S.F., L.S.C.; investigation, S.F., F.A., L.S.C.; resources, R.C.; writing—original draft preparation, S.F.; writing—review and editing, R.C. and J.I.R.; supervision, R.C. and J.I.R. All authors have read and agreed to the published version of the manuscript.

Funding: This study received Portuguese national funds from FCT-Foundation for Science and Technology through projects MARE/UIDB/MAR/04292/2020 granted to MARE (MARE-ISPA) and UIDB/04326/2020, UIDP/04326/2020 and LA/P/0101/2020 granted to CCMAR.

Institutional Review Board Statement: Not applicable.

Informed Consent Statement: Not applicable.

Data Availability Statement: Not applicable.

Conflicts of Interest: The authors declare no conflict of interest.

Abstract

Emerging Trends in the Eastern Cantabrian Small-Scale Fishery [†]

Eneko Bachiller *, Maria Mateo [‡], Estanis Mugerza, Arantza Murillas, Maria Korta and Lucía Zarauz

AZTI, Sustainable Fisheries Management, Basque Research and Technology Alliance (BRTA), Txatxarramendi Ugartea z/g, 48395 Sukarrieta, Bizkaia (Basque Country), Spain; mmateo@azti.es (M.M.); emugerza@azti.es (E.M.); amurillas@azti.es (A.M.); mkorta@azti.es (M.K.); lzarauz@azti.es (L.Z.)
* Correspondence: ebachiller@azti.es
† Presented at the IX Iberian Congress of Ichthyology, Porto, Portugal, 20–23 June 2022.
‡ Presenting author (Oral communication).

Abstract: The general picture of most harbours in the Basque region (eastern Cantabrian coast) has changed during the last few decades, suggesting a decline in small-scale fisheries (SSF) activity which is carried out by vessels with LOA < 15 m. However, little is known in detail about this change, i.e., the recent development of the different fleet segments, or temporal changes regarding landed species. The present study shows that during the last decade (2010–2020) trolling lines targeting albacore in summer and especially handline fishery targeting mackerel in spring was the most important seasonal SSF in the region. Moreover, these fisheries intensified the fishing effort and showed the highest landings, especially for mackerel. In contrast, a decline in vessel numbers as well as fishing efforts and therefore landings was observed for netters, i.e., gillnets and trammel nets. The use of longlines and pots did not show any time trend. Regarding targeted species, the mean fish length landed by both long-liners and netters decreased with time, and so did their fish length-based niche breadth, indicating a lower length range in landed fish. In contrast, while fish diversity landed by hand-liners decreased, probably due to the mackerel fishing intensification, netters targeted a wider variety of small fish species. Technical optimization, probably related to specific market demands, suggests that Basque SSF fleet are shifting to specialized hookers, i.e., seasonal mackerel and albacore fishing, while netters, which are declining in number, are landing a wider range of target species. Given that knowledge on SSF has been trapped in a data-poor cycle, due to the lower importance in data collection when compared to other commercial fleets, understanding such developments might contribute to future management plans on a regional scale.

Keywords: SSF; fleet segmentation; landings; target fish species; fish length; fish diversity; fishing optimization; market demands

Citation: Bachiller, E.; Mateo, M.; Mugerza, E.; Murillas, A.; Korta, M.; Zarauz, L. Emerging Trends in the Eastern Cantabrian Small-Scale Fishery. *Biol. Life Sci. Forum* **2022**, *13*, 10. https://doi.org/10.3390/blsf2022013010

Academic Editor: Alberto Teodorico Correia

Published: 2 June 2022

Publisher's Note: MDPI stays neutral with regard to jurisdictional claims in published maps and institutional affiliations.

Author Contributions: Conceptualization, E.B.; methodology, E.B.; Formal analysis, investigation and resources, E.B., M.M., E.M.; data curation, E.B.; writing-original draft preparation, E.B, M.M., E.M., A.M., M.K., L.Z.; visualization, E.B.; supervision, E.M., A.M., L.Z.; project administration, E.M., L.Z; funding acquisition, E.M., L.Z. All authors have read and agreed to the published version of the manuscript.

Funding: This research received no external funding.

Institutional Review Board Statement: Not applicable.

Informed Consent Statement: Not applicable.

Data Availability Statement: The data presented in this study will be openly available in dryad repository.

Conflicts of Interest: The authors declare no conflict of interest.

biology and life sciences forum

Abstract

Potential Impacts in the Gilthead Seabream Larviculture by Nodavirus [†]

Miguel Ángel García-Álvarez [1], Laura Cervera [1], Carmen González-Fernández [1], Marta Arizcun [2], José Carlos Campos-Sánchez [1,‡], María Ángeles Esteban [1], Elena Chaves-Pozo [2] and Alberto Cuesta [1,*]

[1] Immunobiology for Aquaculture Group, Department of Cell Biology and Histology, Faculty of Biology, Regional Campus of International Excellence "Campus Mare Nostrum", University of Murcia, 30100 Murcia, Spain; miguelangel.garciaa1@um.es (M.Á.G.-Á.); laura.cerveram@um.es (L.C.); carmen.gonzalez1@um.es (C.G.-F.); josecarlos.campos@um.es (J.C.C.-S.); aesteban@um.es (M.Á.E.)

[2] Centro Oceanográfico de Murcia, Instituto Español de Oceanografía, Consejo Superior de Investigaciones Científicas, IEO-CSIC, Carretera de la Azohía s/n, Puerto de Mazarrón, 30860 Murcia, Spain; marta.arizcun@ieo.es (M.A.); elena.chaves@ieo.es (E.C.-P.)

[*] Correspondence: alcuesta@um.es; Tel.: +34-868-884-536

[†] Presented at the IX Iberian Congress of Ichthyology, Porto, Portugal, 20–23 June 2022.

[‡] Presenting author (Poster presentation).

Abstract: The nervous necrosis virus (NNV) leads to viral encephalopathy and retinopathy (VER) disease in more than 170 fish species, mainly from marine habitats. It replicates in the central nervous tissues, reaching up to 100% mortalities after a few days of infection, mainly in the larvae and juvenile stages. This is continuously spreading and affecting more species, both wild and cultured, posing a risk to the development of the aquaculture industry. In the Mediterranean Sea, it mainly affects European sea bass (*Dicentrarchus labrax*) and some grouper species (*Epinephelus* spp.). Interestingly, in the gilthead seabream (*Sparus aurata*), typically resistant to common NNV strains, great mortalities in hatcheries associated with typical clinical signs of VER have been confirmed to be caused by RGNNV/SJNNV reassortants. Thus, we have evaluated the susceptibility of seabream larvae to either RGNNV/SJNNV or SJNNV/RGNNV reassortants, as well as the larval immunity. Based on our results we can conclude that: (i) gilthead seabream larvae are susceptible to infection with both NNV reassortant genotypes, but mainly to RGNNV/SJNNV; (ii) virus replicated and infective particles were isolated; (iii) larval immunity was correlated with larval survival; and (iv) larval resistance and immunity were correlated with age of the larvae. Further investigations should be carried out to ascertain the risks of these new pathogens to Mediterranean larviculture.

Keywords: Nodavirus; reassortants; virus; gilthead seabream; larvae; immunity; aquaculture

Citation: García-Álvarez, M.Á.; Cervera, L.; González-Fernández, C.; Arizcun, M.; Campos-Sánchez, J.C.; Esteban, M.Á.; Chaves-Pozo, E.; Cuesta, A. Potential Impacts in the Gilthead Seabream Larviculture by Nodavirus. *Biol. Life Sci. Forum* **2022**, *13*, 12. https://doi.org/10.3390/blsf2022013012

Academic Editor: Alberto Teodorico Correia

Published: 2 June 2022

Publisher's Note: MDPI stays neutral with regard to jurisdictional claims in published maps and institutional affiliations.

Author Contributions: Methodology and investigation, M.Á.G.-Á., L.C., J.C.C.-S., C.G.-F., M.A., E.C.-P.; software and data curation, M.Á.G.-Á. and A.C.; conceptualization and supervision, E.C.-P. and A.C.; writing—original draft preparation, M.Á.G.-Á.; writing—review and editing, E.C.-P., M.Á.G.-Á., M.A., M.Á.E. and A.C.; project administration and funding acquisition, M.A., E.C.-P., A.C. All authors have read and agreed to the published version of the manuscript.

Funding: This work was funded by Ministerio de Ciencia e Innovación-Agencia Estatal de Investigación [MCIN/AEI/10.13039/501100011033, grant PID2019-105522GB-I00 to AC], Ministerio de Economía y Competitividad and FEDER [grant RTI2018-096625-B-C33 to ECP] and Fundación Séneca, Grupo de Excelencia de la Región de Murcia [19883/GERM/15].

Institutional Review Board Statement: All specimens were handled in accordance with the Guidelines of the European Union Council (2010/63/EU) and the optimal zootechnical protocols of the *Centro Oceanográfico de Murcia* (IEO-CSIC, REGA ES300261040017), the Bioethical Committee of the

IEO-CSIC and the approval of the Ministry of Water, Agriculture and Environment of the Autonomous Community Region of Murcia (Permit number A13200602).

Informed Consent Statement: Not applicable.

Data Availability Statement: Not applicable.

Conflicts of Interest: The authors declare no conflict of interest.

Abstract

Optimizing Size Selectivity and Catch Patterns for Hake (*Merluccius merluccius*) and Blue Whiting (*Micromesistius poutassou*) by Combining Square Mesh Panel and Codend Designs [†]

Elsa Cuende [1,*,‡], Manu Sistiaga [2,3], Bent Herrmann [4,5,6] and Luis Arregi [1]

1. Marine Research, AZTI Basque Research and Technology Alliance (BRTA), 48395 Sukarrieta, Spain; larregi@azti.es
2. Fish Capture Division, Institute of Marine Research, 5817 Bergen, Norway; manu.sistiaga@hi.no
3. Department of Marine Technology, Norwegian University of Science and Technology, 7491 Trondheim, Norway
4. Fisheries and New Biomarine Industry, SINTEF Ocean, Fishing Gear Technology, 9850 Hirtshals, Denmark; bent.herrmann@sintef.no
5. The Norwegian College of Fishery Science, UiT The Arctic University of Norway, 9037 Tromsø, Norway
6. Fisheries Technology, DTU Aqua, Denmark Technical University, 9850 Hirtshals, Denmark
* Correspondence: ecuende@azti.es; Tel.: +34-671-700-519
† Presented at the IX Iberian Congress of Ichthyology, Porto, Portugal, 20–23 June 2022.
‡ Presenting author (Oral communication).

Citation: Cuende, E.; Sistiaga, M.; Herrmann, B.; Arregi, L. Optimizing Size Selectivity and Catch Patterns for Hake (*Merluccius merluccius*) and Blue Whiting (*Micromesistius poutassou*) by Combining Square Mesh Panel and Codend Designs. *Biol. Life Sci. Forum* **2022**, *13*, 37. https://doi.org/10.3390/blsf2022013037

Academic Editor: Alberto Teodorico Correia

Published: 6 June 2022

Publisher's Note: MDPI stays neutral with regard to jurisdictional claims in published maps and institutional affiliations.

Abstract: Gear modifications in fisheries are usually implemented to obtain catch patterns that meet management objectives. In the Basque bottom trawl fishery, gear regulations include the use of a square mesh panel (SMP) placed at the top panel of the extension piece of the trawl to supplement diamond mesh codend selectivity. However, the catch patterns obtained with this combination have raised concern among scientists and authorities. This study combines new data on different SMP and codend designs with existing data from the literature to produce new results that are applied to predict the size selectivity and catch patterns of different gear combinations for a variety of fishing scenarios. A systematic approach based on the concept of treatment trees was outlined and applied to depict the effect of individual and combined gear design changes on size selectivity and catch patterns for hake (*Merluccius merluccius*) and blue whiting (*Micromesistius poutassou*). This approach led to identification of the gear combination with the most appropriate exploitation pattern for these two species and improved the readability and interpretation of selectivity results. The results demonstrated that changes both in SMP and, especially, codend designs have a significant effect on hake and blue whiting size selectivity and catch patterns. We believe the results obtained provide new insight about the potential of gear modification to improve exploitation patterns in trawl fisheries and illustrate the suitability of the methodology used to explore more sustainable gear designs.

Keywords: gear selectivity; catch patterns; treatment trees; square mesh panel; bottom trawl

Author Contributions: Conceptualization: E.C., M.S. and B.H.; Methodology: E.C., B.H., M.S. and L.A.; Software: B.H.; Validation: M.S. and B.H.; Formal analysis: E.C.; Investigation: E.C., M.S. and L.A.; Resources: L.A.; Data curation: E.C.; Writing—original draft: E.C.; Writing—review & editing: E.C., M.S. and B.H.; Visualization: E.C.; Supervision: M.S., B.H. and L.A.; Project administration and Funding acquisition: L.A. All authors have read and agreed to the published version of the manuscript.

Funding: This research was funded by Secretaria General de Pesca (MAPA), by the "Acuerdo Marco".

Institutional Review Board Statement: Not applicable.

Informed Consent Statement: Not applicable.

Data Availability Statement: All relevant data are within the published manuscript.

Conflicts of Interest: The authors declare no conflict of interest.

224

Abstract

Exploring Parasitic Load in European Sardine: Applying Two Methodological Approaches along the Catalan Coast [†]

Marta Caballero-Huertas [1,‡], **Xènia Frigola-Tepe** [1,‡], **Marta Muñoz** [1], **Simonetta Mattiucci** [2] and **Jordi Viñas** [3,*]

[1] Department of Environmental Sciences, Institute of Aquatic Ecology (IEA), Universitat de Girona (UdG), Campus Montilivi, 17003 Girona, Spain; marta.caballero@udg.edu (M.C.-H.); xenia.frigola@udg.edu (X.F.-T.); marta.munyoz@udg.edu (M.M.)

[2] Department of Public Health and Infectious Diseases, Section of Parasitology, Sapienza-University of Rome, 00185 Rome, Italy; simonetta.mattiucci@uniroma1.it

[3] Genetic Ichthyology Laboratory (LIG), Department of Biology, Universitat de Girona (UdG), Campus Montilivi, 17003 Girona, Spain

[*] Correspondence: jordi.vinas@udg.edu

[†] Presented at the IX Iberian Congress of Ichthyology, Porto, Portugal, 20–23 June 2022.

[‡] Presenting author (oral communication), they participated equally in this study.

Abstract: The European sardine, *Sardina pilchardus* (Walbaum, 1792), is a cold-temperate water species from the Clupeid family. This small pelagic has a key functional role in the marine ecosystem along its distributional range in the northeast of the Atlantic Ocean, the Sea of Marmara, the Black Sea and the Mediterranean Sea. Furthermore, sardine is one of the most important commercial fishery resources caught by the purse seine fleet in the Mediterranean Sea. There is reported a decline in annual sardine catches in the Northwestern Mediterranean Sea, as well as smaller mean total length and sexual maturation size of the individuals, mainly attributed to increased water temperature and overfishing, which have a negative impact on energetic body condition, growth and reproduction. However, little is known about the impact of parasitism on sardine health status. In this work we analyzed the incidence of parasites in sardines from the Catalan Coast of the Northwestern Iberian Peninsula, using two approaches: visual inspection by stereo microscope and the UV-press method, based on the autofluorescence of certain parasites, along a complete reproductive cycle. The main parasite found was the nematode *Hysterothylacium* sp., although other species were identified by visual inspection (i.e., *Contracaecum* sp., digenea trematodes). No *Anisakis* spp. were observed under either of the methodologies. Using the former method, we detected 35.3 % of individuals infected by, at least, one nematode parasite with a mean intensity of 1.68 %. With the UV-press technique, we quantified a prevalence of 16.5 %. In both cases, the Southern Catalan Coast presented a larger number of parasitized sardines, which was remarkably higher in spring. Combined methodologies are suggested to more precisely detect parasites, since visual inspection allows a more detailed study of the viscera, but UV-press also allows the study of parasites that may be present in the musculature. Furthermore, genetic validation should be performed in order to accurately determine the parasite species found.

Keywords: *Hysterothylacium* sp.; nematode; *Sardina pilchardus*; UV-press

Funding: We want to thank the funding support of the Spanish Ministry of Science, Innovation and Universities (RTI2018-097544-B-I00, 'ConSarVar', I+D+i Retos de Investigación), as well as the Fons Europeu Marítim i de la Pesca (FEMP) and the Agency for Management of University and Research Grants (AGAUR) of the Generalitat de Catalunya (Ajuts per a la contractació de personal investigador novell (FI-2020)).

Institutional Review Board Statement: Not applicable.

Informed Consent Statement: Not applicable.

Data Availability Statement: Not applicable.

Conflicts of Interest: The authors declare no conflict of interest.

biology and life sciences forum

MDPI

Abstract

Historical Evolution of the Reconstructed Catches of Four Species of the *Pagellus* Genus for Two Large Marine Ecosystems †

Víctor Sanz-Fernández *,‡, Juan Carlos Gutiérrez-Estrada and Inmaculada Pulido-Calvo

Departamento de Ciencias Agroforestales, Escuela Técnica Superior de Ingeniería, Universidad de Huelva, 21007 Huelva, Spain; juanc@uhu.es (J.C.G.-E.); ipulido@uhu.es (I.P.-C.)
* Correspondence: victor.sanz@dcaf.uhu.es
† Presented at the IX Iberian Congress of Ichthyology, Porto, Portugal, 20–23 June 2022.
‡ Presenting author (oral communication).

Abstract: *Pagellus acarne*, *Pagellus bellottii*, *Pagellus bogaraveo* and *Pagellus erythrinus* are sparids distributed throughout Large Marine Ecosystems (LMEs), the Iberian Coastal region (25) and Canary Current region (27). They are target species due to their important commercial value to local and international fleets from three different continents: Africa, Asia and Europe. Given the high exploitation interest of these species, sustainable management of the resource is essential. For this reason, a key element for its implementation is the knowledge of the historical behaviour of catches by geolocalised areas. To this end, marine catches reconstructed in total wet-weight tonnes from 1950 to 2014 from the Sea Around Us database were analysed. A total of 2,058,172.60 tn of species of the *Pagellus* genus were caught for the entire region, of which 83.20% (1,712,552.21 tn) corresponded to the Canary Current area and the remaining 16.79% (345,620.38 tn) to the Coastal Iberian area. The dominant area was Canary Current; its catches were higher than those of the Coastal Iberian area, with an annual average percentage of 78.21%. Overall, the fishery showed a negative trend of −511.37 tn/year. In terms of species, 61.52% of the catches were of *Pagellus bellottii* (1,266,219.36 tn), 20.04% of *Pagellus sp* (not identified at species level, only to genus) (412,482.53 tn), 8.91% of *Pagellus erythrinus* (183,434.67 tn), 6.74% of *Pagellus bogaraveo* (138,717.29 tn) and the remaining 2.78% of *Pagellus acarne* (57,318.74 tn). Our results suggest the existence of important variations in the reconstructed catches of the four species analysed in two large marine ecosystems, showing an overall decreasing behaviour. Canary Current was undoubtedly the region with the highest fishing pressure during the 65 years analysed and *Pagellus bellotii* was the dominant species in the Current Canary region and in the whole region. This multispecies analysis presented could help the development of sustainable management protocols by providing insight into the historical evolution and status of the reconstructed catches for large marine ecosystems.

Keywords: demersal species; LMEs; Iberian Coastal; Canary Current; time series

Citation: Sanz-Fernández, V.; Gutiérrez-Estrada, J.C.; Pulido-Calvo, I. Historical Evolution of the Reconstructed Catches of Four Species of the *Pagellus* Genus for Two Large Marine Ecosystems. *Biol. Life Sci. Forum* **2022**, *13*, 53. https://doi.org/10.3390/blsf2022013053

Academic Editor: Alberto Teodorico Correia

Published: 7 June 2022

Publisher's Note: MDPI stays neutral with regard to jurisdictional claims in published maps and institutional affiliations.

Author Contributions: Conceptualization and formal analysis, V.S.-F. and J.C.G.-E.; methodology, software, data curation, visualization and writing—original draft preparation, V.S.-F.; validation, investigation and writing—review and editing, V.S.-F., J.C.G.-E. and I.P.-C.; resources and supervision, J.C.G.-E. and I.P.-C.; project administration and funding acquisition, J.C.G.-E. All authors have read and agreed to the published version of the manuscript.

Funding: This research was funded by the Government of Spain Ministry of Science, Innovation and Universities with a FPU fellowship, grant number (FPU17/04298).

Institutional Review Board Statement: Not applicable.

Informed Consent Statement: Not applicable.

Data Availability Statement: All data are available from www.seaaroundus.org. Data was downloaded on 11 March and 1 April 2019 with version 47.1 Sea Around Us database.

Conflicts of Interest: The authors declare no conflict of interest.

Abstract

Seasonal Comparison of Length-Weight Ratio of Sea Bass (*Dicentrarchus labrax*), in Two Types of Aquaculture Facilities [†]

África Oliver-Blanco [1], Amadora Rodríguez-Ruiz [2], Lourdes Encina [2], Juan Ramón Cid-Quintero [1] and Carlos Orduna [1,*,‡]

[1] EcoFishUS Research, 41009 Seville, Spain; africa@ecofishus.com (Á.O.-B.); juancid@ecofishus.com (J.R.C.-Q.)
[2] Departamento de Biología Vegetal y Ecología, Universidad de Sevilla, 41089 Seville, Spain; dora@us.es (A.R.-R.); lencina@us.es (L.E.)
* Correspondence: carlos@ecofishus.com
† Presented at the IX Iberian Congress of Ichthyology, Porto, Portugal, 20–23 June 2022.
‡ Presenting author (poster presentation).

Abstract: Sea bass (*Dicentrarchus labrax*) is one of the most important fish species in European aquaculture, and it is farmed in different types of facilities and environmental conditions, which may imply differences in the development and the condition of the farmed fish. In this study, we compare the seasonal length–weight relationships of the fish, in two different types of facilities. Some fish were farmed in the Atlantic Ocean, in cages located offshore; some fish were farmed in ponds, in the Guadalquivir River estuary. Fish collected for sale from both facilities were measured and weighed seasonally for one year. Additionally, environmental factor such as water temperature, salinity, and oxygen concentration were measured. These differences in the length–weight ratio are very important in the use of hydroacoustics and computer vision techniques for estimating fish biomass in aquaculture, because they provide good estimates of length, which must be converted to weight with suitable conversion equations. The results of this study show that fish farmed in both facilities presents differences in the length–weight ratio in the different climatic seasons. Comparing the measurements of the fish from the two facilities, those raised in offshore cages, with lower water temperature and higher salinity, exhibited a shorter standard length but a greater weight compared with their size. In contrast, fish farmed in ponds, with higher water temperatures and lower salinity, exhibited a longer standard length, but a lower weight in relation to their length.

Keywords: length–weight; sea bass; *Dicentrarchus labrax*

Citation: Oliver-Blanco, Á.; Rodríguez-Ruiz, A.; Encina, L.; Cid-Quintero, J.R.; Orduna, C. Seasonal Comparison of Length-Weight Ratio of Sea Bass (*Dicentrarchus labrax*), in Two Types of Aquaculture Facilities. *Biol. Life Sci. Forum* **2022**, *13*, 124. https://doi.org/10.3390/blsf2022013124

Academic Editor: Alberto Teodorico Correia

Published: 17 June 2022

Publisher's Note: MDPI stays neutral with regard to jurisdictional claims in published maps and institutional affiliations.

Author Contributions: Conceptualization, A.R.-R. and L.E.; methodology, A.R.-R., L.E., C.O. and J.R.C.-Q.; software, J.R.C.-Q.; validation, A.R.-R. and L.E.; formal analysis, C.O. and Á.O.-B.; investigation, J.R.C.-Q., Á.O.-B. and C.O.; resources, C.O.; data curation, Á.O.-B.; writing—original draft preparation, C.O.; writing—review and editing, A.R.-R. and L.E.; visualization, A.R.-R. and L.E.; supervision, A.R.-R. and L.E.; project administration, C.O.; funding acquisition, C.O. All authors have read and agreed to the published version of the manuscript.

Funding: This research received no external funding.

Institutional Review Board Statement: Not applicable.

Informed Consent Statement: Not applicable.

Data Availability Statement: Not applicable.

Conflicts of Interest: The authors declare no conflict of interest.

Abstract

Impact of Fisheries on Allis Shad's (*Alosa alosa* L.) Spawning Population Structure in the Mondego River [†]

Ana Filipa Belo [1,*,‡], Catarina Sofia Mateus [1], Bernardo Ruivo Quintella [2,3], Esmeralda Pereira [1], André Moreira [1], Roberto Oliveira [1], Carlos Batista [4] and Pedro Raposo de Almeida [1,5]

[1] MARE—Centro de Ciências do Mar e do Ambiente/ARNET—Aquatic Research NETwork, Universidade de Évora, 7002-554 Évora, Portugal; cspm@uevora.pt (C.S.M.); ecdp@uevora.pt (E.P.); andre.moreira@uevora.pt (A.M.); rlpo@uevora.pt (R.O.); pmra@uevora.pt (P.R.d.A.)

[2] MARE—Centro de Ciências do Mar e do Ambiente/ARNET—Aquatic Research NETwork, Faculdade de Ciências da Universidade de Lisboa, 1749-016 Lisboa, Portugal; bsquintella@fc.ul.pt

[3] Departamento de Biologia Animal, Faculdade de Ciências da Universidade de Lisboa, 1749-016 Lisboa, Portugal

[4] Departamento de Recursos Hídricos e Departamento do Litoral e Proteção Costeira, Agência Portuguesa do Ambiente, I.P., 2610-124 Amadora, Portugal; carlos.batista@apambiente.pt

[5] Departamento de Biologia, Escola de Ciências e Tecnologias, Universidade de Évora, 7002-554 Évora, Portugal

* Correspondence: affadb@uevora.com

† Presented at the IX Iberian Congress of Ichthyology, Porto, Portugal, 20–23 June 2022.

‡ Presenting author (Oral communication).

Citation: Belo, A.F.; Mateus, C.S.; Quintella, B.R.; Pereira, E.; Moreira, A.; Oliveira, R.; Batista, C.; Almeida, P.R.d. Impact of Fisheries on Allis Shad's (*Alosa alosa* L.) Spawning Population Structure in the Mondego River. *Biol. Life Sci. Forum* **2022**, *13*, 126. https://doi.org/10.3390/blsf2022013126

Academic Editor: Alberto Teodorico Correia

Published: 17 June 2022

Publisher's Note: MDPI stays neutral with regard to jurisdictional claims in published maps and institutional affiliations.

Abstract: Allis shad (*Alosa alosa* L.) populations have suffered declines and even regional extinction across their distribution range, mainly due to river impoundment, overexploitation, and pollution. In Portugal, the species is classified as Endangered (EN). This fish is regarded as a valuable delicacy, and commercial fisheries dedicated to this species in Portugal are found in rivers in Minho, Lima, Cávado, Douro, Vouga, Mondego, Tagus, and Guadiana. Official landings state that, in the last 10 years, around 10 t of allis shad were sold annually at the Figueira da Foz fish market. Fisheries are selective since the allowed mesh size for the nets employed does not capture smaller fish. Moreover, larger fish are more desirable, attaining higher prices per kg at fish auctions and thus generating higher revenue for the fishermen. The fishing pressure on this threatened resource in the Mondego is significant and bound to impact its population structure. With the present work, we assess the impact of fishing pressure and gear selectivity on the dimensional structure of the Mondego's shad spawning population, one of the last strongholds for allis shads, and identify possible consequences for the future of the species. The dimensional structure and gender proportion of over 800 shads sampled at Figueira da Foz fish auction from 2015 to 2019 was analyzed. All shads sampled were between 400 mm and 745 mm in total length, with an average of 592 mm. Simultaneously, we used images recorded in the monitoring window at Açude-Ponte Dam's fishway in Coimbra to study the shads reaching the upstream spawning areas in order to check for differences in the dimensional structure possibly related to fishing pressure. The results obtained aim to increase the existing biological knowledge of this population and the human pressure it is subjected to, thus contributing to future management efforts.

Keywords: fishing pressure; dimensional structure; conservation; monitoring

Author Contributions: Conceptualization, A.F.B., B.R.Q., P.R.d.A.; methodology, A.F.B., B.R.Q., P.R.d.A.; formal analysis, A.F.B., B.R.Q., P.R.d.A.; investigation, A.F.B., C.S.M., E.P., R.O., A.M., B.R.Q., P.R.d.A.; resources, C.B., P.R.d.A.; data curation, A.F.B., C.S.M., C.B.; writing–original drafts presentation, A.F.B.; writing–review and editing, A.F.B., C.S.M., E.P., R.O., A.M., B.R.Q., C.B., P.R.d.A.; supervision, B.R.Q., P.R.d.A.; project administration, C.B., P.R.d.A.; funding acquisition, B.R.Q., P.R.d.A. All authors have read and agreed to the published version of the manuscript.

Funding: The present study was supported by the project ANADROMOS–Plano Operacional de Monitorização e Gestão de Peixes Anádromos em Portugal (MAR-01.03.02-FEAMP-0002) funded by the European Fisheries Fund, and by the Portuguese Environment Agency (APA) through the "Coimbra fish pass monitoring program". In addition, this study had the support of FCT through the strategic project UIDB/04292/2020 awarded to MARE and through project LA/P/0069/2020 granted to the Associate Laboratory ARNET". FCT also supported this study through the individual contract attributed to Bernardo R. Quintella (2020.02413.CEECIND and the PhD scholarships attributed to Ana Filipa Belo (SFRH/BD/123434/2016) and Esmeralda Pereira (SFRH/BD/121042/2016).

Institutional Review Board Statement: Not applicable.

Informed Consent Statement: Informed consent was obtained from all subjects involved in the study.

Data Availability Statement: Data is available from correspondence author, upon reasonable request.

Conflicts of Interest: The authors declare no conflict of interest.

biology and life sciences forum 

Abstract

Integrated System of Red Seaweed *Kappaphycus alvarezii* (Rhodophyta, Solieriaceae) and Clownfish *Amphiprion ocellaris* (Perciformes, Pomacentridae): An Experimental Study [†]

Alejandra Filippo Gonzalez Neves dos Santos *,[‡], Nathalia da Silva Lucarevschi, Eshefison Rodrigues Batista and Simone Gomes Ferreira

Laboratory of Applied Ecology, Department of Zootechny and Sustainable Socioenvironmental Development, Fluminense Federal University (UFF), Rua Vital Brazil Filho, 64, Niterói CEP 24230-340, Brazil; nathalialucarevschi@id.uff.br (N.d.S.L.); esthefison2015@gmail.com (E.R.B.); simonegomes@id.uff.br (S.G.F.)
* Correspondence: afilippo@gmail.com; Tel.: +55-21-2629-9518
† Presented at the IX Iberian Congress of Ichthyology, Porto, Portugal, 20–23 June 2022.
‡ Presenting author (Poster presentation).

Abstract: This study aimed to analyse the integrated system of red seaweed *Kappaphycus alvarezii* and clownfish *Amphiprion ocellaris*. Experiments were performed in a 45 L aquarium under the following controlled conditions: temperature 22 °C, dissolved oxygen 5 mg/L, salinity 30, and photoperiod 12 h light. There were 10 fish per aquarium, fed with commercial food. Four treatments were tested (three replicates per treatment): (1) only fish (control); (2) fish with 250 g seaweed; (3) fish with 500 g seaweed; and (4) fish with 750 g seaweed. Weekly seaweed and fish were weighed and measured, and ammonia (NH_3), nitrite (NO_2), nitrate (NO_3^-), phosphate (PO_4^{3-}), and phosphorus (P) were analyzed. The total weight (g) (TW) of fish differed over time (F = 0.00) and between treatments (F = 2.84; p = 0.03). Higher TW ($\pm$0.75 g) occurred in treatment 3 and after 14 days did not significantly differ from the end of the experiment (p < 0.01). The total length of fish (cm) did not differ significantly over time (F = 0.64; p = 0.66) or between treatments (F = 0.34; p = 0.79). The relative growth rate of seaweed (RGR) did not differ over time (p > 0.12). There was a significant interaction between the nutrients present in the water by treatment and time (F = 3.59; p = 0.00). The nutrient removal efficiency (NRE) of NH_3 and NO_2 was higher in treatments 3 and 1, respectively, reaching $\pm$30% at 14 days (p < 0.001), and from the 21st day of the experiment, the values were close to zero, with no significant difference between treatments up to 35 days (p > 0.90 for both). The NRE of NO_3^- increased by 2% at 28 days of the experiment, with no significant difference between treatments (p > 0.07). The NRE of P was $\pm$25% at 14 days and was not different between treatments 1 and 2. The NRE of PO_4^{3-} reached 14% at 21 days for treatments 1, 2 and 3 (p < 0.001). The control showed lower values for the NRE of NH3, NO_3^-, P, and PO_4^{3-} compared to the treatments, but the NRE of NO_2 increased in 14 days. It is concluded that a greater abundance of *K. alvarezii* helps in water quality and contributes to the total weight of clownfish between 14 and 21 days.

Keywords: water quality; sustainability; coral reef fish

Author Contributions: Methodology, A.F.G.N.d.S.; N.d.S.L.; E.R.B.; and S.G.F.; Formal analyses, A.F.G.N.d.S.; Writing—original draft preparation, A.F.G.N.d.S. and N.d.S.L.; and Writing—review and editing, A.F.G.N.d.S. All authors have read and agreed to the published version of the manuscript.

Funding: This research was funded by FAPERJ, grant number E26/202826/2019.

Institutional Review Board Statement: CEUA—UFF n.115416092.

Informed Consent Statement: Informed consent was obtained from all subjects involved in the study.

Data Availability Statement: Not applicable.

Conflicts of Interest: The authors declare no conflict of interest.

Abstract

Husbandry Procedure Effects on Brood-Stock Gilthead Seabream's Heart Rate Housed under Enriched and Bare Environments [†]

María J. Cabrera-Álvarez [1,2,*,‡], Ana Rita Oliveira [1,2], Florbela Soares [3], Ana Candeias-Mendes [3], Pablo Arechavala-Lopez [1,4] and Joao L. Saraiva [1,2]

[1] FishEthoGroup, 8700-1590 Olhao, Portugal; ana@fishethogroup.net (A.R.O.); arechavala@imedea.uib-csic.es (P.A.-L.); joao@fishethogroup.net (J.L.S.)
[2] CCMAR—Centro de Ciências do Mar, Universidade do Algarve, 8005-139 Faro, Portugal
[3] IPMA—Instituto Português do Mar e da Atmosfera, 8700-305 Olhao, Portugal; fsoares@ipma.pt (F.S.); ana.mendes@ipma.pt (A.C.-M.)
[4] IMEDEA (CSIC/UIB)—Institut Mediterrani d'Estudis Avançats (Consejo Superior de Investigaciones Científicas/Universitat de les Illes Balears), 07190 Mallorca, Spain
[*] Correspondence: maria@fishethogroup.net
[†] Presented at the IX Iberian Congress of Ichthyology, Porto, Portugal, 20–23 June 2022.
[‡] Presenting author (oral communication).

Abstract: Husbandry procedures, albeit essential for good welfare, can be stressful for captive individuals. Therefore, being aware of the physiological effects of these procedures and reducing stress during regular maintenance is of pivotal importance to ensure outstanding welfare. Environmental enrichment can be an asset to animal keepers since it has many benefits on captive animals, including reducing stress in many aquatic species. We studied whether structural enrichment had a positive effect on brood-stock gilthead seabream (*Sparus aurata*) during four husbandry procedures. We studied the stress levels of the subjects by measuring their heart rate with an internal bio-logger (DST milli HRT, Star-Oddi) surgically implanted in 18 fish. These fish were distributed in six 3000 L cylindrical tanks, and housed with seven more fish in the tanks, which made a total of 10 fish per tank. Three of the tanks had an environmental enrichment structure consisting of a 1 m^2 floating structure with 9 hanging organic ropes, while the other three tanks had no enrichment. Fish were exposed to their housing setting for five months. After this environmentally enriched/bare period, we carried out feeding, netting, and cleaning each day for three consecutive days, and a formaldehyde bath on the fourth day in logger-implanted fish, and continued recording their recovery for eight more days. We expect the husbandry procedures to evoke a stress response in all the subjects by increasing their heart rate, and the fish housed in enriched environments to have a reduced heart rate and to recover faster from the stressors compared to the fish housed in bare tanks.

Keywords: welfare; stress resilience; environmental enrichment; gilthead seabream; heart rate; bio-loggers; brood-stock; husbandry procedures; precision fish farming

Citation: Cabrera-Álvarez, M.J.; Oliveira, A.R.; Soares, F.; Candeias-Mendes, A.; Arechavala-Lopez, P.; Saraiva, J.L. Husbandry Procedure Effects on Brood-Stock Gilthead Seabream's Heart Rate Housed under Enriched and Bare Environments. *Biol. Life Sci. Forum* **2022**, *13*, 21. https://doi.org/10.3390/blsf2022013021

Academic Editor: Alberto Teodorico Correia

Published: 6 June 2022

Publisher's Note: MDPI stays neutral with regard to jurisdictional claims in published maps and institutional affiliations.

Author Contributions: Conceptualization, M.J.C.-Á., F.S., P.A.-L. and J.L.S.; methodology, M.J.C.-Á., A.R.O. and A.C.-M.; software, M.J.C.-Á.; validation, M.J.C.-Á.; formal analysis, M.J.C.-Á.; investigation, M.J.C.-Á., P.A.-L. and J.L.S.; resources, M.J.C.-Á., P.A.-L. and J.L.S.; data curation, M.J.C.-Á.; writing—original draft preparation, M.J.C.-Á.; writing—review and editing, M.J.C.-Á., A.R.O., P.A.-L. and J.L.S.; visualization, M.J.C.-Á.; supervision, F.S., P.A.-L. and J.L.S.; project administration, F.S., P.A.-L. and J.L.S.; funding acquisition, F.S., P.A.-L. and J.L.S. All authors have read and agreed to the published version of the manuscript.

Funding: This study received Portuguese national funds from the FCT-Foundation for Science and Technology through project UIDB/04326/2020 and doctorate grant UI/BD/151304/2021, and DIVERSIAQUA II (MAR-02.01.01-FEAMP-0175).

Institutional Review Board Statement: The experiment complied with the Guidelines of the European Union Council (Directive 2010/63/EU) and Portuguese legislation for the use of laboratory animals, and was conducted at Centre of Marine Sciences (CCMAR) facilities (Faro, Portugal). CCMAR facilities and their staff are certified to house and conduct experiments with live animals (Group-C licences by the Direção Geral de Alimentação e Veterinária, Ministério da Agricultura, Florestas e Desenvolvimento Rural, Portugal).

Informed Consent Statement: Not applicable.

Data Availability Statement: Not applicable.

Conflicts of Interest: The authors declare no conflict of interest.

Abstract

Are There Differential Roles in the Parental Behaviour of the Chameleon Cichlid *Australoheros facetus*? †

Gonçalo Oliveira ‡, João L. Saraiva, Nuno F. Jesus and Pedro M. Guerreiro *

Centre of Marine Sciences (CCMAR), University of Algarve, 8005-139 Faro, Portugal; a61229@ualg.pt (G.O.); jsaraiva@ualg.pt (J.L.S.); nunofjesus@gmail.com (N.F.J.)

* Correspondence: pmgg@ualg.pt
† Presented at the IX Iberian Congress of Ichthyology, Porto, Portugal, 20–23 June 2022.
‡ Presenting author (Poster presentation).

Citation: Oliveira, G.; Saraiva, J.L.; Jesus, N.F.; Guerreiro, P.M. Are There Differential Roles in the Parental Behaviour of the Chameleon Cichlid *Australoheros facetus*? *Biol. Life Sci. Forum* **2022**, *13*, 34. https://doi.org/10.3390/blsf2022013034

Academic Editor: Alberto Teodorico Correia

Published: 6 June 2022

Publisher's Note: MDPI stays neutral with regard to jurisdictional claims in published maps and institutional affiliations.

Abstract: Cichlids are social fishes well known for their complex behaviour. The chameleon cichlid, *Autraloheros facetus*, is native to South American river drainages and is currently established in several Mediterranean-type drainages in southern Portugal as an invasive species. Their high local recruitment, territorially, and parental care activities are possible advantages in competing with native fish and achieving high reproductive success. The main objective of this work was to characterise the behaviours of the males and females of the species *A. facetus* upon pair formation and their roles during parental behaviour at different stages of offspring development for the purpose of gathering important basic knowledge on fish biology to apply in the control of the species' populations. To attain this initial goal, we used observation techniques and video recording protocols to characterise the specific activities performed by each individual during reproductive and parental stages (pair formation, eggs, attached larvae, and free-swimming larvae), identifying the main tasks, and assessing the time spent on each task by each member of the reproductive pair. The breeding pairs were obtained as a result of the social hierarchy formed in each social group: groups of six individuals of similar size were placed in individual tanks fitted with a bottom biological filter, in which the temperature was increased to 24 °C at an expanded photoperiod. Social behaviours were recorded and characterised, expanded on a previously established ethogram. The results obtained so far allow us to establish a set of aggressive behaviours towards other fish (striking, chasing, biting), nest preparing behaviours (digging, cleaning), caring (caring, fetching), and guarding behaviours towards the offspring (hovering, patrolling). Concerning the rate of occurrence of social behaviours in randomised 5 min periods (frequency of specific behaviours per each 5 min period) show that in the pair formation stage, males present a more aggressive nature with frequent occurrence of striking ($r_{striking}$ = 1) and biting (r_{biting} = 0.31). On the other hand, at the egg stage, the rate of occurrence of parental behaviours shows the dominance of females to prepare the nest ($r_{digging}$ = 0.91) and caring for the eggs (r_{caring} = 3.64), while males are more vigilant ($r_{parental\ hovering}$ = 0.95). At the attached larvae stage, this tendency continues with higher occurrence rates: males focus on vigilance ($r_{parental\ hovering}$ = 1.6) and patrolling ($r_{patrolling}$ = 2), and females care for the recently hatched larvae ($r_{fetching}$ = 4). Finally, at the free-swimming larvae stage, there is a turn and a small decrease in the occurrence of these activities: males oversee nest maintenance ($r_{digging}$ = 1.09), and females patrol the tank ($r_{patrolling}$ = 0.71) and care for the larvae ($r_{fetching}$ = 2.02). These preliminary data suggest differential roles for male and female *A. facetus*, that evolve during parental behaviour, for which further experimental paradigms will be designed to explore underlying proximate causes.

Keywords: ethogram; fish behaviour; fish reproduction; invasive species; video recordings

Funding: Funded by Fundação para a Ciência e Tecnologia (FCT) (FCT-UIDB/04326/2022).

Institutional Review Board Statement: Not applicable.

Informed Consent Statement: Not applicable.

Data Availability Statement: Not applicable.

Conflicts of Interest: The authors declare no conflict of interest.

biology and life sciences forum

Abstract

Effects of River Salinization on Freshwater Fish Behavior—Cerebral Lateralization, Activity, Boldness, and Schooling [†]

Tamara Leite [1,2,3,*,‡], Paulo Branco [1,2], Cristina Canhoto [3], Maria Teresa Ferreira [1,2] and José Maria Santos [1,2]

[1] Forest Research Centre (CEF)—School of Agriculture, University of Lisbon, 1349-017 Lisbon, Portugal; pjbranco@isa.ulisboa.pt (P.B.); terferreira@isa.ulisboa.pt (M.T.F.); jmsantos@isa.ulisboa.pt (J.M.S.)
[2] Associated Laboratory TERRA, 1349-017 Lisbon, Portugal
[3] Centre of Functional Ecology (CFE), Department of Life Sciences, University of Coimbra, 3004-531 Coimbra, Portugal; ccanhoto@ci.uc.pt
[*] Correspondence: tamaraleite@e-isa.ulisboa.pt
[†] Presented at the IX Iberian Congress of Ichthyology, Porto, Portugal, 20–23 June 2022.
[‡] Presenting author (oral communication).

Abstract: Rivers are experiencing increasing salinization due to anthropogenic disturbances, and salinity has been shown to negatively affect freshwater fish behavioral expression, potentially disrupting ecological processes. In this study, the aim was to determine the sublethal effects of secondary salinization (anthropic in origin) on freshwater fish behavior, using a widespread native cyprinid species, the Iberian barbel (*Luciobarbus bocagei*), as the model species. Behavioral trials were performed in a mesocosm setting, focusing on fish cerebral lateralization, routine activity, boldness, and schooling behavior. The impact of salinity stress was assessed by exposing the barbels to three levels of a salinity gradient—Control (no salt added to the water, 0.8–0.9 mS/cm), Low (9 mS/cm), and High concentration (18–19 mS/cm). Behavioral parameters were recorded every three minutes by visual observation. Our results show that, with increased salinity in the flume channels, fish were less active, and formed less cohesive shoals. Moreover, individuals became bolder, since a higher number of attempts to escape their environment was recorded in greater salinity levels. Laterality of the population appeared to become more evident, as fish revealed a tendency in their decision making to turn left more frequently. Behavioral changes in fish caused by salinization stress should be further researched regarding other freshwater species with different tolerances, in addition to their interaction with other environmental stressors. This broader approach would allow us to recognize salinity thresholds, and also understand the true scope of the consequences of salinization for fish species.

Keywords: secondary salinization; behavior; boldness; cerebral lateralization; activity; shoaling cohesion; cyprinids

Citation: Leite, T.; Branco, P.; Canhoto, C.; Ferreira, M.T.; Santos, J.M. Effects of River Salinization on Freshwater Fish Behavior—Cerebral Lateralization, Activity, Boldness, and Schooling. *Biol. Life Sci. Forum* **2022**, *13*, 77. https://doi.org/10.3390/blsf2022013077

Academic Editor: Alberto Teodorico Correia

Published: 9 June 2022

Publisher's Note: MDPI stays neutral with regard to jurisdictional claims in published maps and institutional affiliations.

Author Contributions: T.L.: Methodology, Investigation, Formal analysis, Data curation, Writing—original draft. P.B.: Supervision, Methodology, Conceptualization, Investigation, Writing—review & editing. C.C.: Supervision, Writing—review editing; M.T.F.: Writing—review & editing. J.M.S.: Conceptualization, Methodology, Investigation, Writing—review & editing. All authors have read and agreed to the published version of the manuscript.

Funding: This study was supported by the MARS project (Managing Aquatic ecosystems and water Resources under multiple Stress) funded by the European Commission under Framework Program 7 (contract no. 603378). Tamara Leite was supported by a Ph.D. grant from the FLUVIO–River Restoration and Management programme funded by Fundação para a Ciência e a Tecnologia I. P. (FCT), Portugal (UI/BD/15052/2021). Paulo Branco is supported by a FCT grant (SFRH/BPD/94686/2013) and "Norma Transitória—DL57/2016/CP1382/CT0020". José Maria Santos was funded by a post-doctoral grant (MARS/BI/2/2014) from the MARS project (http://www.mars-project.eu/) and is presently recipient of a FCT researcher contract (IF/00020/2015). The study was partially funded by the project Dammed Fish (PTDC/CTA-AMB/4086/2021). CEF is a research unit funded by Fundação para a Ciência e a Tecnologia (FCT), Portugal (UIDB/00239/2020).

Institutional Review Board Statement: The Institute for Nature Conservation and Forests (ICNF) provided the necessary fishing and handling permits. All research was conducted in accordance with national and international guidelines for animal welfare.

Informed Consent Statement: Not applicable.

Data Availability Statement: Data is available upon request from the authors.

Conflicts of Interest: The authors declare no conflict of interest.

biology and life sciences forum

Abstract

Behaviour of Fish in Bottom-Trawling Gear to Assess the Effectiveness of Cetacean Excluder Devices and Codend Selectivity Modifications [†]

Julio Valeiras [1,*], Nair Vilas [1,‡], Esther Abad [1], José Carlos Fernández-Franco [2], Mateo Barreiro [2], Camilo Saavedra [1] and Eva Velasco [1]

1   Centro Nacional Instituto Español de Oceanografía-CSIC, CO-Vigo, Subida a Radio Faro 50-52, 36390 Vigo, Spain; nair.vilas@ieo.csic.es (N.V.); esther.abad@ieo.csic.es (E.A.); camilo.saavedra@ieo.csic.es (C.S.); eva.velasco@ieo.csic.es (E.V.)
2   OPPF-4 Organización de Productores de Pesca Fresca del Puerto de Vigo, Edificio Ramiro Gordejuela, Puerto Pesquero s/n, 36202 Vigo, Spain; josecarlos@arvi.org (J.C.F.-F.); mateobarreirosim@gmail.com (M.B.)
*   Correspondence: julio.valeiras@ieo.csic.es
†   Presented at the IX Iberian Congress of Ichthyology, Porto, Portugal, 20–23 June 2022.
‡   Presenting author (Poster presentation).

Abstract: The selective retention of fish is a consequence of size and species-dependent fish behaviour during the trawling process. The observation of fish reactions to trawl gear is critical to understanding the behavioural mechanisms responsible for trawl selectivity and to develop future trawl gear for research. In demersal trawling, there is a need to develop more species-selective trawls to minimize discarding and bycatch in multispecies fisheries. This requires observational tools that can operate at depths and light levels encountered by the commercial fleets and with the ability to quantify the herding and capture efficiency by species and age groups of such gears. A range of optical and acoustic observation techniques has been developed over the past few decades to assist in these goals. Work with underwater cameras in fishing gear varies from the simple observation of the presence of certain species during capture to the study of the complex interactions of fishing gear and species during the trawling process. Previous work seems to show that there are different response patterns in the general behaviour of some fish species when entering a trawl. One of these behaviours is that the vertical preference in the trawl cavity differs between species. Several authors have complemented behavioural studies carried out at sea with experimental studies to further explore the interaction between fish and fishing gear. Some authors indicate that the vertical preferences of some fish species change as they move through the net towards the codend. This implies that behavioural selection can potentially change along the horizontal axis of the fishing net. A description of the bycatch (cetaceans, sharks, and skates) and fish behaviour within bottom trawling gear was given through the use of underwater cameras. The objective was to determine the most suitable fishing gear configurations to prevent the escape of the fishing catch during fishing trials carried out to test the operation of dolphin-exclusion devices and selective codends. The behaviours recorded by the cameras were classified into seven types of behaviour: orientation, reaction, resistance, re-entry, final entry, escape attempts, and panic reaction.

Keywords: fish behaviour; fisheries; bycatch; selectivity; cetaceans; technology

Citation: Valeiras, J.; Vilas, N.; Abad, E.; Fernández-Franco, J.C.; Barreiro, M.; Saavedra, C.; Velasco, E. Behaviour of Fish in Bottom-Trawling Gear to Assess the Effectiveness of Cetacean Excluder Devices and Codend Selectivity Modifications. *Biol. Life Sci. Forum* **2022**, *13*, 79. https://doi.org/10.3390/blsf2022013079

Academic Editor: Alberto Teodorico Correia

Published: 13 June 2022

Publisher's Note: MDPI stays neutral with regard to jurisdictional claims in published maps and institutional affiliations.

Author Contributions: J.V. conceived the original idea, performed the analysis and interpretation of data, and wrote the manuscript. N.V., E.A., J.C.F.-F., M.B. and E.V. participate in work at sea and video visualization. C.S. provided critical feedback. All authors have read and agreed to the published version of the manuscript.

Funding: The research received no external funding.

Institutional Review Board Statement: No ethical approval was required as this study is non-experimental.

Informed Consent Statement: Not applicable.

Data Availability Statement: All relevant data are within the published manuscript.

Conflicts of Interest: The authors declare no conflict of interest.

biology and life sciences forum

Abstract

Effects of Environmental Enrichment on the Welfare of Gilthead Seabream Broodstock [†]

Ana Rita Oliveira [1,2,*,‡], María J. Cabrera-Álvarez [1,2], Florbela Soares [3], Ana Candeias-Mendes [3], Pablo Arechavala-Lopez [1,4] and João L. Saraiva [1,2]

[1]　FishEthoGroup Association, 8700-159 Olhão, Portugal; maria@fishethogroup.net (M.J.C.-Á.); arechavala@imedea.uib-csic.es (P.A.-L.); joao@fishethogroup.net (J.L.S.)
[2]　CCMAR—Centro de Ciências do Mar, Universidade do Algarve, 8005-139 Faro, Portugal
[3]　EPPO—Aquaculture Research Station, IPMA—Portuguese Institute of the Sea and the Atmosphere, 8700-305 Olhão, Portugal; fsoares@ipma.pt (F.S.); ana.mendes@ipma.pt (A.C.-M.)
[4]　IMEDEA (CSIC/UIB)—Institut Mediterrani d'Estudis Avançats, 07190 Mallorca, Spain
*　Correspondence: ana@fishethogroup.net
†　Presented at the IX Iberian Congress of Ichthyology, Porto, Portugal, 20–23 June 2022.
‡　Presenting author (oral presentation).

Citation: Oliveira, A.R.; Cabrera-Álvarez, M.J.; Soares, F.; Candeias-Mendes, A.; Arechavala-Lopez, P.; Saraiva, J.L. Effects of Environmental Enrichment on the Welfare of Gilthead Seabream Broodstock. *Biol. Life Sci. Forum* **2022**, *13*, 90. https://doi.org/10.3390/blsf2022013090

Academic Editor: Alberto Teodorico Correia

Published: 13 June 2022

Publisher's Note: MDPI stays neutral with regard to jurisdictional claims in published maps and institutional affiliations.

Abstract: The intensification of aquaculture practices in the last decade has led to the reduction in welfare of farmed fish. Recently, one of the tools that has been considered important to guarantee or improve the welfare of captive fish is the application of environmental enrichment (EE). The physiological state and behaviour of fish can be used as indicators of the welfare of the animal, as well as of the positive impact of the EE in their well-being. In this study, behavioural and physiological indicators were measured to assess the effects of structural environmental enrichment on the welfare of gilthead seabream broodstock. Over the course of 5 months, 60 adult seabreams were distributed in six 3000 L cylindrical tanks. Three of the tanks were enriched with nine hanging organic ropes on 1 m^2 floating structures, while the other three tanks had no enrichment. Fish were filmed regularly before, during, and after feeding, cleaning, and sampling procedures. Operational welfare indicators (OWIs) recently developed for farmed seabream were used and adapted to build an ethogram for the broodstock behaviour analysis. According to our results, fish reared in enriched tanks hardly schooled and presented a more independent swimming activity compared to fish from non-enriched tanks. Moreover, structural enrichment seemed to increase the spatial use of the bottom of the tank, and promoted seabream natural behaviour (hierarchical competitions, foraging, etc.). In addition, fish in enriched tanks presented a higher growth rate, and further studies will determine if such enrichment structures also affect reproductive potential of seabream broodstocks as well as epigenetic effects on offspring.

Keywords: welfare; environmental enrichment; gilthead seabream; broodstock; behaviour; operational welfare indicators

Author Contributions: Conceptualization, A.R.O., P.A.-L., F.S. and J.L.S.; methodology, A.R.O., M.J.C.-Á., A.C.-M.; software, A.R.O.; validation, A.R.O.; formal analysis, A.R.O.; investigation, A.R.O., M.J.C.-Á., P.A.-L. and J.L.S.; resources, A.R.O., M.J.C.-Á., P.A.-L. and J.L.S.; data curation, A.R.O.; writing—original draft preparation, A.R.O.; writing—review and editing, A.R.O., M.J.C.-Á., P.A.-L. and J.L.S.; visualization, A.R.O.; supervision, F.S., P.A.-L. and J.L.S.; project administration, F.S., P.A.-L. and J.L.S.; funding acquisition, F.S., P.A.-L. and J.L.S. All authors have read and agreed to the published version of the manuscript.

Funding: This study received Portuguese national funds from the FCT-Foundation for Science and Technology through project UIDB/04326/2020 and doctorate grant UI/BD/151304/2021, and DIVERSIAQUA II (MAR-02.01.01-FEAMP-0175).

Institutional Review Board Statement: The experiment complied with the Guidelines of the European Union Council (Directive 2010/63/EU) and Portuguese legislation for the use of laboratory animals, was conducted at Estação Piloto Píscicola Experimental de Olhão (EPPO) facilities from IPIMAR (Olhão, Portugal). IPIMAR/EPPO facilities and their staff are certified to house and conduct experiments with live animals (Group-C licences by the Direção Geral de Alimentação e Veterinária, Ministério da Agricultura, Florestas e Desenvolvimento Rural, Portugal).

Informed Consent Statement: Not applicable.

Data Availability Statement: Not applicable.

Conflicts of Interest: The authors declare no conflict of interest.

Abstract

Evaluating Repetitive Mucus Extraction Effects on Mucus Biomarkers, Mucus Cells and Skin-Barrier Status in a Marine Fish Model [†]

Ignasi Sanahuja [1,2,*], Pedro M. Guerreiro [3,‡], Albert Girons [4], Laura Fernández-Alacid [1] and Antoni Ibarz [1]

[1]   Department of Cell Biology, Physiology and Immunology, Faculty of Biology, University of Barcelona, Avda. Diagonal 643, 08028 Barcelona, Spain; fernandez_alacid@ub.edu (L.F.-A.); tibarz@ub.edu (A.I.)
[2]   Aquaculture Program, Institute of Agrifood Research and Technology (IRTA), 43540 La Ràpita, Spain
[3]   CCMAR-Centre for Marine Sciences, University of Algarve, 8005-139 Faro, Portugal; pmgg@ualg.pt
[4]   Ictiovet-Veterinary Services for the Aquaculture, 08025 Barcelona, Spain; albert.girons@ictiovet.com
*   Correspondence: isanahuja@ub.edu
†   Presented at the IX Iberian Congress of Ichthyology, Porto, Portugal, 20–23 June 2022.
‡   Presenting author (poster presentation).

Abstract: Among all the mucosal barriers, the skin and its surrounding mucus are possibly the main defensive tools used by fish against the environment. Due to its less-invasive extraction, the study of its production and functions has gained high interest in the last years. However, there are still many gaps in research, such as the possible alteration of mucus composition or the skin integrity resulting from the sampling process. In the current study, skin mucus extraction impacts were determined by comparing the effects of one-single extraction (SEG; single extraction group) and three successive extractions (REG; repetitive extractions group, separated by 4 days) on mucus properties and on skin epithelial integrity. In terms of analytical evaluation, plasma biomarkers and plasma antibacterial capacity were also determined. With regard skin histology and skin barrier properties, both SEG and REG did not evidence differences with respect to intact skin (ØEG). Interestingly, the repetitive mucus extraction protocol seemed to activate skin mucus turnover, significantly increasing the number of low-size mucus cells (cell area < 100 μm^2) and reducing the number of high-size mucus cells (cell area > 150 μm^2). Repetitive extraction of skin mucus diminished the amounts of soluble protein and glucose in mucus with regard to one-single extraction and increased cortisol exudation. These metabolites remained unaltered in plasma, indicating the different response among both sampling targets. Despite mucus biomarkers modification, the antibacterial capacity against the pathogenic bacterial (*P. anguilliseptica* and *V. anguillarum*) was maintained in both plasma and mucus irrespective of the number of mucus extractions. Overall, the mucus sampling protocol scarcely affected skin integrity and mucus antibacterial properties and only modified metabolites exudation, evidencing a feasible and minimally invasive method for studying fish health and welfare as an alternative or as a complement to plasma. The knolwdege provided here highlighted that this methodology is putatively transferable to farm culture conditions and showed that it is very useful for the study of threatened species aimed at preserving fish welfare.

Keywords: skin mucus; mucus barrier; biomarkers; antibacterial activity

Author Contributions: Conceptualization, I.S., P.M.G., L.F.-A. and A.I.; methodology, I.S., P.M.G., L.F.-A. and A.G.; validation, I.S., P.M.G., L.F.-A. and A.I.; formal analysis, I.S., P.M.G., L.F.-A. and A.I.; investigation, I.S., P.M.G., L.F.-A. and A.I.; resources, P.M.G. and A.I.; data curation, I.S. and L.F.-A.; writing—original draft preparation, I.S.; writing—review and editing, I.S., P.M.G., L.F.-A. and A.I.; funding acquisition, A.I. All authors have read and agreed to the published version of the manuscript.

Funding: This research was funded by means of grants from the Spanish Government: PID2019-106878RB-I00 and IS was granted with a Postdoctoral fellowship (FJC2020-043933-I).

Institutional Review Board Statement: The study was conducted according to the guidelines of the European Union Council (Directive 2010/63/EU) and Portuguese legislation for the use of laboratory animals (Decreto-Lei n.º 113/2013 de 7 de Agosto), and approved by the Institutional Review Board of CCMAR, Adelino V. M. Canário, on 3 April 2021.

Informed Consent Statement: Not applicable.

Data Availability Statement: Not applicable.

Conflicts of Interest: The authors declare no conflict of interest.

biology and life sciences forum

MDPI

Abstract

Into the Wild: A New Approach to the Aquaculture Production of Brown Trout (*Salmo trutta* L.) to Enhance Restocking Success [†]

Andreia Domingues [1,*,‡], Carlos M. Alexandre [1], Catarina S. Mateus [1], Sara Silva [1], Joana Pereira [1] and Pedro R. Almeida [1,2]

[1] MARE-Centro de Ciências do Mar e do Ambiente/ARNET-Rede de Investigação Aquática, Universidade de Évora, 7004-516 Évora, Portugal; cmea@uevora.pt (C.M.A.); cspm@uevora.pt (C.S.M.); ssrs@uevora.pt (S.S.); joana_gomespereira@hotmail.com (J.P.); pmra@uevora.pt (P.R.A.)

[2] Departamento de Biologia, Escola de Ciências e Tecnologias, Universidade de Évora, 7004-516 Évora, Portugal

[*] Correspondence: afdomingues@uevora.pt; Tel.: +351-913-233-492

[†] Presented at the IX Iberian Congress of Ichthyology, Porto, Portugal, 20–23 June 2022.

[‡] Presenting author (Oral communication).

Citation: Domingues, A.; Alexandre, C.M.; Mateus, C.S.; Silva, S.; Pereira, J.; Almeida, P.R. Into the Wild: A New Approach to the Aquaculture Production of Brown Trout (*Salmo trutta* L.) to Enhance Restocking Success. *Biol. Life Sci. Forum* **2022**, *13*, 115. https://doi.org/10.3390/blsf2022013115

Academic Editor: Alberto Teodorico Correia

Published: 17 June 2022

Publisher's Note: MDPI stays neutral with regard to jurisdictional claims in published maps and institutional affiliations.

Abstract: The brown trout (*Salmo trutta* L.) is one of the most iconic native species from European river ecosystems and is also one of the main species of interest for recreational fishing activities (e.g., fly fishing). This species has a significant potential to attract anglers and related investment to the main fishing grounds, which are usually located in poorly developed areas and away from main city centers. Due to its environmental and socioeconomic value, this species is often targeted by management programs directed to the protection and sustainable exploitation of this valuable natural resource. One of the most common actions to enhance the abundance and condition of trout populations is the restocking of wild populations with fish from aquaculture facilities. However, most fish come from fishfarms using production methods such as high densities in the tanks, use of commercial food and standardized feeding methods, lack of environmental stimulus, and domestication of breeding stocks, which usually results in poor fitness and very low survival rates after release. This consequently leads to reduced success of these management actions. To contribute to solving these problems and enhancing the success of restocking actions for the recovery and sustainable enhancement of wild trout populations, we propose a novel approach to the production of this species, by testing and implementing a new protocol that aims to produce wild-reared trout. These fish come from wild breeders and will be produced with the least human contact in conditions that mimic their natural habitat. Taking advantage of a recently remodeled and re-equipped aquaculture facility, located in Central Portugal (Posto Aquícola de Campelo, Figueiró dos Vinhos), we are rearing trout in low densities (10–20 trout/m^3), like the ones observed in natural habitats, using live food (larvae and insects), and subjected to environmental stimuli such as refuges, and water and flow variability, equivalent to those observed in local streams. Accompanied by a pre- and post-restocking monitoring program (e.g., evaluation of trout abundances and habitat in the target stream, dispersion and movement patterns and survival), this study will contribute to enhancing the success of future restocking actions, promoting the sustainable enhancement of wild trout populations, and, thus, increasing the interest of restocked fishing grounds for angling activities and associated incomes.

Keywords: *Salmo trutta*; wild-rearing; restocking and management; recreational fisheries

Author Contributions: Conceptualization, C.M.A., C.S.M. and P.R.A.; methodology, A.D., C.M.A. and P.R.A.; formal analysis, A.D., C.M.A., C.S.M. and S.S.; investigation, A.D., C.M.A., C.S.M., S.S. and J.P.; resources, C.M.A. and P.R.A.; writing-original draft preparation, A.D. and C.M.A.; writing-review and editing, A.D., C.M.A., C.S.M., S.S., J.P. and P.R.A.; supervision, C.M.A. and P.R.A.; project administration, C.M.A. and P.R.A.; funding acquisition, C.M.A., C.S.M. and P.R.A. All authors have read and agreed to the published version of the manuscript.

Funding: The present study was supported by European Funds (EMFF – European Maritime and Fisheries Fund), more specifically, by the Operational Program MAR2020 (project CRER: MAR-02.01.01-FEAMP-0106). Support was also provided by the Portuguese Foundation for Sciences and Technology (FCT) though the strategy plan for MARE (Marine and Environmental Sciences Centre), via project UIDB/04292/2020, and under the project LA/P0069/2020 granted to the Associate Laboratory ARNET. FCT also supported this study through the individual contract attributed to Carlos M. Alexandre within the project CEECIND/02265/2018, and the PhD scholarships attributed to Andreia Domingues (2021.05644.BD) and Sara Silva (2021.05558.BD).

Institutional Review Board Statement: Not applicable.

Informed Consent Statement: Informed consent was obtained from all subjects involved in the study.

Data Availability Statement: Data is available from correspondence author, upon reasonable request.

Conflicts of Interest: The authors declare no conflict of interest.

Abstract

Mating Territory Location Drives Mating Success by Male Wrasses (Labridae) at a Resident Spawning Aggregation Site [†]

Terry J. Donaldson [‡]

Marine Laboratory, University of Guam, UOG Station, Mangilao, GU 96923, USA; tdonaldson@triton.uog.edu; Tel.: +1-671-735-0135

† Presented at the IX Iberian Congress of Ichthyology, Porto, Portugal, 20–23 June 2022.

‡ Presenting author (Oral communication).

Abstract: An increasing number of reef fish species have been shown to form spawning aggregations. These aggregations occur at predictable times and places, with participants utilizing single or mixed mating systems. In a lek-like mating system, males establish temporary courtship territories that they defend against rival males while attracting females to spawn. The location of these territories often contributes to differential mating success. The males holding territories deemed "desirable" by females because of the physical attributes of their location attract more females and secure greater mating opportunities compared to those males that defend territories elsewhere within the spawning aggregation site. Presumably, females favor locations where newly spawned eggs may be carried away from potential predators more effectively. Thus, males holding territories located at the outer edge of the site in an area exposed to water currents running parallel to the reef face have far greater mating success than those males that hold territories found on inner or middle sections of the site where currents are less pronounced. This pattern is consistent across a range of taxa within the family Labridae, co-occurring at a multispecies spawning aggregation site on a coral reef in Guam, Western Pacific.

Keywords: courtship behavior; lek-like mating; reproduction; territoriality

Citation: Donaldson, T.J. Mating Territory Location Drives Mating Success by Male Wrasses (Labridae) at a Resident Spawning Aggregation Site. *Biol. Life Sci. Forum* **2022**, *13*, 137. https://doi.org/10.3390/blsf2022013137

Academic Editor: Alberto Teodorico Correia

Published: 20 June 2022

Publisher's Note: MDPI stays neutral with regard to jurisdictional claims in published maps and institutional affiliations.

Funding: This research was supported by grants awarded to the University of Guam by the U.S. National Science Foundation (OIA-1457769 and OIA-1946352).

Institutional Review Board Statement: Not applicable.

Informed Consent Statement: Not applicable.

Data Availability Statement: The data reported here have not been published previously.

Conflicts of Interest: The author declares no conflict of interest.

biology and life sciences forum

Abstract

The Stress Response in Antarctic Fish: HPI Modulation, Cortisol Profiles, Interrenal Sensitivity, and Gene Expression of *Notothenia rossii* Acclimated to Temperature Challenges [†]

Pedro M. Guerreiro *,‡, Sandra Silva, Bruno Louro, Alexandra Alves, Elsa Couto and Adelino V.M. Canário

Center of Marine Sciences (CCMAR), University of Algarve, 8005-139 Faro, Portugal; scsilva@ualg.pt (S.S.); blouro@ualg.pt (B.L.); alcalves@ualg.pt (A.A.); ecouto@ualg.pt (E.C.); acanario@ualg.pt (A.V.M.C.)
* Correspondence: pmgg@ualg.pt
† Presented at the IX Iberian Congress of Ichthyology, Porto, Portugal, 20–23 June 2022.
‡ Presenting author (Oral Communication).

Citation: Guerreiro, P.M.; Silva, S.; Louro, B.; Alves, A.; Couto, E.; Canário, A.V.M. The Stress Response in Antarctic Fish: HPI Modulation, Cortisol Profiles, Interrenal Sensitivity, and Gene Expression of *Notothenia rossii* Acclimated to Temperature Challenges. *Biol. Life Sci. Forum* **2022**, *13*, 58. https://doi.org/10.3390/blsf2022013058

Academic Editor: Alberto Teodorico Correia

Published: 7 June 2022

Publisher's Note: MDPI stays neutral with regard to jurisdictional claims in published maps and institutional affiliations.

Abstract: The Antarctic Ocean is one of the most extreme marine environments. Antarctic fishes evolved in stable cold thermal conditions (-1.9 °C to 2 °C) for roughly 20 million years, displaying structural and functional features resulting from adaptation or inherited from resilient ancestral species. Climate change forecast models show temperatures may increase at a relevant pace. As fish face a warmer future, their physiological ability to adapt is uncertain. We aimed at evaluating the capabilities of the hypothalamus–pituitary–interrenal (HPI) axis in Antarctic fish and show plasma cortisol profiles, expression of key genes, the sensitivity of the ex vivo interrenal tissue, and the responses to known modulators of the HPI axis in temperature-acclimated fish before and after stress. *Notothenia rossii* were collected from the waters of Admiralty and Maxwell bays, in King George Island, and transferred to an open circuit with ocean-pumped seawater. Upon acclimation, three sets of experiments were performed: (1) eight groups at 2 °C were injected with drugs involved in blockage or stimulation of cortisol release/action (saline, cortisol, dexamethasone, metyrapone, spironolactone, mifepristone) and then kept at control or transferred to 6 °C and sampled after 36 h; (2) fish at 2 °C were exposed to a standard stress test (SST: chasing+netting+1min air exposure), returned to the respective tank and sampled after 1, 4 and 24 h, while one undisturbed group served as control; (3) six groups were acclimated to 2, 5 and 8 °C for 10 days when the control group of each temperature was sacrificed. The other group received SST and was sacrificed 90 min after. Plasma and tissue samples were collected for cortisol and stress-related genes, and the interrenal was used in vitro to determine sensitivity to ACTH in a perfusion system with a continuous flow of oxygenated ringers, and 20 min fractions were collected for 240 min. Cortisol was measured via radioimmunoassay, while glucose and lactate were determined using colorimetric kits; gene expression was evaluated by qPCR. Manipulation of the HPI axis revealed that these fish show similar dynamics to those reported in temperate fish but with lower amplitude. After SST, cortisol peaked at 1–4 h and reduced to basal between 24 and 48 h. Temperature influenced the cortisol response to SST. At higher temperatures, cortisol levels in the non-stressed group were as high as in fish subjected to SST. Interrenal cells showed little response to ACTH in warm conditions, suggesting low sensitivity and/or exhaustion. Liver cortisol receptor genes were downregulated, possibly indicating a peripheral desensitization process that parallels HPI. These results show the ability to respond to stress at cold and mild temperatures but important impairments and substantial allostasis in warm or continuously increasing temperatures.

Keywords: stress response; Antarctic fish; interrenal; cortisol; climate change; physiology

---

Author Contributions: Conceptualization, P.M.G. and A.V.M.C.; methodology, P.M.G., S.S.; formal analysis, P.M.G., S.S., A.A.; investigation, P.M.G., S.S., B.L., A.A., E.C.; resources, P.M.G., A.V.M.C.; data curation, S.S., P.M.G.; writing—original draft preparation, P.M.G., S.S.; writing—review and editing, P.M.G.; supervision, P.M.G., A.V.M.C.; project administration, P.M.G., A.V.M.C.; funding acquisition, A.V.M.C. All authors have read and agreed to the published version of the manuscript.

Funding: This research is supported by FCT through ProPolar and grants PTDC/BIAANM/3484/2014 and UIDB/04326/2021.

Institutional Review Board Statement: Not applicable.

Informed Consent Statement: Not applicable.

Data Availability Statement: Not applicable.

Conflicts of Interest: The authors declare no conflict of interest.

biology and life sciences forum

Abstract

Contribution for the Understanding of Coral Bleaching Events: The Case of *Palythoa caribaeorum* off Porto Belo, South Brazil [†]

Rafael Schroeder [1,2,*,‡], Lucas Gavazzoni [1], Carlos E. N. de Oliveira [3], Pedro H. M. L. Marques [3] and Ewerton Wegener [3]

[1]   Laboratório de Estudos Marinhos Aplicados, Escola do Mar, Ciência e Tecnologia, Universidade do Vale do Itajaí (UNIVALI), Itajaí 88302-901, Brazil; gavazzonilucas@gmail.com
[2]   Centro Interdisciplinar de Investigação Marinha e Ambiental (CIIMAR), 4550-208 Matosinhos, Portugal
[3]   Laboratório de Mergulho Submarino, Escola do Mar, Ciência e Tecnologia, Universidade do Vale do Itajaí (UNIVALI), Itajaí 88302-901, Brazil; kadunemitz@univali.br (C.E.N.d.O.); pedro_marques@univali.br (P.H.M.L.M.); ewerton.wegener@univali.br (E.W.)
*   Correspondence: schroederichthys@gmail.com
†   Presented at the IX Iberian Congress of Ichthyology, Porto, Portugal, 20–23 June 2022.
‡   Presenting author (Poster presentation).

Citation: Schroeder, R.; Gavazzoni, L.; de Oliveira, C.E.N.; Marques, P.H.M.L.; Wegener, E. Contribution for the Understanding of Coral Bleaching Events: The Case of *Palythoa caribaeorum* off Porto Belo, South Brazil. *Biol. Life Sci. Forum* **2022**, *13*, 78. https://doi.org/10.3390/blsf2022013078

Academic Editor: Alberto Teodorico Correia

Published: 9 June 2022

Publisher's Note: MDPI stays neutral with regard to jurisdictional claims in published maps and institutional affiliations.

Abstract: The white encrusting zoanthid *Palythoa caribaeorum* is known to form large colonies, which serve as aggregators of marine fauna. In Porto Belo, south Brazil (27.158° S, 48.553° W), the fish community was more abundant over *P. caribaeorum* grounds and more diverse, as described by traditional ecological descriptors, beta-diversity index, and discriminant analysis. In addition to its important ecological role, it is of great economic relevance, as it is conducive to the establishment of underwater diving trails. In the last 20 years, however, the bleaching of this zoanthid was observed within the study area. As suggested in the literature, we evaluated environmental temperature from three different sources: 1. Meteorological stations (MS) between 1961 and 2019; 2. Satellite images (SI) (1993–2019), and 3. Water temperature (WT) obtained in situ with dive computers, between February and September 2018. Initially, a Pearson correlation coefficient was calculated for temperature values obtained in MS and by SI (between 1993 and 2018) and between temperature values obtained by SI and by dive computers (February to September 2018—WT). Subsequently, the temperatures obtained in MS and in SI were standardized in additive generalized models for location scale and shape. The temperature values obtained in MS and by SI showed perfect correlation (R = 1). In the comparison between temperature values obtained by SI and WT, the correlation coefficient was 0.92, indicating that air, sea surface, and water temperatures showed the same increasing trend. The adjusted temperature values for the weather stations showed an average increase of 1.68 °C between 1961 and 2018 and 0.64 °C between 1993 and 2018 for the values adjusted from SI. The results obtained in a distance-based redundancy analysis showed the spatial distribution of fish species sighted at Porto Belo during the monitoring conducted between February and September 2018 was significant for the preferred temperature of these species. Despite the lack of available estimates for preferential temperature of *P. caribaeorum*, this progressive increase in temperature is one of the main causes for the bleaching of several coral species, as reported in the literature.

Keywords: marine spatial planning; ecosystem impact; marine tourism; habitat modification; fish community

Author Contributions: Conceptualization, formal analysis, investigation, original draft preparation, R.S.; Data acquisition & investigation: L.G., C.E.N.d.O. and P.H.M.L.M.; Data curation & project administration: E.W. All authors have read and agreed to the published version of the manuscript.

Funding: This research received no external funding.

Institutional Review Board Statement: Not applicable.

Informed Consent Statement: Not applicable.

Data Availability Statement: The data that support the findings of this study are available from the corresponding author upon reasonable request.

Conflicts of Interest: The authors declare no conflict of interest.

biology and life sciences forum

Abstract

Should I Stay or Should I Go? Fish Passability in a Rock Weir (Tagus River) under Climate Change Scenarios [†]

Daniel Mameri [1,*,‡], Rui Rivaes [1], Maria Teresa Ferreira [1], Stefan Schmutz [2] and José Maria Santos [1]

[1] Forest Research Centre (CEF) and Associate Laboratory TERRA, School of Agriculture, University of Lisbon, Tapada da Ajuda, 1349-017 Lisboa, Portugal; ruirivaes@isa.ulisboa.pt (R.R.); terferreira@isa.ulisboa.pt (M.T.F.); jmsantos@isa.ulisboa.pt (J.M.S.)

[2] Institute of Hydrobiology and Aquatic Ecosystem Management, Department of Water, Atmosphere and Environment, University of Natural Resources and Life Sciences, Gregor-Mendel-Straße 33/DG, 1180 Wien, Austria; stefan.schmutz@boku.ac.at

* Correspondence: dmameri@isa.ulisboa.pt; Tel.: +351-914201592

† Presented at the IX Iberian Congress of Ichthyology, Porto, 20–23 June 2022.

‡ The presenting author (oral communication).

Abstract: Iberian rivers often face annual periods of flow intermittence during the dry season, when habitat availability for freshwater organisms is drastically reduced. Climate change and the presence of small barriers such as weirs may further exacerbate this lack of suitable habitat, particularly for freshwater and migratory fish that perform seasonal movements to complete their life cycle, by narrowing the submersed area and their possibilities of overcoming these barriers. Using the River2D model, we investigated how the current released flows from a nearby large hydropower plant in the Tagus River affect the passability of native fish species at a downstream rock weir (Pego), equipped with a nature-like fish ramp. Using mean daily flow data from gauging stations, we compared the passability of six fish species under low flow conditions (Q90), considering a historical period (1991–2005), and two flow datasets based on climate change projections until the end of the century (2071–2100) for the Tagus River Basin ("moderate" RCP 4.5 and "extreme" RCP 8.5). Target species included three migratory guilds: (i) anadromous—Allis shad (*Alosa alosa*), twaite shad (*Alosa fallax*) and sea lamprey (*Petromyzon marinus*); (ii) catadromous—European eel (*Anguilla anguilla*) and thlinlip grey mullet (*Chelon ramada*); and (iii) potamodromous species—Iberian barbel (*Luciobarbus bocagei*). Overall, our results show that the passability of all fish species may only be ensured with a minimum flow of 3 m s^{-1} and by using a fish ramp. Furthermore, the passability for all species was found to be significantly lower in the RCP scenarios when compared to the historical period. Our study suggests that climate change is expected to reduce the passability of native fish species in weirs in Iberian rivers, highlighting the importance of considering future flow conditions for a proper management of fish populations in the presence of weirs and other barriers.

Keywords: nature-like fishway; rock weir; climate change; low flows; habitat suitability

Citation: Mameri, D.; Rivaes, R.; Ferreira, M.T.; Schmutz, S.; Santos, J.M. Should I Stay or Should I Go? Fish Passability in a Rock Weir (Tagus River) under Climate Change Scenarios. *Biol. Life Sci. Forum* **2022**, *13*, 81. https://doi.org/10.3390/blsf2022013081

Academic Editor: Alberto Teodorico Correia

Published: 13 June 2022

Publisher's Note: MDPI stays neutral with regard to jurisdictional claims in published maps and institutional affiliations.

Author Contributions: Conceptualization, J.M.S. and M.T.F.; methodology, D.M., J.M.S. and R.R.; software, D.M. and R.R.; formal analysis, D.M., J.M.S. and R.R. and S.S.; investigation, D.M., J.M.S. and R.R.; writing—original draft preparation, D.M.; writing—review and editing, J.M.S., M.T.F., R.R. and S.S.; supervision, J.M.S., M.T.F. and S.S.; project administration, M.T.F.; funding acquisition, M.T.F. All authors have read and agreed to the published version of the manuscript.

Funding: The study was funded by PEGOP under the protocol "Passability of the Pego weir by fish species". The funders had no role in the design of the study; in the collection, analyses, or interpretation of data; in the writing of the manuscript, or in the decision to publish the results. Daniel Mameri was supported by a Ph.D. grant from the FLUVIO—River Restoration and Management program funded by FCT (PD/BD/142885/2018).

Institutional Review Board Statement: Not applicable. This study followed an eco-hydraulic modeling approach using fish habitat suitability data, based on the existing literature and expert judgment.

Informed Consent Statement: Not applicable.

Data Availability Statement: The data presented in this study are available from the corresponding author on reasonable request.

Conflicts of Interest: The authors declare no conflict of interest.

Abstract

Global Patterns of Small Pelagic Fishes' Diversity: Present and Future [†]

Joana Boavida-Portugal [1,*,‡], Miguel B. Araújo [2,3] and Rui Rosa [4,5]

1   MARE-Centro de Ciências do Mar e do Ambiente/ARNET-Rede de Investigação Aquática, Universidade de Évora, 7004-516 Évora, Portugal
2   "Rui Nabeiro" Biodiversity Chair, MED Institute, University of Évora, Largo dos Colegiais, 7004-516 Évora, Portugal; mba@uevora.pt
3   Department of Biogeography and Global Change, National Museum of Natural Sciences, Consejo Superior de Investigaciones Científicas (CSIC), 28006 Madrid, Spain
4   MARE-Centro de Ciências do Mar e do Ambiente/ARNET-Rede de Investigação, Laboratório Marítimo da Guia, Faculdade de Ciências da Universidade de Lisboa, Avenida Nossa Senhora do Cabo, 2750-374 Cascais, Portugal; rrosa@fc.ul.pt
5   Departamento de Biologia Animal, Faculdade de Ciências, Universidade de Lisboa, Campo Grande, 1749-016 Lisboa, Portugal
*   Correspondence: jbp@uevora.com
†   Presented at the IX Iberian Congress of Ichthyology, Porto, Portugal, 20–23 June 2022.
‡   Presenting author (oral communication).

Citation: Boavida-Portugal, J.; Araújo, M.B.; Rosa, R. Global Patterns of Small Pelagic Fishes' Diversity: Present and Future. *Biol. Life Sci. Forum* **2022**, *13*, 108. https://doi.org/10.3390/blsf2022013108

Academic Editor: Alberto Teodorico Correia

Published: 16 June 2022

Publisher's Note: MDPI stays neutral with regard to jurisdictional claims in published maps and institutional affiliations.

Abstract: Many of highly productive mid-latitude marine ecosystems share a prominent aspect in the configuration of their biological community structures. They appear to have a 'wasp-waist' food web, whereby the bottom (planktonic trophic levels) and top (apex and near-apex levels) of the food chain have high species diversity, while intermediate trophic levels are occupied only by few small pelagic fishes (SPF). SPF population dynamics is strongly correlated with temperature fluctuations, and since average global sea surface temperatures are expected to increase up to 1–6 °C by 2100, this warming trend may dictate profound impacts on SPF distribution and abundance. In this study, we used an ensemble of 6 ecological niche models and 21 earth system models to project, for the first time, changes in SPF richness, catch potential and geographic range size (comprising 47 anchovies, 33 herrings and 23 sardine species) by the end-century, by using two extreme scenarios (RCP2.6 and 8.5). We then predicted the latitudinal shifts that major SPF species might undergo due to climate change. Finally, we discussed the ecological and economic impacts potentially induced by climatic change, linking the projections with the global trends in landings since 1950. Our results suggest major effects on fisheries worldwide and highlight the need for adopting precautionary management that can easily adapt to projected changes.

Keywords: climate change; global diversity; catch potential; small pelagic fish; anchovies; herrings; sardines; ecological niche models; ensemble forecast

Funding: This study was supported by the Portuguese Foundation for Science and Technology (FCT) through: (i) the strategic project UID/MAR/04292/2013 granted to MARE, (ii) a PhD grant (SFRH/BD/51514/2011) to JBP and (iii) the Integrated Program of IC&DT Call No 1/SAESCTN/ALENT-07-0224-FEDER-001755 to MBA.

Institutional Review Board Statement: Not applicable.

Informed Consent Statement: Not applicable.

Data Availability Statement: All data used in this manuscript are publicly available; data sources are detailed in the methods.

Conflicts of Interest: The authors declare no conflict of interest.

Abstract

Climate-Change-Proof Riverine Ecosystems for Sustainable Management: The AQUADAPT Project [†]

Ana Filipa Filipe [1,*,‡], José M. Santos [1], Paulo Branco [1], Maria Teresa C de Melo [2], Rodrigo Proença de Oliveira [2], Susana Fernandes [3], Maria Helena Alves [3], Alice Fialho [3], Maria José Moura [3], Luísa Pinto [4], Noémia P. Santiago [4] and Maria Teresa Ferreira [1]

[1] Forest Research Centre and Associated Laboratory TERRA, School of Agriculture, University of Lisboa, 1349-017 Lisboa, Portugal; jmsantos@isa.ulisboa.pt (J.M.S.); pjbranco@isa.ulisboa.pt (P.B.); terferreira@isa.ulisboa.pt (M.T.F.)
[2] Civil Engineering Research and Innovation for Sustainability (CERIS), Instituto Superior Técnico, Universidade de Lisboa, 1049-001 Lisboa, Portugal; teresa.melo@tecnico.ulisboa.pt (M.T.C.d.M.); rodrigopoliveira@tecnico.ulisboa.pt (R.P.d.O.); thiagovmdon@gmail.com (T.N.)
[3] Agência Portuguesa do Ambiente, 2610-124 Amadora, Portugal; susana.fernandes@apambiente.pt (S.F.); helena.alves@apambiente.pt (M.H.A.); alice.fialho@apambiente.pt (A.F.); maria.moura@apambiente.pt (M.J.M.)
[4] EDIA Empresa de Desenvolvimento e Infraestruturas do Alqueva, 7800-522 Beja, Portugal; lpinto@edia.pt (L.P.); nsparada@edia.pt (N.P.S.)
*   Correspondence: affilipe@isa.ulisboa.pt
†   Presented at the IX Iberian Congress of Ichthyology, Porto, Portugal, 20–23 June 2022.
‡   Presenting author (poster).

Citation: Filipe, A.F.; Santos, J.M.; Branco, P.; de Melo, M.T.C.; Proença de Oliveira, R.; Fernandes, S.; Alves, M.H.; Fialho, A.; Moura, M.J.; et al. Climate-Change- Proof Riverine Ecosystems for Sustainable Management: The AQUADAPT Project. *Biol. Life Sci. Forum* **2022**, *13*, 136. https://doi.org/10.3390/ blsf2022013136

Academic Editor: Alberto Teodorico Correia

Published: 20 June 2022

Publisher's Note: MDPI stays neutral with regard to jurisdictional claims in published maps and institutional affiliations.

Abstract: Recent climate scenarios predict dramatic changes for the inland region of Portugal, where the increase in air temperature might reach +5 °C by 2100, accompanied by a strong decrease in precipitation and an increase in extreme events. Such forecasts imply changes in thermal and hydrological patterns in the coming decades, leading to an increase in the frequency, intensity and duration of droughts and floods. Consequently, the sustainable development of Portugal's inland region will inevitably depend on the ability to adapt to such climate-related changes. The project AQUADAPT, funded by 'La Caixa' Foundation, aims to promote the resilience of river ecosystems to climate change, through risk assessment and the construction of adaptation tools. We will develop a high-resolution monitoring and warning system through modelling, forecasting and planning techniques using freshwater fishes as indicators, and test nature-based solutions in degraded areas of protected and agricultural areas. The innovative character of this project lies in the multidisciplinary approach gathering investigation, planning tools and dissemination, and its relevance lies in the construction of replicable products at the national and international context. By bringing together academic partners (ISA-CEF and IST-CERIS), public administration (APA) and companies (EDIA), the project AQUADAPT uses a multidisciplinary-approach gathering investigation, planning tools and dissemination. The gained knowledge of climate and hydrological changes, their impacts and possible natural responses to promote resistance and resilience of ecosystems will allow the construction of scenarios and alternatives for informed decision making by politicians, managers and other stakeholders for the coming decades. This way, the project AQUADAPT will nurture the transformation towards a more sustainable region for people and nature.

Keywords: freshwater biodiversity; intermittent streams; sustainability

Author Contributions: All authors contributed equally to the conceptualization, methodology and writing. All authors have read and agreed to the published version of the manuscript.

Funding: This research was funded by AQUADAPT Project "Climate change-proof riverine ecosystems for sustainable management" funded by 'La Caixa' Foundation (Promove Program, grant number: PD20-00008).

Institutional Review Board Statement: Not applicable.

Informed Consent Statement: Not applicable.

Data Availability Statement: Not applicable.

Conflicts of Interest: The authors declare no conflict of interest.

Abstract

Awareness and Prevention of Aquatic Invasive Alien Species in the Iberian Peninsula by LIFE INVASAQUA: Midterm Outcomes [†]

Francisco J. Oliva-Paterna [1,*,‡], Rosa Olivo del Amo [1], Mar Torralva [1], Pedro M. Anastácio [2], Filipe M. S. Banha [2], Sandra Barca [3], Frederic Casals [4], Fernando Cobo [3], Antonio Guillén [1], Celia López-Cañizares [1], Annie Machordom [5], Rafael Miranda [6], Catherine Numa [7], Francisco J. Oficialdegui [1], Javier Oscoz [6], Anabel Perdices [5], Filipe Ribeiro [8], Jorge R. Sánchez-González [4], Rufino Vieira-Lanero [3] and José Manuel Zamora-Marín [1]

[1] Department of Zoology and Physical Anthropology, University of Murcia, 30100 Murcia, Spain; rosa.olivo@um.es (R.O.d.A.); torralva@um.es (M.T.); antonio.guillenb@usc.es (A.G.); celia.lopez10@um.es (C.L.-C.); oficialdeguifj@gmail.com (F.J.O.); josemanuel.zamora@um.es (J.M.Z.-M.)

[2] Department of Landscape, Environment and Order, University of Évora, 7000-671 Évora, Portugal; anast@uevora.pt (P.M.A.); filipebanha@uevora.pt (F.M.S.B.)

[3] Department of Zoology and Physical Anthropology, University of Santiago de Compostela, 15782 Santiago de Compostela, Spain; sandra.barca@usc.es (S.B.); fernando.cobo@usc.es (F.C.); rufino.vieira@usc.es (R.V.-L.)

[4] SIBIC, Department of Animal Science, University of Lleida, 25198 Lleida, Spain; frederic.casals@udl.cat (F.C.); jorge.r.sanchez.gonzalez@gmail.com (J.R.S.-G.)

[5] CSIC-National Museum of Natural Sciences, 28006 Madrid, Spain; annie@mncn.csic.es (A.M.); aperdices@mncn.csic.es (A.P.)

[6] Department of Environmental biology, University of Navarra, 31009 Pamplona, Spain; rmiranda@unav.es (R.M.); javioscoz@gmail.com (J.O.)

[7] IUCN-Center for Mediterranean Cooperation, 29590 Málaga, Spain; catherine.numa@iucn.org

[8] SIBIC, MARE Marine and Environmental Sciences Centre, 1749-016 Lisboa, Portugal; fmvribeiro@gmail.com

[*] Correspondence: fjoliva@um.es

[†] Presented at the IX Iberian Congress of Ichthyology, Porto, Portugal, 20–23 June 2022.

[‡] Presenting author (oral communication).

Citation: Oliva-Paterna, F.J.; Olivo del Amo, R.; Torralva, M.; Anastácio, P.M.; Banha, F.M.S.; Barca, S.; Casals, F.; Cobo, F.; Guillén, A.; López-Cañizares, C.; et al. Awareness and Prevention of Aquatic Invasive Alien Species in the Iberian Peninsula by LIFE INVASAQUA: Midterm Outcomes. *Biol. Life Sci. Forum* **2022**, *13*, 47. https://doi.org/10.3390/blsf2022013047

Academic Editor: Alberto Teodorico Correia

Published: 7 June 2022

Publisher's Note: MDPI stays neutral with regard to jurisdictional claims in published maps and institutional affiliations.

Abstract: The Iberian society has a limited understanding of the threats posed by invasive alien species (IAS) in aquatic ecosystems. This lack of knowledge and awareness about IAS problems hampers any management policy proposed by public administration and stakeholders, contributing to missing an IAS management strategy. We present the midterm outcomes of the Environmental Governance and Information LIFE project—LIFE INVASAQUA—that will run between 2018 and 2023 in the Iberian Peninsula. The main goal of INVASQUA is to increase the Iberian public and stakeholders' awareness of aquatic IAS problems and to develop tools that will improve an efficient early warning and rapid response (EWRR) framework for new IAS in freshwater and estuarine habitats in the Iberian Peninsula. We focus on new challenges and outcomes of the project to explore some of the problems and solutions encountered in the project implementation.

Keywords: invasive; exotic species; LIFE projects; prevention

Author Contributions: Conceptualization, F.J.O.-P. and R.O.d.A.; writing, F.J.O.-P.; funding acquisition, F.J.O.-P., R.O.d.A., F.R., P.M.A., F.C. (Fernando Cobo), A.P., R.M., C.N. All authors have read and agreed to the published version of the manuscript.

Funding: This project received funds from the LIFE Programme (LIFE17 GIE/ES/000515).

Institutional Review Board Statement: Not applicable.

Informed Consent Statement: Informed consent was obtained from all subjects involved in the study.

Data Availability Statement: Not applicable.

Conflicts of Interest: The authors declare no conflict of interest.

*biology and life
sciences forum*

Abstract

Control of Invasive Plant Species in Wetland Forests (91E0*) [†]

Estêvão Portela-Pereira [1,‡], Paulo Monteiro [2] and Patricia María Rodríguez-González [2,*]

[1] Centro de Estudos Geográficos, Instituto de Geografia e Ordenamento do Território, Universidade de Lisboa, 1600-276 Lisboa, Portugal; estevao@campus.ul.pt

[2] Centro de Estudos Florestais, Instituto Superior de Agronomia, Universidade de Lisboa, 1349-017 Lisboa, Portugal; cef.isa.life@gmail.com

[*] Correspondence: patri@isa.ulisboa.pt; Tel.: +351-213653492

[†] Presented at the IX Iberian Congress of Ichthyology, Porto, Portugal, 20–23 June 2022.

[‡] Presenting author (oral communication).

Abstract: The main objective of the LIFE Fluvial project is the improvement of the conservation status of Atlantic river corridors in the Natura 2000 network, developing a transnational strategy for the sustainable management of river corridor habitats in several Atlantic river basins of the Iberian Peninsula. The project includes seven partners in Northwest Spain (Galicia, Asturias) and one partner in Portugal (Instituto Superior de Agronomia). In Portugal, the preparatory, conservation, monitoring and dissemination actions of the project are focused on the improvement of the state of conservation of habitat 91E0* in the Estorãos River (ZEC Lima River, PTCON0020), with a total intervention area of circa 21 ha, within the property of the Municipality of Ponte de Lima. Special effort has been devoted to the control and removal of invasive plants directly affecting the riparian habitats, and indirectly affecting the aquatic habitats. The major target species addressed in ZEC Lima River are the trees *Acacia melanoxylon, A. dealbata,* invading the riparian zone; *Eucalyptus camaldulensis* plantations in the floodplains of Estorãos river; and the herbaceous species *Phytolacca americana* and *Tradescantia fluminensis.* The restoration measures were designed according to the spatial and temporal scale of threats and applied species-specific methods. This included the tree cut and control of vegetative sprouts (*Eucalyptus*), debarking and cut (*Acacia* spp.), and the uprooting of seedlings of invasive exotic individuals; complete uprooting of the individuals and the destruction of fruits (*Phytolacca americana*); and exposition to sunlight (*Tradescantia fluminensis*). An additional key action included Public Awareness and Dissemination. Throughout the development of the project, knowledge transfer to different target audiences was promoted, and several didactic materials, including an online game for children, were produced. For the general public, the project created a website, in four languages, and different social media pages, TV programs, promotional videos and an annual bulletin, and celebrated several events used for awareness-raising (e.g., World Wetlands Day in 2018, or Natura 2000 day). Notably, for invasive species control and awareness, we promoted training and volunteer actions that engaged students, technicians, NGOs and other relevant stakeholders. The major lessons learned are to be followed up during the After Life period.

Keywords: ecological restoration; habitat conservation; knowledge transfer; riparian vegetation

Author Contributions: Conceptualization, methodology, validation, formal analysis, investigation, resources, data curation, writing—original draft preparation, writing—review and editing, visualization, supervision, E.P.-P., P.M. and P.M.R.-G.; project administration and funding acquisition, P.M.R.-G. All authors have read and agreed to the published version of the manuscript.

Funding: This research was funded by European Commission, LIFE16 NAT/ES/000771.

Institutional Review Board Statement: Not applicable.

Informed Consent Statement: Not applicable.

Citation: Portela-Pereira, E.; Monteiro, P.; Rodríguez-González, P.M. Control of Invasive Plant Species in Wetland Forests (91E0*). *Biol. Life Sci. Forum* **2022**, *13*, 84. https://doi.org/10.3390/blsf2022013084

Academic Editor: Alberto Teodorico Correia

Published: 14 June 2022

Data Availability Statement: Not applicable.

Conflicts of Interest: The authors declare no conflict of interest.

Abstract

Lessons Learned from the Elimination of Plant IAS in Natura 2000 River Corridors of the Iberian Atlantic Region (Galicia, NW Spain) †

Javier Ferreiro da Costa *,‡, Pablo Ramil-Rego, Hugo López Castro, Carlos Oreiro Rey and Luis Gómez-Orellana

Instituto de Biodiversidade Agraria e Desenvolvemento Rural (IBADER), Universidade de Santiago de Compostela, Campus Terra s/n, 27002 Lugo, Spain; ramil.rego@usc.es (P.R.-R.); hugo.lopez@usc.es (H.L.C.); carlos.oreiro@usc.es (C.O.R.); luis.gomez-orellana@usc.es (L.G.-O.)
* Correspondence: javier.ferreiro@usc.es
† Presented at the IX Iberian Congress of Ichthyology, Porto, Portugal, 20–23 June 2022.
‡ Presenting author (oral communication).

Abstract: LIFE FLUVIAL project (LIFE16 NAT/ES/000771) is a transnational project between Spain and Portugal, whose overall objective is the improvement in the conservation of Atlantic Natura 2000 river corridors and associated wetlands, mainly targeting the 91E0* priority habitat. IAS have become one of the main threats to these ecosystems, as they affect the composition and structure of riverine habitats, as well as the decrease in their area. In this conference, we present the results and learned lessons of IAS elimination in several river basins (Miño, Mandeo, Mero, Barcés) of Galicia (NW Spain) by LIFE FLUVIAL, including herb (*Cortaderia selloana, Crocosmia x crocosmiiflora, Delairea odorata*), scrub (*Tamarix gallica,*) and tree (*Eucalyptus* spp., *Acacia dealbata, Robinia pseudoacacia, Populus x canadensis, Salix viminalis*) alien species. An integral strategy has been developed, as they have been removed from the upper and middle basin, as well as from estuaries, so LIFE FLUVIAL has a highly demonstrative character as it can be applied on a larger scale to similar situations, or elsewhere in similar circumstances. The project has refused to use herbicides because of the high fragility of the aquatic environment, nor are they a 100% effective method. Removing plant IAS through manual or light mechanical means has revealed the best practices to improve the conservation status of priority habitat 91E0* and river corridors, carrying out active and continuous management, because many of them have a high resprouting potential. Following this methodology, plant IAS can be completely eliminated, or at least controlled, as they are confined in very specific places from where they cannot disperse. In some cases, the eliminated plants (both herbaceous and woody plants) reached large sizes, which has been a major challenge when it comes to eliminating, handling and removing them. When this happened, the plant material was stacked and dried in an isolated place from which it cannot spread, to finally be removed for treatment by an authorized waste manager. In this regard, the herbaceous IAS have been valued as raw material in the economic system, undergoing treatment to obtain agricultural fertilizers; thus, the project has also contributed to the circular economy.

Keywords: IAS; fluvial corridors; priority habitats; manual treatments; Natura 2000

Citation: Ferreiro da Costa, J.; Ramil-Rego, P.; López Castro, H.; Oreiro Rey, C.; Gómez-Orellana, L. Lessons Learned from the Elimination of Plant IAS in Natura 2000 River Corridors of the Iberian Atlantic Region (Galicia, NW Spain). *Biol. Life Sci. Forum* **2022**, 13, 87. https://doi.org/10.3390/blsf2022013087

Academic Editor: Alberto Teodorico Correia

Published: 14 June 2022

Author Contributions: Conceptualization, P.R.-R.; methodology, J.F.d.C.; field work H.L.C., C.O.R. and L.G.-O.; writing J.F.d.C.; project administration and funding acquisition P.R.-R. All authors have read and agreed to the published version of the manuscript.

Funding: This research was funded by LIFE Programme, grant agreement No. LIFE16 NAT/ES/000771.

Institutional Review Board Statement: Not applicable.

Informed Consent Statement: Not applicable.

Data Availability Statement: The data presented in this study are openly available in https://www.lifefluvial.eu.

Conflicts of Interest: The authors declare no conflict of interest.

*biology and life
sciences forum*

Abstract

Elimination and Control of Flora Exotic Species on Natural Habitats in Natura 2000 Islands of the Atlantic Ocean [†]

Javier Ferreiro da Costa [1,*,‡], José Antonio Fernández Bouzas [2], Pablo Ramil-Rego [1], Hugo López Castro [1], Luis Gómez-Orellana [1] and Carlos Oreiro Rey [1]

[1] Instituto de Biodiversidade Agraria e Desenvolvemento Rural (IBADER), Universidade de Santiago de Compostela, Campus Terra s/n, 27002 Lugo, Spain; ramil.rego@usc.es (P.R.-R.); hugo.lopez@usc.es (H.L.C.); luis.gomez-orellana@usc.es (L.G.-O.); carlos.oreiro@usc.es (C.O.R.)

[2] Maritime-Terrestrial Galician Atlantic Islands National Park, 36202 Vigo, Spain; jose.antonio.fernandez.bouzas@xunta.gal

* Correspondence: javier.ferreiro@usc.es

† Presented at the IX Iberian Congress of Ichthyology, Porto, Portugal, 20–23 June 2022.

‡ Presenting author (oral communication).

Citation: Ferreiro da Costa, J.; Fernández Bouzas, J.A.; Ramil-Rego, P.; López Castro, H.; Gómez-Orellana, L.; Oreiro Rey, C. Elimination and Control of Flora Exotic Species on Natural Habitats in Natura 2000 Islands of the Atlantic Ocean. *Biol. Life Sci. Forum* **2022**, *13*, 88. https://doi.org/10.3390/blsf2022013088

Academic Editor: Alberto Teodorico Correia

Published: 14 June 2022

Publisher's Note: MDPI stays neutral with regard to jurisdictional claims in published maps and institutional affiliations.

Abstract: Islands in the Atlantic Ocean (spread across the Atlantic and Macaronesian biogeographical regions) have been identified as one of the most biodiverse areas in the EU. Thanks to the combination of climatic conditions alongside their edaphic and coastal dynamics, these regions harbor a very high diversity of habitats and species, included in Directives 92/43/EEC and 2009/147/EC, many of which are considered for priority conservation. These island ecosystems generally present shared environmental problems, so they are extremely threatened and consequently present complex biodiversity mosaics that hold joint patterns of fragmentation and vulnerability, nowadays aggravated by global change. To solve this, the LIFE INSULAR project (LIFE20 NAT/ES/001007) is a transnational project between Spain and Ireland which implements a transnational strategy for the integrated restoration of insular habitats in Atlantic Ocean Natura 2000 islands spread across the Atlantic and Macaronesian biogeographical regions, promoting their favourable conservation status and increasing their resilience as the main measure of adaptation to current global changes. In these territories, the main threats and conservation problems that have been identified in the 2013–2018 EU Biogeographical Assessments under Article 17 of the Habitats Directive are the existence of senescent forest plantations (made by fast-growth exotic species with a high invasive potential that are encroaching the surroundings), and the presence of Invasive Alien Species (IAS) of flora. These threats cause an unfavourable conservation status for the natural insular habitats that are targeted by the project, triggering the reduction of their occupied area, as well as severely affecting their structure and functionality. Therefore, during the 2021–2026 period, LIFE INSULAR will carry out the elimination of old forest plantations that negatively affect the natural habitats in the islands and will erase the plant IAS. It will be necessary to carry out active and continuous management because many of the present invasive species have a high resprouting potential. In every case, manual methods will be employed to avoid the use of herbicides because of the high fragility of the insular environment, and these are not 100% effective methods. These operations will enable the subsequent restoration of the targeted habitats by planting and sowing their characteristic plant species.

Keywords: IAS; senescent forest plantations; insular habitats; manual treatments; Natura 2000

Author Contributions: Conceptualization, P.R.-R. and J.A.F.B.; methodology, J.F.d.C.; field work H.L.C., C.O.R. and L.G.-O.; writing J.F.d.C.; project administration and funding acquisition P.R.-R. and J.A.F.B. All authors have read and agreed to the published version of the manuscript.

Funding: This research was funded by LIFE Programme, grant agreement No. LIFE20 NAT/ES/001007.

Institutional Review Board Statement: Not applicable.

Informed Consent Statement: Not applicable.

Data Availability Statement: The data presented in this study will be regularly and periodically updated, as well as openly available, in https://www.lifeinsular.eu/.

Conflicts of Interest: The authors declare no conflict of interest.

Abstract

LIFE FLUVIAL: Improvement and Sustainable Management of River Corridors of the Iberian Atlantic Region [†]

Javier Ferreiro da Costa [1,*,‡], Patricia María Rodríguez-González [2], Pablo Ramil-Rego [1], Estêvão Portela-Pereira [3], Paulo Monteiro [2], Hugo López Castro [1], Carlos Oreiro Rey [1] and Luis Gómez-Orellana [1]

1   Instituto de Biodiversidade Agraria e Desenvolvemento Rural (IBADER), Universidade de Santiago de Compostela, Campus Terra s/n, 27002 Lugo, Spain; ramil.rego@usc.es (P.R.-R.); hugo.lopez@usc.es (H.L.C.); carlos.oreiro@usc.es (C.O.R.); luis.gomez-orellana@usc.es (L.G.-O.)

2   Centro de Estudos Florestais, Instituto Superior de Agronomia (ISA), Universidade de Lisboa, Tapada da Ajuda, 1349-017 Lisboa, Portugal; patri@isa.ulisboa.pt (P.M.R.-G.); cef.isa.life@gmail.com (P.M.)

3   Centre of Geographical Studies, Institute of Geography and Spatial Planning, Universidade de Lisboa, 1600-276 Lisboa, Portugal; estevaohnh@yahoo.com.br

*   Correspondence: javier.ferreiro@usc.es

†   Presented at the IX Iberian Congress of Ichthyology, Porto, Portugal, 20–23 June 2022.

‡   Presenting author (oral communication).

Citation: Ferreiro da Costa, J.; Rodríguez-González, P.M.; Ramil-Rego, P.; Portela-Pereira, E.; Monteiro, P.; López Castro, H.; Oreiro Rey, C.; Gómez-Orellana, L. LIFE FLUVIAL: Improvement and Sustainable Management of River Corridors of the Iberian Atlantic Region. *Biol. Life Sci. Forum* **2022**, *13*, 89. https://doi.org/10.3390/blsf2022013089

Academic Editor: Alberto Teodorico Correia

Published: 14 June 2022

Publisher's Note: MDPI stays neutral with regard to jurisdictional claims in published maps and institutional affiliations.

Abstract: At a global level, different land use change processes (changes in use, presence of invasive alien species, public use activities, intensification in agricultural and livestock activities, loss of compatible and low-intensity traditional uses, climate change, etc.) and phytosanitary problems (black alder disease) are currently threatening factors that generate, with different degrees of intensity, the deterioration and fragmentation of the habitats of river corridors, both in the upper and middle basin, as well as in the estuarine sections. LIFE FLUVIAL (LIFE16 NAT/ES/000771) develops a transnational strategy of sustainable management of river corridor habitats in several Atlantic river basins of the Iberian Peninsula (Spain and Portugal), so the overall objective is the improvement in the conservation status of Atlantic Natura 2000 river corridors and associated wetlands, mainly targeting the 91E0* priority habitat (alluvial forests with Alnus glutinosa and Fraxinus excelsior). Another target habitat is considered in the project, the habitat type 9230 Galician-Portuguese oak woods with Quercus robur and Quercus pyrenaica, which represents continuity with the 91E0* habitat type. To achieve the overall objective, a series of specific objectives are considered, which are designed to combat threats that contribute to habitat degradation: (1) implementation of a transnational model for sustainable management of river corridors for the improvement in their conservation status, through the restoration of the composition, structure and functionality of their types of habitats, as well as improving the connectivity and reduction in fragmentation; (2) control of exotic invasive alien plants; (3) improvement in the plant health of river corridors by the partial removal of dead trees; (4) promoting the dissemination and awareness of the natural values, socio-economic benefits and ecosystem services of river corridors; (5) improvement in training and technical capacity for stakeholders in the management and conservation of river corridors. To achieve these objectives, the project proposes a set of specific conservation actions that constitute a compendium of best practices in relation to the design and execution of viable and effective action measures to improve the conservation status of river corridors and contribute to halting the loss of biodiversity.

Keywords: IAS; fluvial corridors; priority habitats; best practices; Natura 2000

Author Contributions: Conceptualization, P.R.-R. and P.M.R.-G.; methodology, J.F.d.C. and E.P.-P.; field work P.M., H.L.C., C.O.R. and L.G.-O.; writing J.F.d.C.; project administration and funding acquisition P.R.-R. All authors have read and agreed to the published version of the manuscript.

Funding: This research was funded by LIFE Programme, grant agreement No. LIFE16 NAT/ES/000771.

Institutional Review Board Statement: Not applicable.

Informed Consent Statement: Not applicable.

Data Availability Statement: The data presented in this study are openly available in https://www.lifefluvial.eu.

Conflicts of Interest: The authors declare no conflict of interest.

*biology and life
sciences forum*

Abstract

LIFE MIGRATOEBRE: Migratory Fish Recovery and Improved Management in the Final Stretch of the Ebre River (Catalonia, NE Iberian Peninsula) †

Marc Ordeix [1,*,‡], Nati Franch [2], Francesc Vidal [2], Miquel Rafa [3], Karl B. Andree [4] and Enric Gisbert [4]

[1] Center for the Study of Mediterranean Rivers (CERM), University of Vic-Central University of Catalonia, Ter River Museum, Plaça de les dones del Ter, 1, 08560 Manlleu, Spain

[2] Natural Park of Delta de l'Ebre, Ministry of Climate Action, Food and Rural Agenda, Government of Catalonia, 08007 Barcelona, Spain; nfranchv@gencat.cat (N.F.); fvidale@gencat.cat (F.V.)

[3] Territory and Environment Area, Catalunya-La Pedrera Foundation, 08008 Barcelona, Spain; miquel.rafa@fcatalunyalapedrera.com

[4] Unit of Aquatic Cultures, Center of Sant Carles de la Ràpita, IRTA—Institute of Agrifood Research and Technology, 43540 La Ràpita, Spain; karl.andree@irta.es (K.B.A.); enric.gisbert@irta.cat (E.G.)

*   Correspondence: marc.ordeix@uvic.cat
†   Presented at the IX Iberian Congress of Ichthyology, Porto, Portugal, 20–23 June 2022.
‡   Presenting author (oral communication).

Citation: Ordeix, M.; Franch, N.; Vidal, F.; Rafa, M.; Andree, K.B.; Gisbert, E. LIFE MIGRATOEBRE: Migratory Fish Recovery and Improved Management in the Final Stretch of the Ebre River (Catalonia, NE Iberian Peninsula). *Biol. Life Sci. Forum* **2022**, *13*, 135. https://doi.org/10.3390/blsf2022013135

Academic Editor: Alberto Teodorico Correia

Published: 20 June 2022

Publisher's Note: MDPI stays neutral with regard to jurisdictional claims in published maps and institutional affiliations.

Abstract: The aim of this project is to promote the recovery of ecological connectivity within 10–20 years in the lower Ebre River and Delta, and a healthy and sustainable population of diadromous fish, including European sturgeon (*Acipenser sturio*), European eel (*Anguilla anguilla*), twaite shad (*Alosa fallax*) and sea lamprey (*Petromyzon marinus*), among other aquatic native species. It is focused on long-term sustainable investments, adapting all present river obstacles to allow fish migration, increasing the river spawning habitat availability and the distribution and growth areas of migrating fish. The main activities of the LIFE MIGRATOEBRE Project (LIFE13 NAT/ES/000237) are: (1) Build a fish ramp at Xerta's weir (located at 58 km from the Mediterranean sea; in 2023), and a fish ramp at Ascó's weir (located at 104 km from the sea; built in 2017), and monitor them regularly. (2) Apply ship locks fish-friendly improved management at Xerta's weir (located at 56 km from the sea), and monitor it regularly. (3) Undertake a pilot project of European sturgeon restocking through an experimental release in the lower Ebre (in 2023). (4) Develop a communication campaign and a community involvement plan for students, the general public, farmers, fishermen, anglers, electric companies, tourism stakeholders and regional and local authorities. This included the production of a great temporary exhibition and a network of volunteers. The potential spawning and feeding areas for European sturgeon, twaite shad and sea lamprey were identified and cartographed. Extensive monitoring of the target fish population of the lower Ebre river was carried out (2000 fish were marquet with PIT tags). Between 2017 and 2020, 150 fish (*Chelon labrosus*, *Mugil cephalus*, *Alosa fallax* and *Anguilla anguilla*) were monitored by acoustic telemetry (using 47 Vemco receivers located along the river and at in the lagoons of the Ebre delta). Obtained results indicate that the Ebre is optimal for the recovery of the target species of this project, but at the same time it would be very important to carry out a control of invasive species. This LIFE project started on 1 July 2014 and, after several extensions, it will finish on 30 June 2024. More information can be found at: www.migratoebre.eu.

Keywords: LIFE projects; fish migration; rivers; ultrasonic telemetry

Funding: This research was funded by EU Life Programm, grant number LIFE13 NAT/ES/000237.

Institutional Review Board Statement: Not applicable.

Informed Consent Statement: Not applicable.

Data Availability Statement: Not applicable.

Conflicts of Interest: The authors declare no conflict of interest.

Abstract

Biodiversity in Neotropical Fishes [†]

Claudio Oliveira [‡]

Instituto de Biociências, Universidade Estadual Paulista Julio de Mesquita Filho (UNESP), Botucatu 18618-689, Brazil; claudio.oliveira@unesp.br
† Presented at the IX Iberian Congress of Ichthyology, Porto, Portugal, 20–23 June 2022.
‡ Presentation type: Plenary talk.

Abstract: With more than 6200 fresh species, the Neotropical Water Fish Fauna (NFF) constitutes the most diverse continental vertebrate fauna on Earth. This diversity is quite impressive when we realize that it is concentrated in less than 0.5% of the total area of the Earth's surface. The diversity of this ichthyofauna has been extensively explored through extensive morphological work by renowned ichthyologists who produced the basic knowledge available today. The use of tools for generating and interpreting morphological data has progressed considerably in recent decades, as has the number of researchers working in this field. Thus, only considering the families of the three main orders of NFF (Siluriformes, Gymnotiformes and Characiformes), 853 new species were described in the last 10 years. With the advancement of DNA sequencing techniques, a new class of data began to be incorporated into NFF biodiversity studies, allowing for a new and more complete view of the group. New phylogenies, such as the proposal for Charciformes, ordered the groups into monophyletic families, and, thus, we advanced in the description of species in these families. Furthermore, the development of data analysis methods for species identification, associated with large DNA barcode generation programs, allowed an unprecedented expansion in our knowledge of NFF. As an example, we can mention the genus *Tetragonopterus*, which just over 15 years ago contained only 2 species and today has 14 species, and the genus *Neoplecostomus*, which had only 3 species and now has 16 species. In both cases the combination of morphological and molecular data was fundamental for a better definition of species in these groups. Currently, with new generation sequencing techniques, our knowledge of fish groups is improving so that large monophyletic or polyphyletic groups are being redescribed as smaller units in order to make possible taxonomic revisions that can effectively better evaluate the diversity of these groups. Our results and experiences will be presented at this congress.

Keywords: neotropical freshwater fishes; systematics; taxonomy; phylogenomics; species richness

Citation: Oliveira, C. Biodiversity in Neotropical Fishes. *Biol. Life Sci. Forum* **2022**, *13*, 28. https://doi.org/10.3390/blsf2022013028

Academic Editor: Alberto Teodorico Correia

Published: 6 June 2022

Funding: This research was funded by Fundação de Amparo à Pesquisa do Estado de São Paulo—FAPESP grants 2020/13433-6, 2018/20610-1, 2016/09204-6, 2014/26508-3 and Conselho Nacional de Desenvolvimento Científico e Tecnológico—CNPq proc. 306054/2006-0 (CO).

Institutional Review Board Statement: Not applicable.

Informed Consent Statement: Not applicable.

Data Availability Statement: Not applicable.

Conflicts of Interest: The author declares no conflict of interest.

Abstract

Impact of Climate Change on Sharks [†]

Rui Rosa [1,2,‡]

[1] MARE—Marine and Environmental Sciences Centre, Laboratório Marítimo da Guia, Faculdade de Ciências, Universidade de Lisboa, Av. Nossa Senhora do Cabo, 939, 2750-374 Cascais, Portugal; rrosa@fc.ul.pt

[2] Departamento de Biologia Animal, Faculdade de Ciências, Universidade de Lisboa, Campo Grande, 1749-016 Lisboa, Portugal

[†] Presented at the IX Iberian Congress of Ichthyology, Porto, Portugal, 20–23 June 2022.

[‡] Presentation type: Key-note talk.

Abstract: The global ocean has been shielding our planet from abrupt climate change by absorbing a large portion of the anthropogenically emitted carbon dioxide and the excess heat trapped in the atmosphere, leading to ocean acidification and warming. Additionally, oxygen loss in the ocean (also known as deoxygenation) is being exacerbated by rising global temperatures. This complex 3-way interaction ("deadly trio") will definitely shape populations' fitness and ecosystems' health in the ocean of tomorrow. Here, I discussed the differential impacts of the "deadly trio" on the marine biota, with a special emphasis on sharks.

Keywords: sharks; climate change; warming; acidification; hypoxia; deoxygenation

Funding: The author would like to thank Fundação para a Ciência e a Tecnologia (FCT) for funding MARE—Marine and Environmental Sciences Centre (UIDB/04292/2020).

Institutional Review Board Statement: Not applicable.

Informed Consent Statement: Not applicable.

Data Availability Statement: Not applicable.

Conflicts of Interest: The author declares no conflict of interest.

Citation: Rosa, R. Impact of Climate Change on Sharks. *Biol. Life Sci. Forum* **2022**, *13*, 41. https://doi.org/10.3390/blsf2022013041

Academic Editor: Alberto Teodorico Correia

Published: 6 June 2022

biology and life sciences forum

Abstract

Environmental Life History of Neotropical Fish through the Chemistry of Calcified Structures [†]

Esteban Avigliano [‡]

Instituto de Investigaciones en Producción Animal (INPA-CONICET-UBA), Universidad de Buenos Aires, Buenos Aires C1427CWO, Argentina; estebanavigliano@conicet.gov.ar
† Presented at the IX Iberian Congress of Ichthyology, Porto, Portugal, 20–23 June 2022.
‡ Presentation type: Plenary talk.

Abstract: The neotropical inland waters hold about 30% of the world's fish species. South America, in particular, brings together two of the largest basins in the world: the Amazon and Plata Basin. These are long migratory corridors of thousands of kilometers for numerous species of commercial importance. Poor management practices have brought many fisheries to the brink of collapse, including species for which little is known about their biology or environmental history. In the last decade, the microchemistry of calcified structures has been a valuable tool to reveal different aspects of the biology of fishes. Calcified structures such as otoliths keep an environmental record as fish grow, thus providing information on nursery areas, population structure, and migrations. This tool is especially powerful when the environmental variability (e.g., water) is known. The use of bioindicators such as bivalves, which keep track of environmental variation over time, is particularly promising for improving otolith-chemistry-based interpretations. The usefulness of different analytical techniques of hard structures (chemical analysis in one and two dimensions) in relation to different geographical scales of evaluation is discussed. Advances in the environmental mapping of radiogenic isotopes ($^{87}Sr/^{86}Sr$) and trace elements in five countries traversed by the Plata Basin (Argentina, Brazil, Paraguay, Uruguay, and Bolivia) and their extension to the southern Atlantic and Pacific basins (Patagonia, Argentina, and Chile) are presented. The results of these techniques have made it possible to reveal the natal origin and reveal the environmental life history of various neotropical fish species, including cross-border migrations and complex population structures.

Keywords: otolith chemistry; microchemistry; life history

Citation: Avigliano, E. Environmental Life History of Neotropical Fish through the Chemistry of Calcified Structures. *Biol. Life Sci. Forum* **2022**, *13*, 56. https://doi.org/10.3390/blsf2022013056

Academic Editor: Alberto Teodorico Correia

Published: 7 June 2022

Publisher's Note: MDPI stays neutral with regard to jurisdictional claims in published maps and institutional affiliations.

Funding: This research received no external funding.

Institutional Review Board Statement: Not applicable.

Informed Consent Statement: Not applicable.

Data Availability Statement: Data are available on request from the author.

Conflicts of Interest: The author declares no conflict of interest.

Abstract

Long-Distance Migrations: Orientation and Navigation of Anguillid Eels [†]

Caroline Durif [‡]

Institute of Marine Research, Austevoll Research Station, 5392 Storebø, Norway; caroline.durif@hi.no
† Presented at the IX Iberian Congress of Ichthyology, Porto, Portugal, 20–23 June 2022.
‡ Presentation type: key-note talk.

Abstract: Anguillid eels grow in freshwater but spawn in the open ocean. Almost all of them undertake long migrations, consisting of several thousands of kilometers, between their feeding and their distant oceanic spawning areas. The cues that guide eels over long distances to the spawning area are unknown. The Earth's magnetic field is one, if not the only, reliable cue that can guide them between these areas. To test whether the use of magnetic cues is compatible with what we know about the life-history and migration of eels, the patterns of magnetic inclination and intensity along the migratory routes of five anguillid species were investigated. Regardless of the species and the differing routes between life stages, larvae of those species always drift along paths of increasing magnetic inclination and intensity, while adults follow reverse gradients. This is consistent with an imprinting/retracing hypothesis. The proposed navigation mechanism suggests that larvae imprint the target magnetic intensity or inclination isoline value upon hatching, and then years later retrace the magnetic gradient until they reach the target isoline value which they can follow to find their conspecifics for reproduction. Such a mechanism does not require a high level of precision to find a specific area but does require imprinting of the magnetic gradient experienced during the early life of the eel. There is already evidence for the imprinting of a magnetic compass direction in glass eels as well as yellow and silver eels. Knowledge about the orientation cues and biological mechanisms used by marine organisms to navigate and orient are important for taking appropriate management steps that are likely to help the conservation of vulnerable or endangered species.

Keywords: *Anguilla* sp.; compass orientation; magnetic inclination; geomagnetic field; secular variation; restocking

Citation: Durif, C. Long-Distance Migrations: Orientation and Navigation of Anguillid Eels. *Biol. Life Sci. Forum* **2022**, *13*, 65. https://doi.org/10.3390/blsf2022013065

Academic Editor: Alberto Teodorico Correia

Published: 8 June 2022

Funding: This research received no external funding.

Institutional Review Board Statement: Not applicable.

Informed Consent Statement: Not applicable.

Data Availability Statement: The data presented here are available at NMDC at http://doi.org/10.21335/NMDC-1870239141, reference number NMDC-1870239141.

Conflicts of Interest: The author declares no conflict of interest.

biology and life sciences forum

MDPI

Abstract

Global Challenges to Feed the Blue Revolution in a Sustainable Way †

Luisa M. P. Valente ‡

CIIMAR-Centro Interdisciplinar de Investigação Marinha e Ambiental and IBAS, Instituto de Ciências Biomédicas Abel Salazar, Universidade do Porto, 4050-313 Porto, Portugal; lvalente@icxbas.up.pt
† Presented at the IX Iberian Congress of Ichthyology, Porto, Portugal, 20–23 June 2022.
‡ Presentation type: Key-note talk.

Abstract: The global demand for food is expected to double by 2050, and trends of healthy eating and ethical consumerism are driving the innovation and consumption trends. Moreover, sustainability messaging is particularly relevant in the seafood category. There will be increasing pressure to reduce marine ingredients in aquafeeds to protect ecosystems at risk and replace them with a wide range of alternatives that include plant sources, invertebrates, microbial biomass, algae, and also agrifood byproducts that may simultaneously contribute towards a circular economy concept. However, the nutritional and functional value of each new feed ingredient has to be thoroughly evaluated to assure good growth and prevent major impacts on fish health status, flesh nutritional value, and safety. A holistic approach towards sustainable farming has to be tackled by addressing the One Health approach envisaging (a) the protection of the environment by improved feed utilization reducing the loss of nutrients into the water; (b) an increased production of robust fish able to cope with global challenges; (c) an increased contribution with safe and healthier food items for consumers, especially for sensitive populations such as elderly people, pregnant women, and children. Aquaculture needs to respond to consumer expectations, positioning itself as a true superfood with a prominent place in protein, healthy, and "clean" diets and improve its perception and image among consumers.

Keywords: one health; aquafeeds; alternative ingredients; sustainable aquaculture; healthy diet; fish consumption

Citation: Valente, L.M.P. Global Challenges to Feed the Blue Revolution in a Sustainable Way. *Biol. Life Sci. Forum* **2022**, *13*, 92. https://doi.org/10.3390/blsf2022013092

Academic Editor: Alberto Teodorico Correia

Published: 15 June 2022

Funding: This research received external funding from FCT (UIDB/04423/2020, UIDP/04423/2020).

Institutional Review Board Statement: Not applicable.

Informed Consent Statement: Not applicable.

Data Availability Statement: Not applicable.

Conflicts of Interest: The authors declare no conflict of interest.

biology and life sciences forum

Abstract

An Updated Assessment on Climate Change Impacts and Adaptation in the Oceans [†]

Elena Ojea [‡]

Future Oceans Lab, CIM-Universidade de Vigo, Torre Cacti, Campus Lagoas Marcosende, 36310 Vigo, Spain; elenaojea@uvigo.es

† Presented at the IX Iberian Congress of Ichthyology, Porto, Portugal, 20–23 June 2022.

‡ Presentation type: key-note talk.

Abstract: According to the latest IPCC report on impacts and adaptation, ocean and coastal systems are already reaching tipping points, where habitat-forming species such as corals and seagrasses will reach non-reversible shifts, even below 1.5 °C warming. Marine species are responding to incremental temperature changes in the oceans by shifting poleward at a pace of 60 km/decade. However, extreme events such as marine heat waves challenge the conditions at which many species can thrive. These changes have significant impacts in human dependent communities, with effects ranging from the capacity of fishing fleets to continue harvesting, to the nutritional intake of marine fisheries in coastal communities. Here, I present an overview of new evidence for anthropogenic climate change impacting upon the oceans and the human dependent communities, focusing on the solutions space and adaptation pathways for the oceans over the next few decades. A combination of nature-based solutions, socio-institutional adaptation and technical interventions are needed to address the impacts of climate change in the oceans. The reach of the adaptation portfolio highly depends on mitigation efforts, where lower emission scenarios allow for a more effective adaptation. Nature-based solutions in the oceans include ecosystem-based management, adaptive fisheries management, restoration and the conservation of habitats and ecosystems. Under higher-emission scenarios, technological solutions and infrastructure interventions are needed; for example, for restoring coral reefs and for adapting to a rise in sea level. In any case, engaging in such options requires a profound transformative change of ocean governance, shifting the policy focus to equity and justice.

Keywords: adaptation; ocean and coastal systems; IPCC; nature-based solutions; adaptation pathways

Citation: Ojea, E. An Updated Assessment on Climate Change Impacts and Adaptation in the Oceans. *Biol. Life Sci. Forum* **2022**, *13*, 131. https://doi.org/10.3390/blsf2022013131

Academic Editor: Alberto Teodorico Correia

Published: 20 June 2022

Funding: This research was funded by ERC CLOCK project grant number 679812.

Institutional Review Board Statement: Not applicable.

Informed Consent Statement: Not applicable.

Data Availability Statement: Not applicable.

Conflicts of Interest: The author declares no conflict of interest.

MDPI
St. Alban-Anlage 66
4052 Basel
Switzerland
Tel. +41 61 683 77 34
Fax +41 61 302 89 18
www.mdpi.com

Biology and Life Sciences Forum Editorial Office
E-mail: blsf@mdpi.com
www.mdpi.com/journal/blsf

9 783036 564791